The Minerals, Metals & Materials Series

Gabrielle Gaustad · Camille Fleuriault ·
Mertol Gökelma · John A. Howarter ·
Randolph Kirchain · Kaka Ma ·
Christina Meskers · Neale R. Neelameggham ·
Elsa Olivetti · Adam C. Powell ·
Fiseha Tesfaye · Dirk Verhulst ·
Mingming Zhang
Editors

REWAS 2019

Manufacturing the Circular Materials Economy

TMS

Springer

Editors
Gabrielle Gaustad
Alfred University
Alfred, NY, USA

Mertol Gökelma
Norwegian University of Science and Technology
Trondheim, Norway

Randolph Kirchain
Massachusetts Institute of Technology
Cambridge, MA, USA

Christina Meskers
Umicore
Olen, Belgium

Elsa Olivetti
Massachusetts Institute of Technology
Cambridge, MA, USA

Fiseha Tesfaye
Åbo Akademi University
Turku, Finland

Mingming Zhang
ArcelorMittal Global R&D
Schererville, IN, USA

Camille Fleuriault
Gopher Resource
Eagan, MN, USA

John A. Howarter
Purdue University
West Lafayette, IN, USA

Kaka Ma
Colorado State University
Fort Collins, CO, USA

Neale R. Neelameggham
IND LLC
South Jordan, UT, USA

Adam C. Powell
Worcester Polytechnic Institute
Worcester, MA, USA

Dirk Verhulst
Reno, NV, USA

ISSN 2367-1181 ISSN 2367-1696 (electronic)
The Minerals, Metals & Materials Series
ISBN 978-3-030-10385-9 ISBN 978-3-030-10386-6 (eBook)
https://doi.org/10.1007/978-3-030-10386-6

Library of Congress Control Number: 2018964934

This Springer imprint is published by the registered company Springer Nature Switzerland AG
The registered company address is: Gewerbestrasse 11, 6330 Cham, Switzerland

Preface

More than ten years ago, in October 2008, I participated in my first REWAS event, giving two presentations on my doctoral dissertation work examining aluminum recycling. At that time REWAS was called the Global Symposium on Recycling, Waste Treatment, and Clean Technology and the conference was in Cancun, Mexico. Much has changed in the decade since that conference but one thing remains the same: the looming sustainability challenges the world faces. Resource consumption continues to rise driving increased waste, impacts of climate change are growing, and energy use is increasing exponentially as the world continues to develop. Fortunately, the minerals, metals, and materials science community has embraced these challenges as opportunities to drive ground-breaking work in these fields. These engineers, scientists, educators, and entrepreneurs are contributing to significant gains in our shared goal of a sustainable materials industry.

The focus of this year's REWAS conference is on *Manufacturing the Circular Materials Economy*. Our current linear economy involves a one-way street where materials are extracted, manufactured, used, and then thrown away. The idea behind a circular economy is to de-couple development and growth from resource consumption which have been historically correlated. Unlocking the potential for circularity in the materials life-cycle can actually enable economic opportunities. All this will require ingenuity in the materials science and manufacturing sectors as well as require trans-disciplinary work with the sustainability and industrial ecology communities. We are highlighting this work in REWAS 2019 in five main thematic sessions.

Disruptive Material Manufacturing: A Systems Perspective

The landscape of materials manufacturing has the potential for dramatic change as new design techniques (e.g., materials genome) and new technologies (e.g., additive manufacturing) begin to scale to industrial production levels. What is the environmental impact of additive manufacturing? How will disruptive technologies change the landscape of producing materials? What is the circularity potential for inputs and outputs of these new production routes? How can these systems be optimized? How does the massive scale-up of clean energy technologies affect process flow sheets and recycling? Many of these questions are being addressed by the proceedings from this session which is a partnership with the Additive Manufacturing Committee.

Secondary and Byproduct Sources of Materials and Minerals

As scarcity and criticality concerns grow, attention has turned to ore alternative sources of important materials, metals, and minerals. Circular economy techniques, industrial symbiosis, and urban mining are sustainability strategies for obtaining these materials from industrial byproducts, end-of-life wastes, and other secondary sources. This includes recycling of electronic waste, batteries of all chemistries, and agricultural byproducts. What extraction technologies will be needed to enable this material recovery? What are the economic and environmental impact implications of such alternative routes? What kinds of material flow analysis and/or metric standardization is needed for tracking circularity at multiple scales? As many of these issues are tackled via pyrometallurgy and hydrometallurgy, these TMS committees were supporting partners in this session.

Rethinking Production

Besides disruptive technologies, there are additional sustainability benefits with high potential in the production sector. In partnership with the Materials Characterization Committee, this session asks, How do we decrease emissions in production? How can we achieve sustainable process design? How will clean energy technologies be manufactured? How do we measure and quantify embodied energy (how do existing methods fall short and how do we align internationally)? How can we enable cost-effective and efficient collection and reprocessing of wastes? What are technologies and strategies for managing mixed materials? How can trace, tramp, and other unwanted contaminants be removed from secondary streams? What are opportunities for direct use of end-of-life products to make new products with minimal or no reprocessing? How do we reach zero-waste production?

Education and Workforce Development

Transitioning knowledge from the research and academic sectors into applied work is critical to realizing sustainability. Papers from this session address novel educational approaches like blended learning, flipped classrooms, and MOOCs, as well as approaches for integrating sustainability into traditional disciplinary curriculums like materials science. Technology transition, applied learning, and workforce development initiatives also will be highlighted. This session was in partnership with the TMS Education Committee and the Professional Development Committee.

In partnership with the Light Metals Division, REWAS 2019 also is showcasing work in the Cast Shop Technology symposium. Proceedings manuscripts from this joint session are featured in the *Light Metals 2019* publication, with abstracts appearing here on the next page. This session focuses on some of the unique challenges the aluminum industry faces in its transition to sustainability. Sustainable operations, life-cycle assessment, recycling impacts and awareness, charge materials, and other environmental issues relevant to cast shops are the focus of this session.

Looking back at the innovations since REWAS 2008 leaves me excited and hopeful for the next decade. Many thanks to the excellent team of organizers, committee chairs, reviewers, and session chairs who made this proceedings publication possible and most importantly, the researchers, authors, and presenters participating in REWAS 2019.

Gabrielle Gaustad
Lead Organizer

REWAS 2019: Cast Shop Recycling Technologies Abstracts

A Method for Assessment of Recyclability of Aluminum from Incinerated Household Waste
Mertol Gökelma, Ingrid Meling, Ece Soylu, Anne Kvithyld, Gabriella Tranell

Aluminum is widely used in daily household consumable goods such as food and drink packaging materials, storage containers, etc. The disposal of such goods into household waste means that this waste stream contains a significant amount of aluminum. Domestic waste is commonly sent to incinerator plants where the organics are combusted while the metallic content stays in the bottom ash, which is subsequently separated into various metal streams. Because of the importance of aluminum in the circular economy, there is a need for efficient recovery procedures for this metal source. This paper discusses the recyclability and the recovery rate of aluminum from the bottom ash through remelting with a molten salt. The remelting experiments were performed under a 50–50 wt% NaCl:KCl mixture with a 2 wt% CaF_2 addition to promote metal coalescence. The oxide thickness and trace element content of the starting metal and the composition of the resulting metal were characterized as these parameters largely determine the recovery rate and recyclability of these secondary metal streams. The laboratory results showed the coalescence efficiency up to 99.5% and the material yield up to 92%. High deviation in oxide content based on the oxide layer thickness measurements was observed which crucially affects the metal losses in recycling.

Aluminum Alloys in Autobodies: Sources and Sinks
Ayomipo Arowosola, Gabrielle Gaustad, Leslie Brooks

Emissions from the transportation industry combined with increasing consumption of materials have inspired the automotive industry to use lightweight materials in autobodies. A wide diversity of materials is being used, for example, aluminum, magnesium, and plastic composites. This lightweighting approach is a proven sustainability strategy improving fuel economy and thereby reducing greenhouse gas emissions. However, increasing the number of differing types of materials in cars is actually complicating recycling operations. Critical metals used as alloying additions are dissipatively lost in the recycling process and can also negatively impact recycling rates by accumulating as tramp elements. This work combines compositional characterization of automotive materials, material flow analysis, and techno-economic assessment to better understand this problem and inform solutions. Results show that both technical solutions like sensor-based sorting and operational solutions like compositionally based blending can decrease material losses, thereby reducing the negative impacts and inching closer to a circular economy.

Isothermal Hot Pressing of Skimmed Aluminium Dross: Influence of the Main Processing Parameters on In-house Molten-Metal Recovery
Varužan Kevorkijan

The Isothermal Hot Pressing (IHP) of skimmed aluminum dross, considered in this study, was performed under laboratory conditions using a cylindrical pressing model made from high-temperature stainless steel. The pressing model was inserted into an electrical furnace with a protective argon atmosphere. The temperature of the pressing was within the interval 650–900 °C, while the applied pressure varied between 5 and 50 bars. The laboratory results showed that when using IHP it is possible to reduce the remaining aluminum content in "pressed dross skulls" below 10%.

LIBS Based Sorting—A Solution for Automotive Scrap
Georg Rombach, Nils Bauerschlag

The demand for automotive sheet shows the highest growth rates in the aluminum industry. Currently, the main alloys are from 5.xxx and 6.xxx series with a variety of approx. 40 different specifications. Such a mixed scrap quality cannot be used in an efficient way in cast houses, due to the divergent Mg and Si concentrations. Additionally, the ratios between 5.xxx and 6.xxx scrap are unpredictable. Today, the only technical possibility to distinguish between these alloys is sensor-based sorting using LIBS. Such a sorting machine has to produce two sorting fractions, which both fulfill the requirements of the new wrought alloys. In this case a high sorting efficiency is mandatory. First results show a purity of the sorted fraction above 95% as well as a high recovery of the sorted material also above 90%.

Manufacturing of Hydrogen on Demand Using Aluminum Can Scrap with Near Zero Waste
Jed Checketts, Neale R. Neelameggham

During the 1990s Powerball Industries, UT had demonstrated producing Hydrogen on demand machinery—where encapsulated sodium metal is reacted with water to release hydrogen gas under pressure. The hydrogen on demand has several applications. Here we discuss the ongoing research of making use of aluminum (recycled or newly produced) with anhydrous sodium sulfate—which will make sodium metal, sulfur oxides, and aluminum oxide. The thermodynamic analysis shows suitable physicochemical conditions for the reactions. Preliminary economics using aluminum scrap—recycled aluminum is shown.

Positive Material Identification (PMI) Capabilities in the Metals Secondary Industry: An Analysis of XRF and LIBS Handheld Analyzers
Leslie Brooks, Gabrielle Gaustad

Recycling is a critical part of obtaining a more circular economy. In the metals secondary industry, traditional equipment (a magnet, file, and/or grinding wheel) used to identify and sort materials at their end of life can aid in grouping metals (i.e., Al+Mg alloys, ferrous, high-temperature alloys, etc.), but they are incapable of identifying the alloy's elemental composition; a necessity for preventing downcycling and maximizing secondary utilization rates. Handheld analyzers that utilize X-ray fluorescence (XRF) and spectroscopy (LIBS) technology may offer technological assistance that is helpful for achieving this level of analysis, often referred to as Positive Material Identification (PMI). This work tests the performance of these units under the challenging conditions present in yards (contaminated, unpolished, rugged scraps). These instruments, with their increasing safety settings, ruggedness, ease of point-click use, and quick read times (for both XRF and LIBS) have significant potential, especially with ability to ID metal faster than cognitive recognition. Additionally, as unit costs of these instruments continue to decrease and the range of varying types of metal entering yards continues to widen, the return on investment becomes more immediate. However, extreme fluctuations of reported elemental compositions are being seen even when measurements have been taken in the same place consecutively; indicating that in their current state, they can inform content of material but are not necessarily reliable for reporting accurate and precise compositional percentages.

The Vertical Floatation Decoater for Efficient, High Metal Yield Decoating and Delacquering of Aluminum Scrap
Robert De Saro, Sam Luke

The Vertical Floatation Melter (VFD) has undergone pilot testing in removing organics from scrap aluminum. The VFD uses a vertical cone in which the scrap is dropped into the top and products of combustion at 1000 °F are introduced into the bottom, flowing countercurrent to the scrap. The gases float the scrap in the cone resulting in very high convective heat transfer which leads to rapid decoating with minimal metal loss.

Turnings, fines, Twitch, and UBC have been processed through a single unit. Decoating times are about one minute. Measured energy use varies from 78 to 854 Btu/lbm; measured preheat temperatures are 850 °F.

Decoat efficiency appears good with visually no oxidation present and no evidence of organics. Samples have been sent to a lab to confirm this.

Contents

About the Editors

Gabrielle Gaustad recently became the Dean of the Inamori School of Engineering at Alfred University. For the prior ten years she was an Associate Professor in the Golisano Institute for Sustainability at the Rochester Institute of Technology. She holds a Ph.D. in Material Science and Engineering from MIT. The Gaustad group conducts research quantifying the economic and environmental trade-offs for materials at their end of life with a focus on recycling, resource recovery, and promoting a circular economy. Methodologies include a variety of systems modeling techniques such as dynamic material flow analysis, optimization, simulation, systems dynamics, economic modeling, process-based cost modeling, and life-cycle assessment, as well as traditional material characterization such as TGA, PSD, SEM, XRD, XRF, EDS, and ICP-MS. Specific projects include implications of material scarcity and criticality for clean energy technologies, aluminum and steel recycling technologies and compositional analysis, and environmentally benign and economically efficient recycling of lithium ion batteries, particularly those containing nanomaterials.

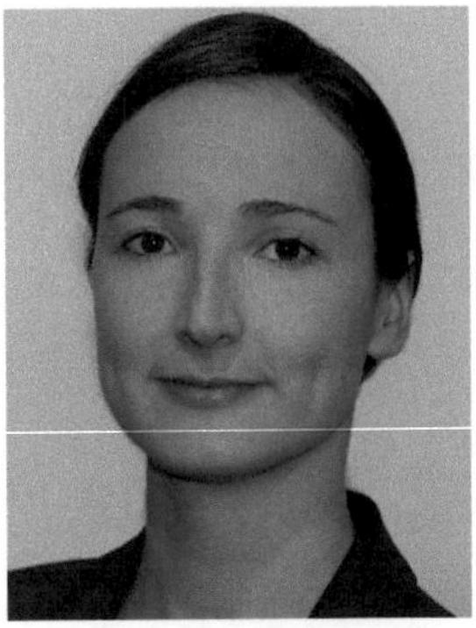

Camille Fleuriault is R&D Metallurgist at Gopher Resource in Eagan, Minnesota, USA. She is developing innovative and environmentally friendly recycling processes for the secondary metals industry. In particular, she is focusing on waste management and energy efficiency of high-temperature systems. She holds a B. S. in geological engineering and an M.Eng. in Mineral Engineering from the National School of Geological Engineering in Nancy, France, and a M.Sc. in Metallurgical Engineering from Colorado School of Mines, USA. She is *JOM* advisor for the TMS Pyrometallurgy Committee.

Mertol Gökelma is a postdoctoral researcher in the Department of Materials Science and Engineering at Norwegian University of Science and Technology (NTNU), Trondheim, Norway. He finished his B.Sc. degree at Dokuz Eylul University in Metallurgical Engineering and Materials Science, Turkey, and his M. Sc. and Ph.D. in Metallurgical Engineering at RWTH Aachen University, Germany. He worked as a research assistant for four years at the Institute of Process Technology and Metal Recycling (IME), RWTH. His research interests focus on process metallurgy of nonferrous metals and he has been involved in different R&D projects focused on recycling of magnesium black dross, recycling and refining of precious metals, powder synthesis of titanium alloys, and metallothermic reduction of oxides. The main focus of his research is recycling and refining of aluminum as well as the behavior of nonmetallic inclusions in Al-melts.

John A. Howarter is an Associate Professor in Materials Engineering at Purdue University with a joint appointment in Environmental and Ecological Engineering. His research interests are centered on synthesis, processing, and characterization of sustainable polymers and nanocomposites, value recovery through recycling and reprocessing of waste materials, and sustainable materials that can enable improved design for the environment.

John is the Chair of the TMS Public and Governmental Affairs committee and serves on the TMS Board of Directors. Since 2014 he has served as the chapter advisor for the Purdue University Material Advantage student organization. John earned a B.S. from The Ohio State University in 2003 and Ph.D. from Purdue

University in 2008, both in Materials Engineering. From 2009 to 2011 he was a National Research Council postdoctoral scholar in the Polymers Division of the National Institute of Standards and Technology in Gaithersburg, Maryland.

Randolph Kirchain's research and teaching aim to improve materials-technology decisions by characterizing the economic and environmental impact of those decisions. That impact may derive from changes in the performance of the products into which those materials are transformed and/or in the systems in which they are produced, used, and eventually discarded. Dr. Kirchain has authored over 200 publications in refereed journals and conferences. He has been awarded the American Iron and Steel Institute's Top Technical Achievement Award, the General Motors Technical Achievement Award, and the TMS Recycling Technology Award. Currently, Dr. Kirchain serves as the co-director of the MIT Concrete Sustainability Hub.

Kaka Ma received her B.S. in Materials Physics from University of Science and Technology of China in 2006 and earned her Ph.D. in Materials Science and Engineering from the University of California (UC), Davis in December 2010. She joined Colorado State University as a tenure-track assistant professor in August 2016 after several years of postdoctoral research at UC Davis and UC Irvine.

Dr. Ma's research interests sit at the interface of materials science, mechanical engineering, and sustainability. She is interested in fabrication and characterization of advanced materials that contain nanoscale microstructural features for high performance such as high specific strength, high ductility, improved reliability and lifetime. Her research also performs mechanical testing at small scales and in localized regions such as boundaries and interface using nanoindentation and nanoscratch techniques. She aims to discover new processing-structure-properties correlation and to manufacture advanced materials for next-generation structural, electronic, and energy components. Her previous research investigated nanostructured or ultrafine grained materials ranging from thermal barrier coatings, aluminum alloys to metal matrix composites. Her recent research activities focus on multiscale hierarchical structured materials, functionally graded materials, low-work-function electride materials, additive manufacturing (AM) and sustainability issues associated with metal AM. Dr. Ma serves on the editorial board for the journal *Materials Science and Engineering A* and is a reviewer for several other journals, including *Nature Communications*, *Metallurgical and Materials Transactions A*, *Journal of Materials Science*, *Surface and Coatings Technology*, and *Journal of Alloys and Compounds*.

Christina Meskers is senior manager Open Innovation at Umicore's Group Research and Development where she is responsible for the development, management, and governance of the global portfolio of bilateral and multilateral research partnerships and (education) programs as well as the University—RTO—industry partner network. Furthermore she develops and implements the open innovation strategy. Christina serves on the Industrial Advisory Boards of among others the Sustainable Materials (SuMa) and Sustainable and Innovative Resource Management (SINReM) international M.Sc. programs and within EIT Raw Materials innovation community serves as member of the Education Committee and is vice-chair of Innovation Hub West steering committee.

With over 15 years' experience in the raw materials sector, her current focus is on the contribution of recycling and extractive technologies to the transition to sustainable cities and e-mobility. Christina co-authored the United Nations' International Resource Panel report "Metal Recycling—Opportunities, Limits, Infrastructure" (2012), among other publications. In 2008 she was a recipient of the Young Leaders Professional Development Award of The Minerals, Metals & Materials Society (TMS) Extraction and Processing Division. She was lead organizer of Sustainable Materials Production and Recycling 2010, REWAS 2013, and REWAS 2016 and connected these symposia to current societal and industrial trends. After volunteer roles in the TMS Recycling Committee, Materials and Society Committee, and Public and Governmental Affairs Committee she currently serves as Vice-chair of the Extraction and Processing Division.

Neale R. Neelameggham is "The Guru" at IND LLC, involved in international technology and management consulting in the field of critical metals and associated chemicals, thiometallurgy, energy technologies, soil biochemical reactor design, lithium ion battery design, and agricultural uses of coal. He was a visiting expert at Beihang University of Aeronautics and Astronautics, Beijing, China and a plenary speaker at the Light Metal Symposium in South Africa on the topic of low carbon dioxide emission processes for magnesium.

Dr. Neelameggham has more than 38 years of expertise in magnesium production and was involved in process development of the startup company NL Magnesium through to the present US Magnesium LLC, UT until 2011. He and Brian Davis authored the ICE-JNME award-winning (2016) article "21st Century Global Anthropogenic Warming Convective Model." He is presently developing "stored renewable energy in coal" Agricoal™ for greening arid soils and has authored an e-book *Eco-stoichiometry of Anthropogenic CO2 That Returns to Earth* on a new discovery of quantification of increasing CO2 returns to Earth.

Dr. Neelameggham holds 16 patents and patent applications, and has published several technical papers. He has served in the Magnesium Committee of the TMS Light Metals Division (LMD) since its inception in 2000, chaired it in 2005, and in 2007 he was made a permanent co-organizer for the Magnesium Technology Symposium. He has been a member of the Reactive Metals Committee, Recycling Committee, and Titanium Committee, and was a Program Committee Representative for LMD.

Dr. Neelameggham was the inaugural chair, when in 2008, LMD and the Extraction and Processing Division created the Energy Committee, and he has been a coeditor of the Energy Technology symposium through the present. He received the LMD Distinguished Service Award in 2010. While he was the chair of Hydrometallurgy and Electrometallurgy Committee he initiated the Rare Metal Technology symposium in 2014. He is coeditor for the 2019 symposia on Magnesium Technology, Energy Technology, Rare Metal Technology, REWAS 2019, and Solar Cell Silicon.

Elsa Olivetti is the Atlantic Richfield Associate Professor of Energy Studies in the Department of Materials Science and Engineering at the Massachusetts Institute of Technology (MIT). Her research focuses on improving the environmental and economic sustainability of materials using methods informed by materials economics, machine learning, and techno-economic analysis. She has received the NSF Career award for her experimental research focused on beneficial use of industrial waste materials. Dr. Olivetti received her B.S. degree in Engineering Science from the University of Virginia. Her Ph.D. in Materials Science and Engineering from MIT was focused on development of cathode materials for lithium ion batteries.

Adam C. Powell joined the faculty at Worcester Polytechnic Institute (WPI) in August 2018 as an Associate Professor in the Mechanical Engineering Department. His field is materials processing, and his research focuses on validated mathematical modeling of metal process development for clean energy and energy efficiency. His research group is developing new projects whose goals are to reduce vehicle body weight, lower solar cell manufacturing cost with improved safety, reduce or eliminate environmental impact of aerospace emissions, and improve grid stability with up to 100% renewables.

Dr. Powell's research has resulted in 67 publications across materials classes: metal extraction/refining and product development, thin films, ceramic coatings, polymer membranes, batteries, and electromagnetic propulsion. He is the author of nine open source computational tools in materials processing, microstructure and thermodynamics modeling.

Dr. Powell is fluent in Japanese, and as a University of Tokyo Foreign Collaborative Researcher, gives technical talks in Japanese to industry, government, and academic audiences.

Fiseha Tesfaye is a Senior Researcher and project manager working in the Johan Gadolin Process Chemistry Centre (PCC) of Åbo Akademi University, Finland. He received his Master's degree in materials processing technology in 2009 from Helsinki University of Technology and Ph.D. degree in metallurgy in 2014 from Aalto University, Finland. During his Ph.D. period, he focused his research on the electrochemical investigation of the thermodynamic properties of sulfide and intermetallic materials.

After a postdoctoral position in the Laboratory of Inorganic Chemistry at Åbo Akademi University from 2015 to 2017, which focused on the sulfosalts and sulfates characterizations, he attracted a large research project related to thermodynamic investigation of complex inorganic material systems in the renewable energy and metals production processes. From September 2017 onward, his research activities have been focused mainly on the sulfate-oxide systems database development with the FactSage software package, as well as rigorous theoretical and experimental investigations for promoting improved recovery of values from waste streams. Dr. Tesfaye was also appointed as a Visiting Research Scientist at Seoul National University, South Korea, for 6 months between March and August 2018. Dr. Tesfaye's current research interests are also within the scope of materials science, recycling technology, circular economy, and metallurgical engineering.

Dr. Tesfaye is a regular contributor and member of TMS, and is the 2018 TMS Young Leaders Professional Development Award winner. He has served on TMS committees including Recycling and Environmental Technologies, Energy, and Young Professionals, and was a member of the scientific committee of Materials San Diego 2018. His personal achievements include significant improvement of experimental research applying the solid-state EMF technique for thermodynamic studies, as well as contribution of new experimental thermodynamic data of several chalcogenide materials. Dr. Tesfaye has published more than 35 peer-reviewed articles in his research areas.

Dirk Verhulst spent the last 40 years in practical process metallurgy research on both sides of the Atlantic, bringing a number of projects from the laboratory to the pilot scale, and a few to industrial implementation. He is presently an independent consultant in process metallurgy and energy efficiency. Until the end of 2008, he was Director of Research at Altairnano in Reno, Nevada. He participated in the development of the Altair Lithium–ion Battery and was involved in the design and procurement of the manufacturing plant for the ceramic materials. Over the period 2003–2008, he worked extensively on the Altair Hydrochloride TiO_2 Pigment Process and the operation of its pilot plant. The complex flow sheet included both hydrometallurgical and pyrometallurgical steps. Optimization of energy use was a key factor to make this new approach competitive. From 1995 to 2000, he was Senior Development Engineer in BHP's Center for Minerals Technology at the same location in Reno. It is at BHP that the development of the hydrochloride TiO2 pigment process was initiated. Other BHP projects included novel processes for nickel, cobalt, zinc, and copper. Prior to 1995, he worked for 17 years in the research department of Umicore in Hoboken, Belgium. He was active in lead refining and in the hydrometallurgy of minor metals (indium, tellurium, selenium), but was mostly involved in the introduction of electric furnaces in lead smelting and slag cleaning. He tackled mathematical models and lab-scale experiments, ran pilot plants, and participated in the start-up of industrial operations. He has a doctor of engineering science degree in extractive metallurgy from Columbia University, and a chemical engineering degree from the Free University of Brussels. He has written and presented publications in the areas of hydrometallurgy, pyrometallurgy, nanomaterials, and environmental science. He holds several patents and patent applications.

Mingming Zhang is a lead research engineer at ArcelorMittal Global R&D at East Chicago, Indiana. Dr. Zhang has over 15 years of research experience in the field of mineral processing, metallurgical, and materials engineering. He obtained his Ph.D. degree in Metallurgical Engineering from The University of Alabama and his master's degree in Mineral Processing from the General Research Institute for Non-ferrous Metals in China. Prior to joining ArcelorMittal, he worked with Nucor Steel in Tuscaloosa, Alabama where he was a metallurgical engineer leading the development of models for simulating slab solidification and secondary cooling process. Dr. Zhang has conducted a number of research projects involving mineral beneficiation, thermodynamics and kinetics of metallurgical reactions, electrochemical processing of light metals, metal recycling, and energy-efficient and environmentally cleaner technologies. He has published over 50 peer-reviewed research papers and he is the recipient of several U.S. patents. Dr. Zhang also serves as editor and reviewer for a number of prestigious journals including *Metallurgical and Materials Transactions A and B*, *JOM*, *Journal of Phase Equilibria and Diffusion*, and *Mineral Processing and Extractive Metallurgy Review*. Dr. Zhang has made more than 20 research presentations at national and international conferences including more than 10 keynote presentations. He was the recipient of 2015 TMS Young Leaders Professional Development Award. He has served as conference/symposium organizer and technical committee chair in several international professional organizations including The Minerals, Metals & Materials Society (TMS), the Association for Iron & Steel Technology (AIST), and the Society for Mining, Metallurgy & Exploration (SME).

Part I
REWAS 2019: Disruptive Material Manufacturing—Scaling and Systems Challenges

From Recycled Machining Waste to Useful Powders for Metal Additive Manufacturing

Blake Fullenwider, Parnian Kiani, Julie M. Schoenung and Kaka Ma

Abstract To fulfill the growing demand for alternative and sustainable feedstock production for metal additive manufacturing, a novel dual-stage ball milling strategy was proposed to effectively convert recycled stainless-steel machining chips to powder with desirable characteristics for metal additive manufacturing. A theoretical analysis was performed to evaluate the impact of ball size on the chips-to-powder evolution and the consequent powder morphology. To verify the viability of using the ball milled powder created from machining chips in metal additive manufacturing, single tracks have been successfully deposited via laser engineered net shaping deposition and compared to the single tracks made from gas atomized powder using identical deposition conditions. The microstructures of these single tracks exhibited adequate adhesion to the substrate, a uniform melt pool geometry, continuity, and minimal splatter. Minimal differences in grain structure were observed between the single tracks made from ball milled powder and those made from gas atomized powder.

Keywords Metal additive · Manufacturing · Stainless steel · Sustainability · Ball milling

Powder metallurgy processing techniques, such as additive manufacturing, thermal spray, spark plasma sintering, and hot isostatic pressing, are widely used to fabricate bulk samples from metal powders. Metal additive manufacturing, one of the advanced powder metallurgy techniques, has attracted extensive research interest in recent decades, because of its capability to create near-net-shape parts in one step [1]. Additive manufacturing (AM) is regarded as a more sustainable process compared to conventional processing such as casting, as it significantly reduces the need of subtractive machining processes and thereby results in less material waste and less use of hazardous cutting fluids [2–4]. The properties of the bulk components

B. Fullenwider · K. Ma (✉)
Department of Mechanical Engineering, Colorado State University, Fort Collins, CO 80523, USA
e-mail: Kaka.Ma@colostate.edu

P. Kiani · J. M. Schoenung
Department of Chemical Engineering and Materials Science, University of California, Irvine, CA 92697, USA

G. Gaustad et al. (eds.), *REWAS 2019*, The Minerals, Metals & Materials Series,
https://doi.org/10.1007/978-3-030-10386-6_1

created through AM processes depend on both the processing method and the feedstock powder properties [5]. Despite the progresses in processing optimization and product property improvement that have been achieved, metal AM still faces several challenges due to low feedstock utilization efficiency and the availability of ideal feedstock powders [3]. Gas atomized (GA) powders are the most common feedstock for current metal additive manufacturing techniques because of their spherical morphology and controllable particle size distribution [1]. However, one environmental challenge of using GA powder in metal AM is the high energy consumption required to produce the powder through atomization, leading to high costs and limited availability in alloy compositions [6, 7]. To fulfill the growing demand for alternative and sustainable feedstock production for metal additive manufacturing, the present work aimed to explore a mechanical milling strategy to fabricate powders from recycled machining waste chips. In addition, the feasibility of using the powders created from machining chips in metal AM was proved by successful deposition of single tracks and multiple layers via laser engineered net shaping (LENS®).

304L stainless steel was selected as the model material due to its wide use in AM for various structural materials [8]. The desirable characteristics of feedstock powder for AM include a near-spherical morphology and particle sizes of 38–150 μm. The machining chips used in the current study were provided by AK steel (West Chester Township, Butler County, Ohio, OH). The individual chips exhibit a length of 5–20 mm, as shown in Fig. 1a. The surface of the chips contains serrations (Fig. 1c) due to the previous machining operation. To effectively convert the machining chips with dimension of several millimeters to powders with particle size on the micron scale, both modeling and experimental work were conducted. A theoretical analysis combining Gusev's model [9] and Hertz's model [9, 10] was performed to evaluate the impact force and stress on the particle, as well as the resultant maximum deformation depth into the particle. Two different types of balls are used as the milling media: Φ-20 balls (ball diameter = 20 mm) and Φ-6 balls (ball diameter = 6 mm). Experimentally, various ball milling procedures were implemented to investigate the effect of ball diameter on the powder morphology evolution and particle size refinement. A novel strategy of changing ball diameters during the ball milling was applied.

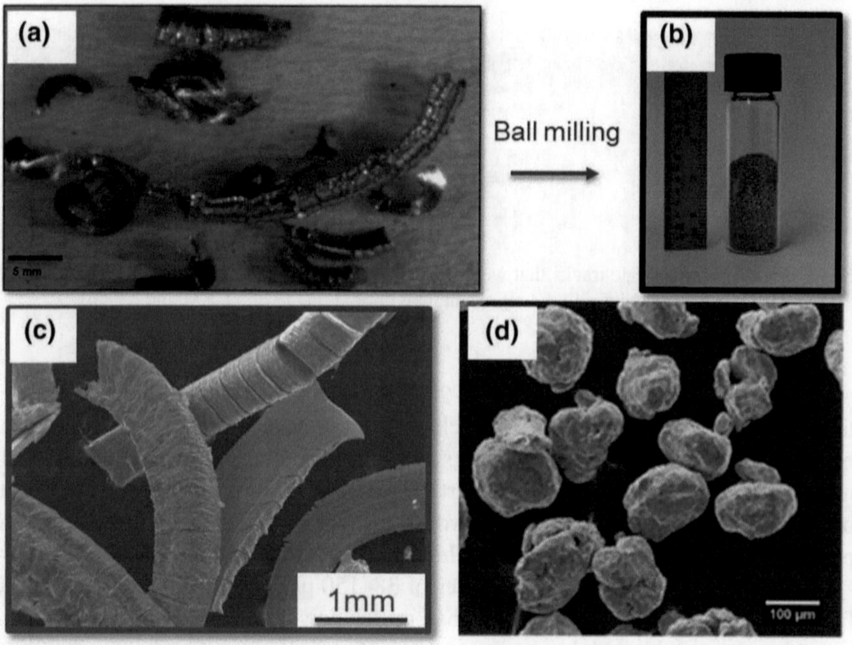

Fig. 1 **a** Optical image of the machining chips; **b** picture of the powders created by ball milling of machining chips; **c** SEM image of the machining chips; **d** SEM image of the powders created by ball milling of the machining chips

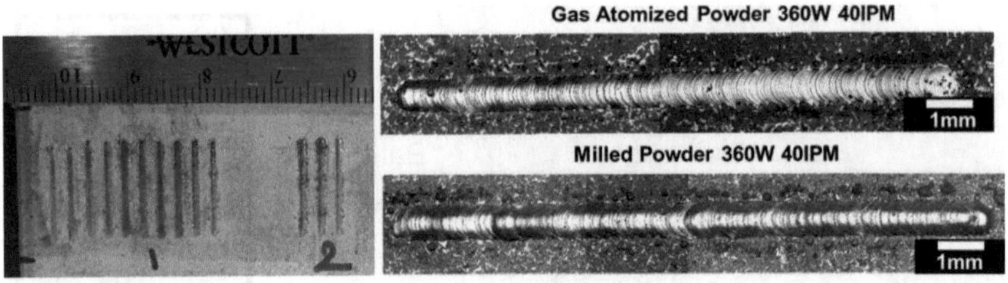

Fig. 2 Picture of the single tracks that were deposited using the ball milled powder created from the recycled machining chips and the representative optical image of ST-BM and ST-GA

The modeling results show that Φ-20 balls effectively reduce the powder particle size while Φ-6 balls effectively change the powder morphology to spherical or near-spherical. Based on the findings from the modeling, a novel dual-stage ball milling strategy was proposed to convert the machining chips to powders. Figure 1a, b exhibits the success of generating powders from the recycled machining waste. The powders created from the machining chips via dual-stage ball milling exhibit near-spherical morphology with particle sizes of 38–150 μm (Fig. 1d), which is suitable for metal additive manufacturing. The ball milled powders created from the machining chips also exhibit a higher hardness than GA powder, based on nanoindentation testing.

In addition, single tracks (ST) have been successfully deposited via LENS® using the ball milled powder created from the recycled machining chips. Single tracks were also made from GA powder (ST-GA) using identical deposition conditions. Figure 2 shows the single tracks that were deposited on the same substrate and the representative optical image of the single tracks made from ball milled powder (ST-BM) and ST-GA. The microstructures of these single tracks exhibited adequate adhesion to the substrate, a uniform melt pool geometry, continuity, and minimal splatter. Minimal differences in grain structure were observed between ST-BM and ST-GA. However, the average nanoindentation hardness of ST-BM is approximately 21% higher than that of ST-GA. The present research has discovered a sustainable approach to fabricate powders from recycled machining chips and has proved it is feasible to utilize these powders as feedstock in metal additive manufacturing.

Acknowledgements The present work is financially supported by NSF-CBET#1605392.

References

1. Frazier WE (2014) Metal additive manufacturing: a review. J Mater Eng Perform 23(6):1917–1928
2. Bourhis FL, Kerbrat O, Hascoet J-Y, Mognol P (2013) Sustainable manufacturing: evaluation and modeling of environmental impacts in additive manufacturing. Int J Adv Manuf Technol 69(9):1927–1939
3. Ford S, Despeisse M (2016) Additive manufacturing and sustainability: an exploratory study of the advantages and challenges. J Clean Prod 137:1573–1587
4. Huang SH, Liu P, Mokasdar A, Hou L (2013) Additive manufacturing and its societal impact: a literature review. Int J Adv Manuf Technol 67(5):1191–1203
5. Tan JH, Wong WLE, Dalgarno KW (2017) An overview of powder granulometry on feedstock and part performance in the selective laser melting process. Addit Manuf 18:228–255
6. Ma K, Smith T, Lavernia EJ, Schoenung JM (2016) Environmental sustainability of laser metal deposition: the role of feedstock powder and feedstock utilization factor. Procedia Manuf 7:198–204
7. Anderson IE, White EMH, Dehoff R (2018) Feedstock powder processing research needs for additive manufacturing development. Curr Opin Solid State Mater Sci 22(1):8–15
8. DebRoy T et al (2018) Additive manufacturing of metallic components—process, structure and properties. Prog Mater Sci 92:112–224
9. Gusev AI, Kurlov AS (2008) Production of nanocrystalline powders by high-energy ball milling: model and experiment. Nanotechnology 19(26):265302
10. Hertz H, Jones DE, Schott GA (1896) Miscellaneous papers. Macmillan, London

Recycling in Supply Chains for Tomorrow's Low-Carbon Industries

Adam C. Powell

Abstract Metal production and recycling technology changes are urgently needed to confront multiple simultaneous grand challenges in society. Energy production and distribution and transportation are undergoing disruptive transformations, and industry will follow. Unlike prior disruptions, such as the Internet and biotech, this will up-end material and energy flows around the planet. Due to both heightened awareness of climate impacts and plunging costs of sustainable technology, fossil energy and metals technologies which took 250 years to reach planet-wide industrial ubiquity must be swept aside and replaced in 2–3 decades. This talk will present energy industry scenarios and discuss new supply chains which will need to emerge in order to support these scenarios, the role which recycling must play in order to make these new supply chains themselves sustainable, and the timescales in which these changes need to happen, in order to meet targets for preventing large-scale climate disruption.

Keywords Recycling · Greenhouse emissions · Solar · Wind · Electric vehicles

Introduction

Dramatic reduction in CO_2 and other greenhouse emissions from all sectors is required to avoid severe climate disruption. The 2014 Intergovernmental Panel on Climate Change (IPCC) report puts it as follows:

> In the majority of stringent mitigation scenarios (430-530 ppm), the share of low-carbon electricity supply (comprising [renewable energy], nuclear and [carbon dioxide capture and storage] (CCS)) increases from presently about 30% to more than 80% by 2050. In the long term (2100), fossil-based electricity generation without CCS is phased out entirely in these scenarios [1, p. 560].

A. C. Powell (✉)
Worcester Polytechnic Institute, Worcester, MA, USA
e-mail: acpowell@wpi.edu

G. Gaustad et al. (eds.), *REWAS 2019*, The Minerals, Metals & Materials Series,
https://doi.org/10.1007/978-3-030-10386-6_2

This work considers three scenarios for the industrial change required to achieve these energy goals. The first two, by Bloomberg New Energy Finance (BNEF) [2] and DNV GL [3] are forecasts based on current technology and development trends. The third, by Mark Jacobson at Stanford, suggested that 100% electricity production by water hydropower, wind, and solar (WWS) is possible by 2050, despite the intermittent nature of wind and solar generation [4]. Though there are disagreements with assumptions in this scenario [5], it represents an upper bound to changes in material flows required to reduce greenhouse emissions. These studies include electrification of vehicles and building heating as added loads on the grid, and consider regional land availability and weather patterns in WWS availability.

However, these analyses make only limited mention of materials required for these changes. DNV GL mentions the cobalt and lithium needs of batteries for transportation and grid storage, based on analysis by Olivetti et al. [6]. But silicon and rare earths for solar, wind, and vehicles, and aluminum and magnesium for lightweight vehicles, are not well described. In particular, recycling can dramatically reduce energy consumption and emissions consumption in supplying these materials but has not entered into the above scenario analysis.

This study will therefore estimate some material production requirements for a renewable energy future envisioned by the above scenarios, with a focus on the role of recycling in the new supply chains. In particular, it will focus on PV use of silicon, wind turbine use of neodymium, and electric transportation use of neodymium, copper, aluminum, and magnesium.

Energy Scenarios

Scenarios by various research groups and organizations fall around two groups: one with nuclear and CCS and the other without. In terms of emissions reduction, Jacobson is the most aggressive and Bloomberg New Energy Finance (BNEF) the least so, with DNV GL in between.

Scenario 1: Bloomberg New Energy Finance

Bloomberg New Energy Finance (BNEF) is the investment advising wing of Bloomberg focused on the clean energy sector. Their forecast published in June 2018 looks at changes in technology and concludes that wind and solar will reach 48% of electricity production by 2050 [2]. This is based on technological forecasts of cost structure changes, and detailed assessments of the lowest cost energy mixes in several countries (Fig. 1).

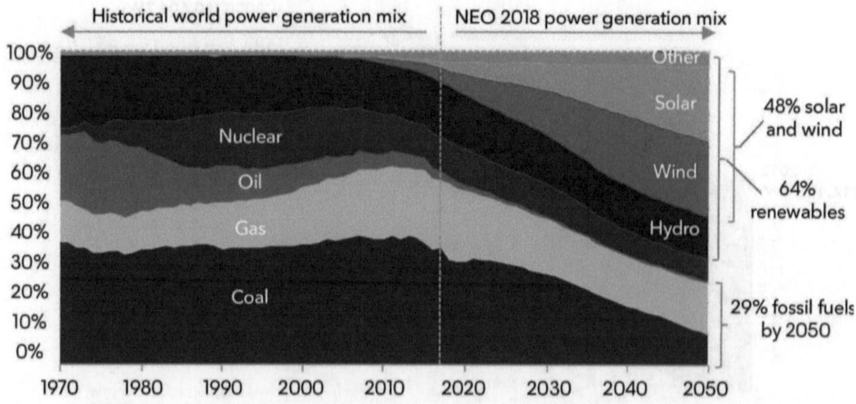

Fig. 1 Historical and projected worldwide electricity generation percentages by fuel [2]

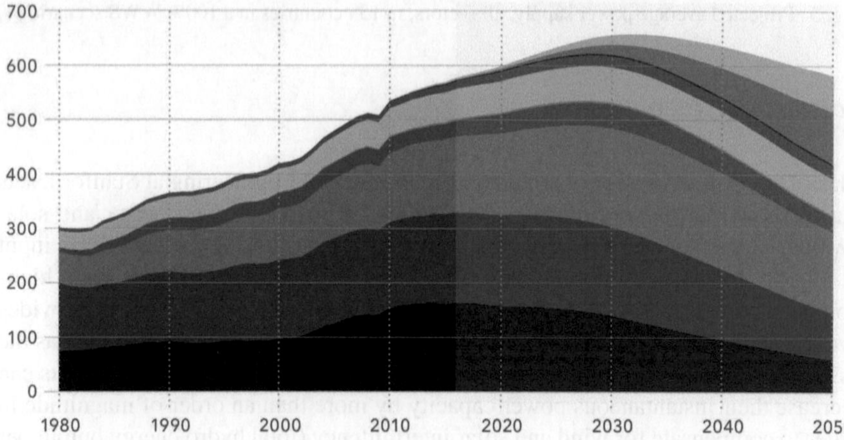

Fig. 2 Historical and projected primary energy use worldwide, all sectors [3], comprising (from bottom) coal, oil, natural gas, nuclear, geothermal, biomass, hydro, solar thermal, solar PV, and wind

Scenario 2: DNV GL

DNV GL is the world's largest ship registrar and classification society, formed by the merger of Del Norske Veritas (Norway) and Germanischer Lloyd (Germany). The company provides consulting services in shipping, logistics, and energy, and published their energy industry forecast in September 2018 [3]. Like BNEF, their approach is based on techno-economic forecasting. The report concludes that electric vehicles will comprise most of the light-duty and more than half of heavy-duty vehicle fleets, and renewables will constitute 50% of total energy in all sectors, and 85% of electricity production (Fig. 2).

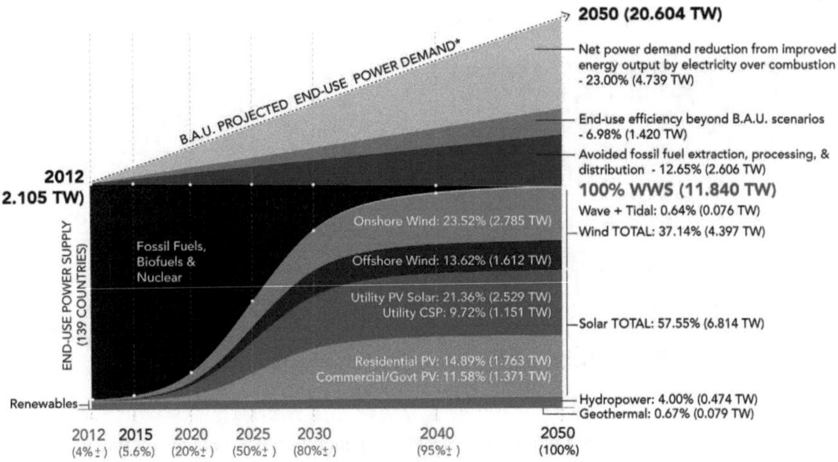

Fig. 3 Projected average power supply, all sectors, in 139 countries in a 100% WWS scenario [4]

Scenario 3: Jacobson

Mark Jacobson, Professor of Civil and Environmental Engineering at Stanford, sets out to show that the world can make a transition to 100% wind, water and solar (WWS) energy sources by 2050, with no fossil fuel, biofuel, or nuclear energy input at all [4]. To balance intermittency of renewables, he proposes large-scale high-voltage direct current (HVDC) grid connections, such that west coast solar provides evening power for the east coast, etc. There are issues with his analysis, such as the assumption, unstated until after publication [7], that hydroelectric power plants can increase their instantaneous power capacity by more than an order of magnitude in order to compensate for wind and solar intermittency (total hydro energy output per year does not change). But his is the only scenario known to the author which reaches emissions required for the 1.5° threshold of IPCC SR15 [8] (Fig. 3).

Summary

Table 1 shows the total energy supply and primary energy mixture in each of the three above scenarios. Though they use different methodologies, they all project very large increases in solar and wind electricity and electric transport within 20–30 years. In particular, solar energy from rises 1.5% of total electricity in 2017 to 24–58% by 2050, and from less than 1 EJ in 2017 to 33, 99, or 216 EJ, respectively, in the three scenarios. Electric vehicles grow from less than 2% of world markets in 2017, to 55% in 2040 [3], 50% in 2037 [4], or 100% by 2025–2030 [5]. These trends will require

Table 1 Total energy supply and primary energy mix in 2050 according to each scenario

Scenario	BNEF[a]	DNV GL	Jacobson
Total energy, EJ	138 EJ	586 EJ	373 EJ
Solar (%)	24	17	58
Wind (%)	24	12	37
Hydro (%)	10	4	4
Geothermal etc. (%)	3	1	1
Biomass (%)	1	11	
Nuclear (%)	7	5	
Natural gas (%)	17	25	
Petroleum (%)	1	15	
Coal (%)	11	10	

[a]BNEF indicates energy use only for electricity production

dramatic changes to the industrial landscape and metal supply chains, as described in the next section.

Materials Requirements for Energy Scenarios

The rapid increase in solar PV, wind turbine, and electric vehicle production required to meet these scenarios requires significant increases in solar silicon, rare earth metal, and copper production. Vehicle lightweighting is more complex. The effect of these energy and transportation trends on materials use is discussed here.

Solar Silicon

A steady state of 33×10^{18} J/year of solar energy production (BNEF estimate for 2050), with a typical capacity factor of 26%, requires about 4 TW of installed solar capacity. With module lifetime of 25 years, this requires about 160 GW/year of solar electricity production at steady state. The DNV GL and Jacobson scenarios require proportionately more: about 480 and 1000 GW/year, respectively. That said, the near-term ramp envisioned by Jacobson will require 2–3 times that production rate.

As solar energy deployment has accelerated upward, silicon use per watt has been slowly trending downward, as shown in Fig. 4 [9]. Kerfless wafer production methods such as 1366 Technologies' Direct Wafer® or Crystal Solar's Direct Gas to Wafer™ can reduce silicon use to 1.5 or 0.5 g/watt, respectively. Though other technologies such as perovskites or other thin films may displace silicon, at this point, projected silicon PV price declines make it seem reasonable to focus on this technology, with long-term use of 0.5–1.5 g/W.

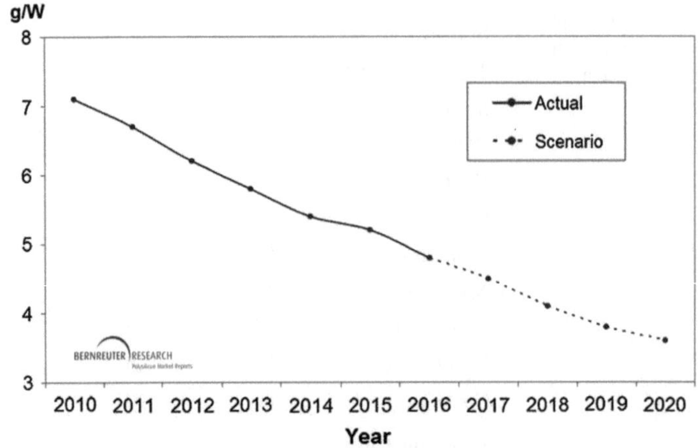

Fig. 4 Silicon PV material intensity with projection to 2020

Table 2 Required solar silicon production for various energy and silicon intensity scenarios, kT/year

Si intensity/Energy scenario	BNEF 33 EJ/year	DNV GL 99 EJ/year	Jacobson 216 EJ/year[a]
0.5 g Si/W	80	240	500–1500
1.5 g Si/W	240	720	1500–4500

[a]Jacobson low/high range reflects steady state versus approx. near-term ramp production requirement

Based on these assumptions, solar silicon production may or may not need to increase substantially from today's 450 kT/year as outlined in Table 2. One can therefore conclude that except in the Jacobson scenario, a solar silicon shortage is not likely in the coming decades.

Wind Energy

Wind already plays a very important role in the grids of many countries, from Denmark and the UK to Texas and the plains states, and continues to grow rapidly. Projected generation of 33, 70, and 138 EJ/year in the BNEF, DNV GL, and Jacobson scenarios, respectively, with 60% capacity factor in new large turbines, leads to 1.7, 3.7, and 7.2 TW of generation capacity. For 30-year generator lifetime, steady state requires replacement of roughly 60 GW/year, 120 GW/year and 240 GW/year, respectively.

Onshore wind has mostly used doubly fed inductive generator (DFIG) technology, while much larger offshore turbines nearly all use rare earth permanent magnet generators with Nd(Pr)FeB magnets. Based on the use of 0.5 T Nd(Pr) per MW, with about one-third of wind generation being offshore, this leads to 10–40 kT/year Nd and NdPr production. Current production is around 35 kT/year for all uses [10], and China is expected to be a net importer of rare earths by 2025, so additional production will need to come online in order to meet demand. This is compounded by motor vehicle use as described below.

Motor Vehicle Materials

An increase in total vehicle passenger-miles per year mainly in the developing world could counter the trend of ride sharing with higher intensity of use (passenger-miles per vehicle-year, e.g., "Peak Car" in the US [11]), likely leading in no major change in vehicle production rate. That said, electrification will require materials changes for batteries and motors, particularly lithium and cobalt for batteries and neodymium and praseodymium for motors. Higher intensity of use makes the cost of energy consumption more significant relative to depreciation, increasing the motivation for electrification—and also for lightweighting. Thus, aluminum, magnesium, and carbon fiber are likely to see considerable growth in this market.

As mentioned above, Olivetti et al. discussed the rising use of lithium and cobalt for batteries [6]. And production of 100 M vehicles/year will almost certainly need to use low- or zero-cobalt batteries, and possibly non-lithium batteries. That said, if roughly 1 kg Nd(Pr) per vehicle motor remains the norm, production of 100 kT/year—plus demand requirements for wind energy described above—will require large increases in NdPr metal production and overall rare earth mining, much more so if the ramp rate approaches that of Jacobson's scenario. In addition, copper use in electric vehicles is roughly 2–4 times as high as in internal combustion engines [12].

For vehicle lightweighting, the Ford-Magna multi-material lightweight vehicle [13] project presents two interesting potential cases for future lightweight vehicles. The Mach I vehicle uses 396 kg iron + steel, 368 kg Al, 16 kg magnesium and 57.6 kg composites for an 1195 kg Vehicle. Mach II uses at least 60 kg magnesium, particularly in closures (62% Mg alloys) and the chassis (70% Mg), in a 761 kg vehicle. 100 M vehicles/year using these body types could thus require up to 37 M T/year aluminum and 6 M T/year magnesium, both substantial increases from today's production rates.

Recycling in the New Energy Supply Chain

Primary production of metals, particularly iron and steel, produced 7% of the world's GHG emissions in 2014 [14]. Table 3 summarizes production rate, energy, and emissions intensities for steel and other metals considered here. The Solar Si energy

Table 3 Production rate and energy and emissions intensity of major metals in the energy economy

	Steel	Al	Solar Si	Mg	RE
Current world production (tons/year)	1.5B	50M	450k	900k	35k
Energy intensity (kWh/kg)	6.4	20	100–300	40–100	12
Emissions intensity (T CO_2e/T metal)	2.3	3–12	>3	7–25	256
Recycling/primary energy use	~50%	5%	Unknown	5%	100%

range is for Elkem metallurgical and Siemens processes [15]; Mg for electrolytic and Chinese Pidgeon production; rare earth emissions numbers are from Vogel and Friedrich "medium case scenario" [10] (currently most magnet recycling reoxidizes the rare earth metal, requiring reduction). Considering emissions, energy use, and cost, recycling is very compelling.

Olivetti and Cullen discuss five approaches, or levers, to reduce industrial energy consumption and emissions for various materials: lifetime extension, dematerialization, manufacturing efficiency, substitution, and recovery [16]. At this point, the technologies are too new, and changing too rapidly, to realistically discuss lifetime extension, though solar and wind plants aim for roughly the same 25–50 year lifetime as fossil fuel plants. Likewise for substitution: it is difficult to foresee major changes. Dematerialization can take multiple forms, such as reducing the number of vehicles, i.e, "Peak Car" mentioned above [11], or mass reduction, e.g, the Magna-Ford multi-material lightweight vehicle as described above [13].

That leaves manufacturing efficiency and recovery as the two primary means to reduce industrial energy consumption and emissions. For this proceedings, recovery/recycling is the primary consideration, and the energy savings described above present significant motivation, particularly for aluminum and magnesium. (For rare earths, the primary motivations are cost and supply risk.)

That said, the rapid growth and long lifetime of these energy generation and efficiency technologies (solar PV, wind turbines, electric vehicles) make recovery/recycling relatively insignificant in the short term, during rapid growth. For example, rare earth production has been growing at about 12%/year, so for products with 15-year lifetimes, that's roughly fivefold increase during the product lifetime, and a *maximum* recycling rate of 20% with perfect recovery. Solar and wind growth rates are higher, and lifetimes considerably longer, such that recycling could only start to significantly reduce energy and emissions >10 years after reaching market saturation. This could be as early as 2050–2060 for Jacobson's scenario, or as late as 2100 in the BNEF and DNV GL scenarios. Recycling thus has an important role to play in the long-term steady state, but cannot play a major role during ramp-up.

Conclusions

The above analysis considers materials requirements for reaching the solar, wind, and electric vehicle production volumes in three scenarios for energy system transformation published in 2018. Results are summarized as follows:

- Except in the most aggressive growth scenario with 1.5 g/W long-term material intensity, solar silicon capacity will likely not need to grow significantly in order to reach these targets.
- For wind energy and especially electric vehicles, barring a widespread change away from rare earth permanent magnet motors or generators, Nd and NdPr production will need to increase by about a factor of 4 to address these markets, over today's total.
- For vehicle lightweighting, aluminum and magnesium production will need to increase significantly: aluminum by about a factor of 2, and magnesium by up to a factor of 5.
- For all of these materials changes, due to rapid growth and long lifetime of vehicles and generation plants, in the short term (30–70 years), manufacturing efficiency will be more important than recycling for reducing material production energy consumption and emissions.

Note that in addition to solar and wind generation and electric vehicles, Jacobson requires long-range energy transmission by an HVDC grid as mentioned above. This could in turn require production of very significant amounts of copper and/or aluminum, with additional energy use and emissions not discussed here.

References

1. Bruckner T et al (2014) Energy systems (Chapter 7). In: Edenhofer O et al (eds) Climate change 2014: contribution of working group III to the fifth assessment report of the IPCC. Cambridge University Press
2. Bloomberg New Energy Finance (2018) New energy outlook 2018, Bloomberg, June 2018. https://about.bnef.com/new-energy-outlook/
3. DNV GL (2018) Energy transition outlook, 10 September 2018. https://eto.dnvgl.com/2018/download
4. Jacobson MZ et al (2017) Joule 1:108. https://doi.org/10.1016/j.joule.2017.07.005
5. Clack CTM et al (2018) Proc Natl Acad Sci USA 114(26):6722. https://doi.org/10.1073/pnas.1610381114
6. Olivetti EA, Ceder G, Gaustad GG, Fu X (2017) Joule 1:229. https://doi.org/10.1016/j.joule.2017.08.019
7. Spector J (2018) Mark Jacobson drops Lawsuit against critics of his 100% renewables plan, 26 February 2018. Greentech Media
8. Myles Allen et al (2018) Global warming of 1.5 °C. IPCC special report 15, 8 October 2018. http://www.ipcc.ch/report/sr15/
9. Bernreuter Research (2016) The Polysilicon market outlook 2020, 24 November 2016. http://www.bernreuter.com/en/shop/polysilicon-market-reports/market-outlook/report-details.html

10. Vogel H, Friedrich B (2018) In light metals. In: O. Martin (ed) TMS. https://doi.org/10.1007/978-3-319-72284-9_197
11. Nicola, Bermann (2018) Peak car and the end of an industry. BNEF
12. Hobson P (2018) Electric cars no game changer for copper in short term. Reuters
13. Skszek T, Conklin J, Wagner D, Zaluzec M (2015) Multi-material lightweight vehicles. Presented at US Department of Energy Vehicle Technologies annual merit review, 11 June 2015. https://www.energy.gov/node/1087038
14. Davis et al (2018) Science 360:1419. https://doi.org/10.1126/science.aas9793
15. Gløckner R, de Wild-Scholten M (2012) Energy payback time and carbon footprint of Elkem solar silicon. In: Proceedings of the European PV conference and exhibition, pp 4661–4666. https://doi.org/10.4229/27theupvsec2012-6cv.4.12
16. Olivetti EA, Cullen JM (2018) Science 360:1396. https://doi.org/10.1126/science.aat6821

The Role of Manufacturing Variability on Environmental Impact

Alexander van Grootel, Jiyoun Chang and Elsa Olivetti

Abstract Additive manufacturing (AM) especially metal additive manufacturing (MAM) is expected to disrupt many industries. Besides being very flexible and allowing bespoke parts with little to no setup time, AM technology is able to fabricate parts with geometries which were previously impossible to create. This allows for dramatically better designs by making the product lighter or more efficient. However, despite these numerous and significant benefits, the uptake of functional additive manufactured parts is slow. A major barrier to expedited uptake of this technology is process control. It is not certain what the most important process parameters or the ideal process windows are and how this changes for different process/material combinations. As of yet, there is not a set process to certify an AM part or process. This makes quality assurance prohibitively longwinded and expensive. Furthermore, to ensure safety under such uncertain conditions, a high safety factor and therefore thicker parts must be used. As a result, uncertainty is also tied to increased material consumption and therefore higher environmental impact. We need to better understand the nature of variability in AM in order to alleviate some of these problems. This manuscript presents several examples of the influence of variability in manufacturing and its potential impact on environmental performance.

Keywords Environmental impact · Manufacturing · Variability

Introduction

Manufacturing plays a critical role in our economy and has a significant impact on the environment. About a third of global energy production is dedicated to manufacturing

---

A. van Grootel
Technology and Policy Program,
Massachusetts Institute of Technology, Cambridge, MA, USA

J. Chang · E. Olivetti (✉)
Department of Materials Science & Engineering, Massachusetts Institute of Technology, Cambridge, MA, USA
e-mail: elsao@mit.edu

G. Gaustad et al. (eds.), *REWAS 2019*, The Minerals, Metals & Materials Series,
https://doi.org/10.1007/978-3-030-10386-6_3

materials into products [1]. Researchers have argued that in order to transition to a more environmentally sustainable economy our reliance on material consumption must fall [2]. One effort to reduce environmental impact from material consumption is dematerialization; reducing the amount of material used while maintaining the same functionality [3].

In order to make effective decisions regarding dematerialization, it is necessary to understand how manufacturers operate and consume materials in production processes. One important reality that manufacturers deal with is variability in their processes and raw materials. This variability inevitably results in variance in the manufactured products, which can be of considerable concern to the manufacturers. Business tools and practices have arisen to help manage variability in processes and products across factories and entire enterprises. Total Quality Management [4, 5] and Six Sigma [6–8] aim at developing strategies or methods to successfully control the quality variation of products and achieve profitable manufacturing practices. Process control, the field that is dedicated to managing and reducing this variability, has a lot of depth and is in close collaboration with industry. Indeed, the concept of manufacturing variability appears to be very well established and completely ingrained in the world of manufacturing.

While the relation between manufacturing and environmental impact is indisputable, and while the importance of managing variability in manufacturing is well accepted, no study has investigated the relation between manufacturing variability and environmental impact. In this paper, we make a contribution to this area by investigating the relationship between manufacturing variability and the environmental impact from material consumption. We aim to demonstrate to the industrial ecology community that instead of viewing mechanical properties and part dimensions as deterministic values, we should instead be viewing them through the framework of distributions and risk levels.

It is worth pointing out that variability has been studied in various different contexts by the sustainability community. For example, the impact of variability in life cycle analysis has been studied [9], and the role of variance in material composition in recycling has been analyzed as well [10]. In this paper, we specifically focus on the type of variability that makes two products of the same model differ in their geometries and mechanical properties. While two products may seem the same, differences will exist and these differences can be significant enough to warrant either the company or a certifying body to put practices in place to deal with the discrepancies. These practices ultimately have environmental implications that we investigate in this contribution.

The rest of the paper is organized as follows: first, we briefly discuss the concept and sources of manufacturing variability and how variability can be linked to increased material consumption. Second, we draw from several examples in industry to illustrate how the issue of manufacturing variability can manifest itself in different scenarios, and we tie this to material use and environmental impact. Third, we discuss how manufacturing variability might influence new technologies, suggesting that processes should be seen as stochastic instead of deterministic. And finally, we close by highlighting some points of discussion and potential future avenues of research.

Manufacturing Variability Is Linked to Environmental Impact

The link between manufacturing variability and environmental impact is through overdesign. In many situations, a dominant strategy to deal with variance in a process is to overshoot the design constraints. As mechanical properties will always be stochastic to some degree, this overshoot helps to ensure that parts will fall out of spec only a small fraction of the time. However, overdesign often requires more material to be used, and this extra material has environmental implications. In this section, we will step through each of these connections, first discussing the source of manufacturing variability, then drawing connections to overdesign, material use, and eventually environmental impact.

Manufacturing Variability and Overdesign as a Strategy

No two fabricated parts are the same in manufacturing. Differences in the final properties of a manufactured part arise due to variations in the environment, materials, or the machine itself [11]. For example, the temperature or humidity in a factory changes over the course of a day, which impacts the behavior of an injection molding machine. The flow paths of the molten plastic will be slightly different each time plastic is injected into a cavity, which has some effect on the internal stresses, properties, and geometries. The materials used for the injection molding are not completely homogenous because the feedstock itself has had its own set of variations in the environment, materials, or machines. In this way, variations may propagate through the supply chain. The distribution of the properties of interest will depend on the process, the part being produced, the processing conditions, and the standard operating procedures used by the machine operators. Because of these differences, one must consider properties as distributions rather than deterministic values.

One of the tools to manage manufacturing variance is overdesign: the practice of overshooting a design constraint on average in order to ensure only a small proportion of fabricated parts fall outside of the design specification. An example is provided in Fig. 1a. Instead of being deterministic, the properties of a part are portrayed using distribution as shown by the blue line. If this distribution is centered exactly on the design constraint, a large portion of parts would fall out of spec, as is shown by the blue area in Fig. 1a. Instead, the entire curve is shifted as we show in Fig. 1b. As a result, only a small portion of all the parts will fall below the acceptable threshold as shown in Fig. 1c. The amount by which the curve is shifted over is what we consider

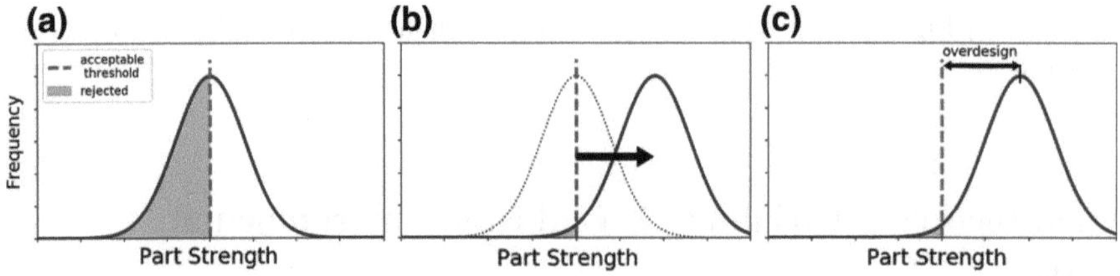

Fig. 1 Schematic of overdesign. **a** No overdesign, causing a large proportion of parts to fall out of spec, **b** the curve is shifted right to create overdesign, **c** the overdesign leads to less parts being out of spec

to be overdesign. It is worth pointing out here that adding this buffer only makes sense when there is effectively only a single constraint on the performance variable. For a property like part, strength designers may not care that the part is significantly stronger than promised. However, there are some instances where a property has both an upper and a lower bound to it. An example may be the dimension of a hole; the diameter of the hole can neither be too big nor too small and therefore the idea of overdesigning does not make sense. Also, note that the scenario could be inverted; there may be an upper bound on an attribute like the amount of contaminants.

Overdesign is just one of the strategies that can be deployed to manage manufacturing variability. Another strategy is to do 100% inspection together with rework and scrapping. However, inspection is not always easy. Reliably establishing the ultimate tensile strength of a part requires the part to be destroyed. It might be possible to test the strength of a part up to its design constraint, and to approve it if it is able to withstand these forces, but this will likely deform and damage the part. Furthermore, there are significant costs associated with the potential rework or scrapping of out-of-spec parts [12]. Of course, in some cases overdesign can be used in conjunction with inspection, rework, and scrapping. In most cases, the strategy of overdesign appears to be the most effective way to manage manufacturing variability.

The buffer for overdesign is typically decided by weighing the risk level that decision makers are willing to tolerate against the cost of mitigating this risk. The policy can either be set internally by a company, or externally, through standards and certification frameworks. Overdesign in certification frameworks can take on various forms. In some cases, a safety factor is added (e.g., [13]), which follows the concept of overdesign directly and essentially sets a higher target for mechanical requirements to account for variation in mechanical properties. For example, one might say that instead of a steel bar needing to withstand 10 kN, it must withstand 13 kN (safety factor of 1.3) to account for variability in the strength. Assuming an average tensile strength of 400 MPa, this turns a bar with 25 mm^2 of cross-sectional area into a bar of 32.5 mm^2. In other cases, the preferred framework is that of design allowables [14]. Here, the mechanical properties that are allowed to be assumed during design depend on the distribution of these properties. A material that has very high spread in its properties will have a lower allowable strength than a material that has a tighter distribution but the same average strength. For example, instead of assuming a tensile strength of 400 MPa for a steel bar, only a tensile strength of 300 MPa can be assumed. For our bar that has to withstand 10 kN, this would mean that instead of 25 mm^2 of cross-sectional area, 32.5 mm^2 of area would be required. Ultimately, these different approaches frame the same concept in different terms.

Why Does Overdesign Cause Increased Material Consumption?

Increased amounts of materials used in a part mean more embodied energy in that part. For many products, a substantial proportion of the lifetime energy can be accredited to the embodied energy in the materials [15, 16]. In addition, the increased weight of a final product due to more materials can also influence the environment impact of the use phase of that product. For example, fuel consumption of a vehicle heavily depends on its weight.

For any design constraint that varies with the amount of material used, it is possible to overdesign by adding more of that material. This includes properties such as the stiffness, strength, and thermal capacity of a part among many others. In reality, the effect of overdesign will be subtler and more complex than simply scaling the thickness of all parts at the end of a project's design. Variability will be taken into consideration from the onset of the project [14]. Significant testing will be conducted early on during the design process to account for variability and to establish the strengths that can be assumed for the design. Or to put it in the language of a different framework, safety factors will be applied from the beginning of the project and parts will be designed to meet the requirements of a more conservative design specifications. Ultimately, the results will be similar, and variability will lead to more material being used in the product.

The link between manufacturing variability and environmental impact can therefore be established under a few conditions. A simple schematic is shown in Fig. 2. If manufacturing variability is sufficiently high then potentially the idea of safety factors is adopted, and overdesign is framed as the need to shift the design specifications by a specified factor. Alternatively, the idea of design allowables is adopted and the effective strength that can be assumed by designers is adjusted based on the variability of the mechanical properties. In both cases, the result is overdesign which implies that more material will likely be used thereby leading to higher environmental impact.

Variability Plays an Important Role in Many Different Industries

Manufacturing variability plays an important role in many industries. In this section, we discuss some examples and explore the opportunities that reductions in variability could present by looking at typical levels of manufacturing variability and considering some standard practices in these industries.

Concrete

Concrete has a very high CO_2 emissions footprint [17], is a critical construction material, and the variance in the strength of concrete is notably high [18]. We will discuss variability in terms of the coefficient of variation (CV), which is defined as the standard deviation divided by the average of a distribution. A CV of 0.14 is rated to be "fair" by industry standard [19]. In the compliance requirements of Section 26.12.3.1 in the standard developed by the American Concrete Institute (ACI) [20], the ACI requires the average of three consecutive tests to exceed the specified strength, with an allowable failure rate of one in a hundred tests. Which translates into $Safety\ Factor = zscore \times \frac{CV}{\sqrt{n}} = 2.33 \times \frac{0.14}{\sqrt{3}} = 18.8\%$. In other words, under these assumptions, almost 19% extra concrete is required in order to overcome issues with manufacturing variability.

In 2016, the total CO_2 attributed to concrete was around 1.45 Gt CO_2 [17]. Assuming that all this concrete has a CV of 0.14 and that for all this concrete, the above compliance requirements were followed, we can extrapolate our calculations to suggest that around 0.27 Gt of CO_2 is being put into the atmosphere every year to make up for the high variability in concrete. In reality, not all of the world's concrete will follow this standard, and not all concrete has a coefficient of variation of 14%. However, this figure presents itself as a reasonable estimate of the total opportunity that can be captured by reductions in variability for concrete.

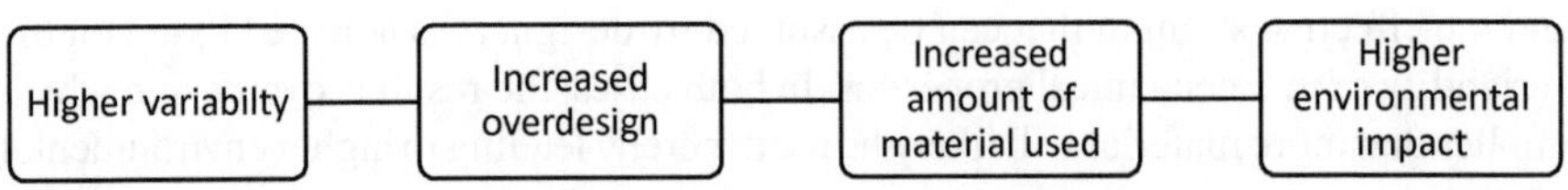

Fig. 2 A schematic of the connections between manufacturing variability and environmental impact. If the variability of a process is high, then there is a need for overdesign to account for this additional variability. More material is needed to satisfy these additional design requirements, which leads to more environmental impact

Integrated Circuits

Integrated Circuits (IC) is an industry familiar with variability in manufacturing processes. There is clear motivation for this; most of the innovation in this industry is driven by the desire to create increasingly smaller transistors, which requires increasingly more accurate and precise manufacturing operations. The trend has been powerful enough to shrink feature sizes from tens of micrometers in the 1970s, to tens of nanometers in today's state of the art [21]. With billions of transistors per chip, each individual transistor must be very reliable otherwise too many chips may have too many defected transistors. To ensure extremely high reliability for each transistor, each feature is made larger (overdesigned) relative to what current technologies may be able to produce on average.

IC manufacturers have to determine tolerable failure rates. This becomes a trade-off between the size of each transistor, and therefore speed and power draw, and the reliability of each transistor, which manifests as the yield of the chips. Interestingly, the yield of chips has been falling as the transistors have shrunk. Transitioning from 350 to 90 nm in feature size, the yield went from around 90–50% [22]. With current technologies around 14 nm, yield likely has dropped even further although the exact numbers remain difficult to obtain. These statistics give an interesting insight to the IC industry. The industry is willing to accept high levels of rejects just for the chance to build very small transistors, and this risk tolerance seems to have shifted over time. It also highlights a further dimension from a sustainability perspective. In this industry, there is a trade-off between yield and the amount of material required per part. With 350 nm technology Intel produced 31 million transistors on a chip with a die size of 242 mm^2 [23] while with 90 nm technology Intel produced 125 million transistors on 112 mm^2 [24]. This is an 8.7-fold decrease in area per transistor. Assuming that the kind of material used is comparable and the thickness remains the same, even if yield decreased from 90 to 50% materials will still be saved. However, there might be a point where yield may be so low that it no longer leads to materials savings.

Nevertheless, process control in the IC industry should be seen as a success story. The field pays close attention to process capabilities and has very sophisticated statistical tools and design strategies to continue progress [25]. This relentless push towards smaller transistors is a form of dematerialization that has been enabled by advances in process control.

Fiber Composites

Fiber composite manufacture provides an interesting example because the composites industry is starting to see many applications where lightweighting is of importance, such as in the transportation industry. However, fiber composites have notoriously high variation in their mechanical properties [26]. Many regulatory frameworks revolve around establishing the effective strength of a part [27], with a reliability of

99% or 95% depending on how critical the component is. Other regulatory frameworks often require an additional safety factor (which is set depending on the variability) when operating with fiber composites (e.g. [13]). We explore how the variability of fiber composites affects the aerospace industry in a separate contribution [28], and find that overdesign is frequently around 30% and can lead to an additional 100 kton of CO_2 output over the life of a Boeing 787 mostly due to the additional fuel requirements associated with the added weight.

In this example, manufacturing variability does not just relate to material consumption but also to the use phase. With fiber composites gaining increasing traction in automotive [29], it is important to remember how manufacturing variability might impact the ultimate lightweighting potential of fiber composites in applications beyond aerospace.

Promises made by early composite advocates included saving 50% of a structure's weight compared to using an aluminum frame [30]. It is difficult to determine how much weight savings occurred in reality, but estimates place it closer to 20% of the structure's total weight [31, 32]. This remains true despite three decades of developments in the composites field. Part of the shortcomings may be explained by the difficulties in manufacturing composites. The composites production process proved to be more problematic to control than anticipated [30] and these challenges were not considered in initial optimistic predictions. As a result, additional safety factors must be applied when using composites [13]. These safety factors vary depending on the specific process, materials, and applications, but discussion with industry experts places this value around 1.3. The additional requirements of safety factors were not foreseen by early composite advocates, but it severely undermines the weight-saving potential of this technology.

The lesson to draw from these developments is that the predicted weight savings of 50% were an unrealistic extrapolation that did not account for realities on the production floor. Perhaps when isolated from other factors and when ignoring variability in properties, it is possible to achieve 50% weight savings. However, once the technology reached the factory floor the results were more sobering. We should remember this lesson when we consider upcoming processes and materials.

Every industry has to deal with manufacturing variability every day. This variability has implications for the environmental impact of various products and materials as we have introduced here. Different industries have different approaches and standards when managing manufacturing variability, but the underlying concepts of viewing properties and geometries as distributions rather than deterministic values can be seen throughout all these examples.

Effective Performance Is More Important than Average Performance

As illustrated in the previous section, in industry, it is not just about a deterministic measure of a property or dimension but rather about its distribution and the risk level companies are willing and able to tolerate. From industry's perspective, reproducibility is vital and the framing falls around effective performance rather than average performance. During research this is not always the norm. Instead, researchers gravitate towards the average performance of a few select samples or even single samples. It is worth discussing how and at what point during the journey from lab to production floor we should transition from talking about averages and point values to talking about distributions.

Differences Between Research Labs and Production Floors

The different focus between research labs and production floors is understandable. For novel processes and materials, it is unreasonable to demand the volume of samples necessary to make accurate estimates of variability, especially when the researchers are still exploring the process space. However, we can still be skeptical of statements that speak about improvements in properties if the processability of this new technology has not been addressed. Before we extrapolate from the average of a select few samples, we should at the very least consider whether process control may be difficult.

Manufacturing variability is often not reported at early stages of development, but even small changes in variability can be impactful. Depending on the distribution of properties (process and material dependent) and on the reliability (derived from external or internal policies), a safety factor is determined. This safety factor can grow rapidly as shown in Fig. 3. If a failure rate of 1% is accepted, then a coefficient of variance of around 0.1 already requires a 20% overdesign. If a new material were to claim a 25% higher strength to weight ratio but made the coefficient of variance increase from 0.01 to 0.10, then the majority of the benefits might be nullified due to the need for a safety factor.

Upcoming Technologies—Additive Manufacturing

There are some technologies currently in development which show high promise for effective lightweighting or to reduce environmental impact. However, the processability of these technologies is not always given due consideration. One such technology is Additive Manufacturing (AM).

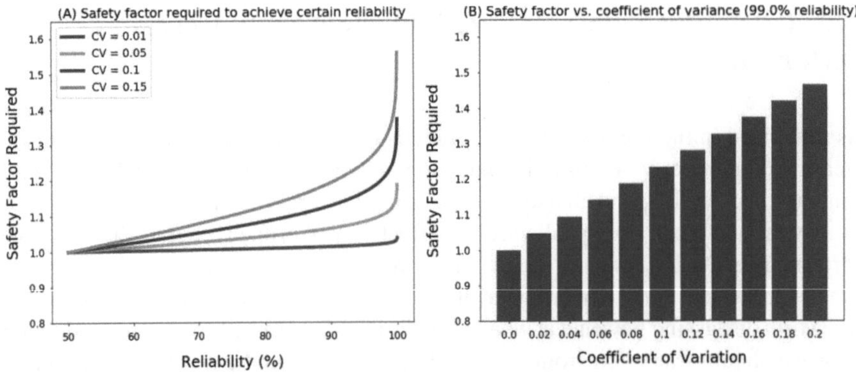

Fig. 3 **a** Relations between safety factor required to meet a certain reliability for various coefficient of variation scenarios. **b** Safety factor required to achieve 99% reliability at various coefficients of variation. Both plots assume a normal distribution

AM, the practice of building a part layer by layer, is being hailed as a disruptive technology that can fundamentally change the way we approach fabrication. One of the main benefits of AM is its ability to fabricate nearly any kind of geometry, which opens up the potential for substantial weight savings. However, one key obstacle that impedes the commercialization of AM is process control and quality assurance [33, 34]. AM is a highly sensitive process, and we do not yet understand exactly how the process works [35]. Due to the difficult nature of this new process will we see special safety factors for AM parts just like what was observed for fiber composites? How will this impact the lightweighting and dematerialization potential of AM? Are there ways to improve processability and thereby increase the dematerialization potential? So far these questions remain unanswered. Nevertheless, studies are coming out which claim impressive dematerialization ratios on the order of 50% for some parts (e.g., [36]) but gloss over issues surrounding manufacturing variability. More prudence may be required lest we repeat the overestimates we observed with fiber composites.

Discussion and Future Work

We have established a link between manufacturing variability and environmental impact through the concept of overdesign. Many questions remain, and in this section, we highlight a few future research areas that may provide answers to these questions.

Estimating Variability of a New Process

If variability is worth considering during the development of new processes, a natural question is how and when this variability should be estimated. Estimating the variance of a distribution is difficult, and requires a large number of samples to do reliably. Additionally, it is not well known whether a lab-scale process is a reasonable proxy for its ultimate mass production version. Both of these are strong arguments to delay estimating variance until later in the process' development. However, delaying too long is similarly unfavorable. A process is most flexible early in its development, which implies that if it is possible to make a process more consistent and less variable, it would be easiest to do so early in the development of the process. Furthermore, if the goal of developing a technology is to save material or to make a part lighter, it is important to know how we should compare different options.

Expecting hundreds of lab-built samples to establish the variance of a distribution is unreasonable. Instead of creating many samples and building a distribution empirically, it may be possible to get a better understanding of what the variability may be by doing sensitivity analyses and investigating the processability of a new technology. At the very least this may help identify areas that need improvement or ways in which the variability can be controlled, and may also help in setting more reasonable expectations of how the new technology may ultimate preform once it has been adjusted to account for variability.

Reducing Variability

If it were possible to reduce variability, then many opportunities to save materials would open up. For example, we estimate that the around 19% of concrete is used to counteract the effects of variance.

There are tools and methods that can be used to reduce variability. So far these methods mainly focus on business practices such as Six Sigma, and the results have been strong [6, 8]. There is also increasing interest in process-level solutions. For example, in situ process monitoring in AM [35]. Here, the process parameters are adjusted in process based off of readings taken during the process, and might reasonably result in reduced variance between parts. Furthermore, the trend of increasingly more precise fabrication techniques, like we see in integrated circuits, shows that there are ways to reduce variability at least in some circumstances.

While studies on variability for various materials and processes exists, there are few that attempt to understand and highlight the root causes of variability on a process level. Having a better understanding of underlying sources of variability, and potential ways to alleviate them may prove an effective way to reduce our material consumption.

Risk Neutrality

Our framing implies that one strategy to reduce the size of a safety factor is to accept a higher risk of failure. For example, one might argue that instead of requiring a reliability of 99%, only 90% is required. With constant variance, this would lower the need for overdesign thereby saving material. Setting risk tolerances is a complicated task that is highly context dependent and we are not advocating that this is the correct strategy. However, in some scenarios, it may be insightful to get a better understanding of how the level of reliability is decided upon.

In the case of integrated circuits, it appears that manufacturers are accepting increasingly higher odds that there will be a chip-disabling defect and that the chip will need to be discarded as this is the only way in which very small chips can be made at all. The scenario is very different when talking about setting risk tolerances for a concrete structure. Here, the result of failure might be catastrophic and should not be taken lightly.

Conclusions

In this work, we connect manufacturing variability to environmental impact and argue that instead of leaving manufacturing variability as a concern at the very end of a process's development, it should be considered and investigated earlier. At the very least, researchers should be aware that deterministic values will never be realized, and that a more realistic representation is that of a distribution and an associated risk level that decision-makers are willing to tolerate. As a result, effective performance of a part, which is adjusted to account for the distribution's variance and the risk level, is more important than average performance.

These points add another consideration for potential light-weight materials and dematerialization strategies; average performance is not sufficient and the manufacturing variability should also be considered. This work suggests opportunities for dematerialization that the industrial ecology may not have recognized before; significant material savings may be achieved by reducing variability in certain material systems.

Acknowledgements This publication was made possible with the support of the Government of Portugal through the Portuguese Foundation for International Cooperation in Science, Technology, and Higher Education, and was undertaken in the MIT Portugal Program.

References

1. EIA (2016) International energy outlook. In: International Energy Outlook, pp 1–2
2. UNEP (2011) Decoupling natural resource use and environmental impacts from economic growth
3. Cleveland CJ, Ruth M (1999) Indicators of dematerialization and the materials intensity of use. J Ind Ecol 2(3):15–50
4. Ng PK, Goh GGG, Eze UC (2010) The influence of total quality management, concurrent engineering and knowledge management in a semiconductor manufacturing firm. In: 2010 IEEE international conference on industrial engineering and engineering management (IEEM), pp 240–244
5. Kaynak H (2003) The relationship between total quality management practices and their effects on firm performance. J Oper Manag 21(4):405–435
6. Gupta V, Jain R, Meena ML, Dangayach GS (2018) Six-sigma application in tire-manufacturing company: a case study. J Ind Eng Int 14(3):511–520
7. Prasad AG, Saravanan S, Gijo EV, Dasari SM, Tatachar R, Suratkar P (2013) Six sigma-based approach to optimise the diffusion process of crystalline silicon solar cell manufacturing. Int J Sustain Energy 35(2):190–204
8. Gijo EV, Scaria J (2014) Process improvement through six sigma with beta correction: a case study of manufacturing company. Int J Adv Manuf Technol 71(1–4):717–730
9. Gustavsson L, Sathre R (2006) Variability in energy and carbon dioxide balances of wood and concrete building materials. Build Environ 41(7):940–951
10. Noshadravan G, Gaustad A, Kirchain R, Olivetti E (2017) Operational strategies for increasing secondary materials in metals production under uncertainty. J Sustain Metall 3(2):350–361
11. Hardt DE (1993) Modeling and control of manufacturing processes: getting more involved. ASME J Dyn Syst Meas Control 115(2B):291–300
12. Tang K (1988) Economic design of product specifications for a complete inspection plan. Int J Prod Res 26(2):203–217
13. ASTM (2015) ASTM F3114-15. Astm i:1–5
14. DoD (1997) Department of defense handbook composite materials handbook. In: Polymer matrix composites guidelines for characterization of structural materials, vol 1
15. Sartori I, Hestnes AG (2007) Energy use in the life cycle of conventional and low-energy buildings: a review article. Energy Build 39(3):249–257
16. Del Pero F, Delogu M, Pierini M, Bonaffini D (2015) Life cycle assessment of a heavy metro train. J Clean Prod 87(1):787–799
17. Andrew RM (2018) Global CO_2 emissions from cement production. Earth Syst Sci Data 1–52
18. Obla K (2010) Sources of concrete strength variation—Part II of concrete quality series. In: Tech talk concrete in focus, pp 21–23
19. Cook JE, Parnes J, Akers DJ, Barringer WL, Brown JL, Graf A (2011) Evaluation of strength test results of concrete. Test 1–20
20. ACI (2014) Building code requirements for structural concrete (ACI 318-14), vol 11
21. Kingon AI, Maria JP, Streiffer SK (2000) Alternative dielectrics to silicon dioxide for memory and logic devices. Nature 406(6799):1032–1038
22. Ozdemir S, Sinha D, Memik G, Adams J, Zhou H (2006) Yield-aware cache architectures. In: Proceedings of annual international symposium on microarchitecture, MICRO. pp 15–25
23. Slater M (1995) Intel boosts pentium pro to 200 MHz. Microprocess Rep 9(17)
24. Boggs D et al (2004) The microarchitecture of the Intel® Pentium® 4 processor on 90 nm technology. Intel Technol J 08(1–18):119–130
25. Kuhn K et al (2008) Managing process variation in Intel's 45 nm CMOS technology. Intel J Technol 12(45):77–85
26. Mesogitis TS, Skordos AA, Long AC (2014) Uncertainty in the manufacturing of fibrous thermosetting composites: a review. Compos Part A Appl Sci Manuf 57:67–75
27. US Department of Defense (2002) Composite materials handbook. In: Polymer matrix composites materials usage, design, and analysis, vol 3

28. van Grootel A, Chang J, Olivetti E. Economic and environmental cost of variability in manufacturing: the case of carbon fiber reinforced polymer composite in the aerospace industry (in progress)
29. Friedrich K, Almajid AA (2013) Manufacturing aspects of advanced polymer composites for automotive applications. Appl Compos Mater 20(2):107–128
30. Vosteen LF, Hadcock RN (1994) Composite chronicles: a study of the lessons learned in the development, production, and service of composite structures
31. Hale J (2008) Boeing 787 from the ground up 06. Boeing
32. Quilter A (2004) Composites in aerospace applications. Inf Handl Serv Inc 1–5
33. Bonnín Roca J, Vaishnav P, Fuchs ERH, Morgan MG (2016) Policy needed for additive manufacturing. Nat Mater 15(8):815–818
34. FAA (2016) Summary report: joint federal aviation administration—air force workshop on qualification/certification of additively manufactured parts, New Jersey
35. Everton SK, Hirsch M, Stravroulakis P, Leach RK, Clare AT (2016) Review of in-situ process monitoring and in-situ metrology for metal additive manufacturing. Mater Des 95:431–445
36. Huang R et al (2016) Energy and emissions saving potential of additive manufacturing: the case of lightweight aircraft components. J Clean Prod 135:1559–1570

Manufacturing Materials Optimization Research at The REMADE Institute

Pradeep Rohatgi, Alan A. Luo and Magdi Azer

Abstract Manufacturing accounts for about 25% of the energy consumption in the United States. To help reduce the energy consumption and emissions, the US Department of Energy (DOE) has supported "The REMADE Institute", a public–private partnership launched in 2017. The objective of REMADE is to increase manufacturing energy efficiency and reduce embodied energy in materials (metals, polymers, electronic waste, and fibers). This talk provides an overview of REMADE's five nodes and the objective of its Manufacturing Materials Optimization (MMO) node.

Keywords Energy efficiency · Manufacturing efficiency · Materials optimization · Manufacturing optimization

Manufacturing accounts for about 25% of the energy consumption in the United States. To help reduce the energy consumption and emissions, the US Department of Energy (DOE) has supported "The REMADE Institute", a $140 million public–private partnership launched in 2017. The objective of REMADE is to increase manufacturing energy efficiency and reduce embodied energy in materials (metals, polymers, electronic waste, and fibers), with the following missions:

- Enable the early stage applied research and development of key industrial platform technologies that could dramatically reduce the embodied energy and carbon emissions associated with industrial-scale materials production and processing.
- Eliminate and/or mitigate technical and economic barriers that prevent greater material recycling, recovery, remanufacturing, and reuse.

P. Rohatgi · A. A. Luo (✉) · M. Azer
The REMADE Institute, Rochester, NY, USA
e-mail: luo.445@osu.edu

P. Rohatgi
University of Wisconsin at Milwaukee, Milwaukee, WI, USA

A. A. Luo
The Ohio State University, Columbus, OH, USA

M. Azer
University of Illinois at Urbana-Champaign, Champaign, IL, USA

G. Gaustad et al. (eds.), *REWAS 2019*, The Minerals, Metals & Materials Series,
https://doi.org/10.1007/978-3-030-10386-6_4

This talk provides an overview of REMADE's five nodes including:

- Systems Analysis and Integration;
- Design for Recovery, Reuse, Remanufacturing, and Recycling (Re-X);
- Manufacturing Materials Optimization;
- Remanufacturing and End-of-life Reuse; and
- Recycling and Recovery.

The objective of the Manufacturing Materials Optimization (MMO) node is to develop technologies to reduce in-process losses, reuse scrap materials, and utilize secondary feedstocks in manufacturing. Specifically, MMO research will optimize the use of materials in manufacturing to control materials properties, improve process efficiency, and increase the use of more cost-effective alternative feedstocks. MMO will also develop and implement tools, technologies, and methods to facilitate precise control of materials manufacturing in order to (1) process secondary feedstocks at cost and energy parity with primary feedstocks; (2) improve manufacturing yields for primary feedstocks; and (3) decrease the embodied energy.

There are several opportunities for reducing embodied energy in materials. These include using lower embodied energy materials and secondary feedstocks, as well as increasing the recycling rates for these materials. Opportunities in reducing embodied energy exist in using functionally gradient materials, manufacture of net-shaped parts, reduction of defects, and increasing yields during manufacturing. Currently, casting yields are in the range of 30–60% and these need to be increased through use of improved thermal physical property data and improved stimulation techniques. If yield in steel castings can be improved from 50 to 60%, there would be significant savings in energy.

In steels, the in-process losses due to melt oxidation need to be reduced. It is necessary to develop in-process sensors for in situ analysis of melts and to take corrective actions in real time available to reduce defective products. Development and manufacture of ultrahigh performance materials requiring lower factors of safety will reduce the weight of material needed, hence reduce the embodied energy. Recycling of steel is very extensive and it can be further increased by increasing the tolerance of tramp elements like copper by reducing their detrimental effects. In the United States, the total steel production is 120 MT whereas the shipment is only 100 MT representing 16.6% or 20 MT yield loss. A reduction in yield loss by 10% will save 2 MT of steel or savings in corresponding energy and emissions. In steel foundry, 2 million tons of molten metal is used and produce 1 million tons of finished castings. The melt loss is 10% or 0.2 million tons. If the melt loss is reduced to 6%, 4 million tons of steel will not need to be produced implying saving of significant amount of energy and emissions. More than 85 million tons of scrap is consumed annually with a recycling rate of 92% in the US. However, there is a need for logistical and technological improvement in steel recycling, recovery, and processing of scrap, including improvement in contaminant removal and recovery.

The aluminum industry is facing a similar situation in the use of secondary alloys. Current aluminum casting industry uses limited amount of secondary alloys. This is because that primary aluminum is generally viewed as higher quality, containing

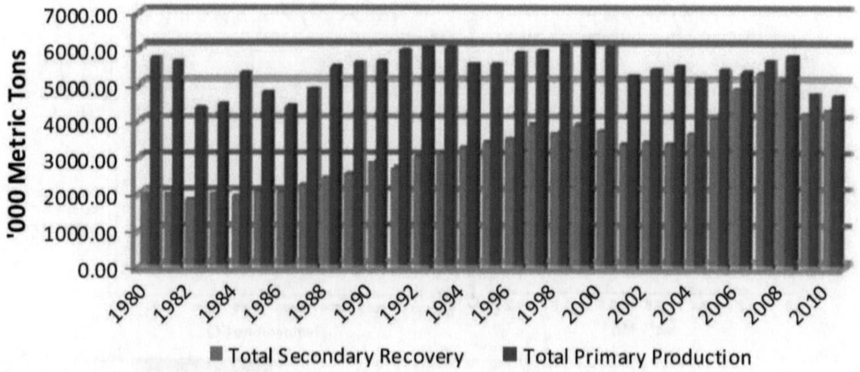

Fig. 1 Primary aluminum production and secondary aluminum recovery from new and old scrap in North America, including metal recovered from net exported scrap [2]

fewer entrained oxides and having better-controlled chemical composition, than secondary aluminum [1]. Despite the recent increase of secondary aluminum production in North America, the current use of secondary aluminum is still lower than primary production, as shown in Fig. 1 [2].

One example project selected for negotiation by REMADE is "Increasing melt efficiency and secondary alloy usage in aluminum die casting" being carried by The Ohio State University (OSU), Alcoa USA Corporation (Alcoa) and North American Die Casting Association (NADCA). In the die casting industry, molten aluminum processing is crucial to energy efficiency and casting quality (chemistry and cleanliness). There are many materials involved in aluminum melt processing, including primary and secondary alloys, refractory, and flux materials. The fundamental thermodynamic and kinetic reactions among these materials in melt processing determine the energy efficiency and throughput (melt recovery) of die casting operations. Such reactions are especially complex with secondary alloys which bring in more impurities and inclusions in the process. Currently, most of the process control and optimization in molten aluminum processing is conducted via traditional "trial-and-error" and "design of experiment" methods, with limited use of analytical tools available to the industry. Therefore, there is significant room for improvement in terms of energy efficiency and melt recovery in aluminum die casting. The project goals are: (1) develop fundamental thermodynamic and kinetic models for aluminum melt processing involving secondary alloys, refractory and flux materials; (2) design improved secondary alloys more tolerant of impurities (such as Fe and Zn); (3) develop new/improved refractory and flux materials for higher melt recovery rates; and (4) provide new/improved melt treatment techniques (such as fluxing and filtration) for better control of alloy chemistry and cleanliness.

The technical approach is to use fundamental thermodynamic and kinetic modeling [3] to guide the composition control of secondary alloys and optimize melting and melt treatments of such alloys for significantly improved melt efficiency and increased use of secondary alloys in aluminum die casting. This approach is a holis-

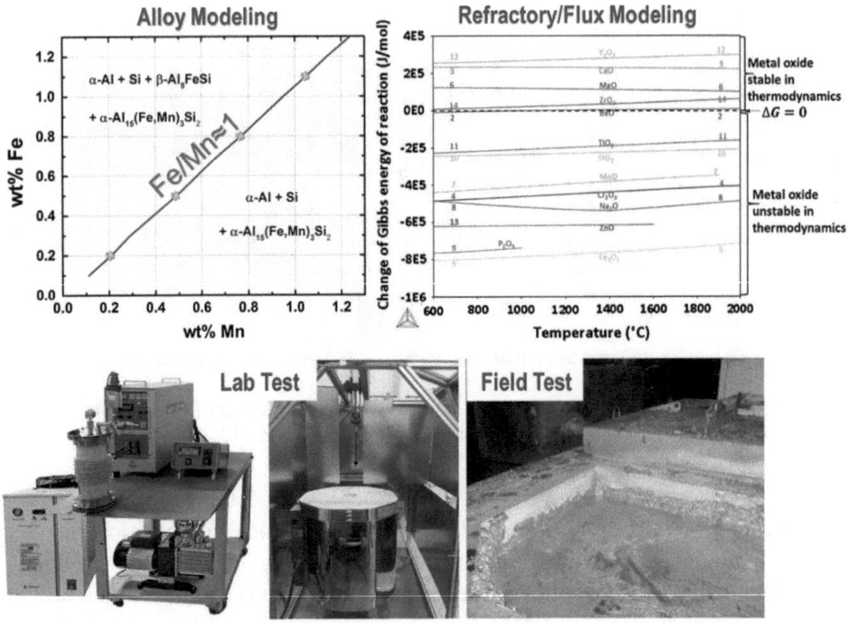

Fig. 2 Technical approach of an example REMADE project "Increasing melt efficiency and secondary alloy usage in aluminum die casting"

tic modeling supported by experiments and validation at both lab-scale (about 30 lb. melt) and pilot/production scales (500–3,000 lb. melt), with technical tasks outlined in Fig. 2.

Another example project at REMADE is on new approaches for removing contaminants in secondary materials. Many metals require high levels of purity to meet performance requirements. The presence of contaminants seriously can erode the value of recycled metals. An initiative led by The Ohio State University, Alcoa and CompuTherm intends to advance new methods for contaminant removal.

References

1. Twarog D, Apelian D, Luo A (2016) High integrity casting of lightweight components. North American Die Casting Association, Arlington Heights, IL, USA
2. The Aluminum Association (2011) Aluminum: the element of sustainability—a North American Aluminum Industry Sustainability Report, September 2011
3. Luo AA (2015) Material design and development: from classical thermodynamics to CALPHAD and ICME approaches. CALPHAD 50:6–22

Sustainable Nitrogen-Based Fertilizer Production from Sun, Air, and Water

Dorottya Guban, Martin Roeb, Josua Vieten, Hanna Krüger, Stephan Petersen, Klaus Hack, Tatjana Jantzen, Martin Habermehl and Markus Hufschmidt

Abstract In the DüSol research project, the technology of sustainable fertilizer production is developed and demonstrated on the basis of solar thermal redox cycle processes. The focus is on the unexplored step of solar thermal air separation for the production of nitrogen. For this reaction, corresponding materials are identified by thermodynamic calculations and qualified and optimized on a laboratory scale. In combination with material development, a prototype reactor is designed based on computer-aided calculation tools. In a test campaign in the new high-performance simulator for concentrated solar radiation SynLight at the Technology Center in Jülich, this reactor is being tested and the solar thermal nitrogen production demonstrated. These experimental works go hand in hand with the overall process simulation and optimization, which lead to a comprehensive economic analysis.

Keywords Fertilizer production · Thermochemistry · Air separation · Solar reactor · Concentrated sunlight

Introduction

There is an ever-growing demand for ammonia production that already reached globally 200 million tons per year by 2018 and is forecasted to increase to over 350 million tons per year by 2050 [1]. The application segment is dominated by the fertilizer industry, since the most important fertilizer and the world's most widely produced chemical is urea. Ammonia is synthesized via the Haber–Bosch process, for which the required hydrogen and nitrogen are currently provided by using fossil fuels. The process consumes about 2% of the world's commercial energy, which corresponds

D. Guban · M. Roeb · J. Vieten · H. Krüger
German Aerospace Center (DLR), Cologne, Germany

S. Petersen (✉) · K. Hack · T. Jantzen
GTT-Technologies, Kaiserstrasse 103, 52134 Herzogenrath, Germany
e-mail: sp@gtt-technologies.de

M. Habermehl · M. Hufschmidt
aixprocess, Aachen, Germany

G. Gaustad et al. (eds.), *REWAS 2019*, The Minerals, Metals & Materials Series,
https://doi.org/10.1007/978-3-030-10386-6_5

to approx. 2 trillion kWh. This work proposes a novel approach to produce ammonia from the raw materials water and air only by utilizing solar energy directly—without the detour of electricity, which is inevitably associated with energy conversion losses.

The proposed ammonia production route consists of two coupled solar-heated thermochemical cycle processes, which aim to specifically remove oxygen from gases. This applies both to gases that have atomically bound oxygen (water vapor), as well as for oxygen-containing gas mixtures (air). The applied redox material is oxidized in a two-stage process with the corresponding gas, and then thermally reduced. The energy required for this is provided by concentrated solar radiation that has the potential to play a major role in the future global energy mix.

In the first cycle, H_2O is used for oxidation and hydrogen is produced. This process has already been successfully tested by DLR on a pilot scale [2]. The second innovative cycle uses air as the oxidant. The air is deprived of oxygen and thereby pure nitrogen is produced. Both gases together, N_2 and H_2, are subsequently converted to ammonia in the well-established Haber–Bosch process.

The focus of the current work is the unexplored solar thermal air separation for the production of nitrogen for the Haber–Bosch process, which requires N_2 with oxygen contamination below 10 ppm. During thermochemical reduction–oxidation (redox) cycles, the redox material (typically metal oxide) is thermally reduced at high temperature

$$MO_x \xrightarrow{Q_{heat}=\Delta h_O} MO_{x-1} + \frac{1}{2}O_2 \tag{1}$$

while heat is converted to chemical energy in an amount equal to the enthalpy of the reduction reaction, and with the release of gaseous oxygen. During the reduction step, this stored energy can then be converted back to heat by the reverse reaction to drive the air separation step for N_2 production. The absorbed O_2 during the oxidation step could later be released and utilized for fertilizer production, e.g., in the Ostwald process for ammonium nitrate synthesis.

Perovskites are promising redox materials that are described with the general formula $AMO_{3-\delta}$, where the A site typically features an alkali, alkaline earth or rare earth metal cation, whereas the M site is occupied in most cases by transition metal cations. The occurrence of an oxygen non-stoichiometry δ ($\delta = 0$–0.5) in $AMO_{3-\delta}$ perovskites, as well as the close structural relationship between perovskites and their defect-ordered reduced form $A_2M_2O_5$ (brownmillerite) allow for fast redox kinetics. Furthermore, the partial reduction that is described by Eq. 2 ensures the stability of these materials since during the redox cycle, oxygen moves through the lattice without a decomposition of the crystal structure.

$$AMO_3 \xleftrightarrow{pO_2,T} AMO_{3-\delta} + \frac{\delta}{2}O_2 \tag{2}$$

Another advantage of perovskites is their tuneable composition, since the A and M sites can be occupied by a number of ions. The possible formation of a given composition is limited by the Goldschmidt tolerance factor [3]:

$$t = \frac{r_A + r_O}{\sqrt{2}(r_M + r_O)} \tag{3}$$

In most cases, the perovskite forms the desired ideal cubic lattice until $1 < t < 1.02$. The tolerance factor can be controlled by selecting the required alkali metal on the A site, while the B site can be occupied by a number of transition metals with the possibility of adding other metals in small quantities. With this method, the reduction enthalpy can be tuned in the range of $\Delta H^0 = 100\text{–}500$ kJ/mol^{-1}.

The selection of the metal ions was based on the rule that the oxygen affinity of the material should be low enough to allow thermal reduction in air at moderate temperatures, but high enough to enable full reoxidation of the reduced form to its initial state in air. A measure of the reducibility of oxide is its reduction enthalpy (ΔH_{red}), a lower ΔH_{red} results in a lower reduction onset temperature.

In this work, the feasibility of using perovskites as redox materials for solar thermochemical air separation is demonstrated. The choice of the applied redox material $SrFeO_{3-\delta}$ is based on our previous results [4]. Furthermore, the raw materials used to make $SrFeO_{3-\delta}$, strontium carbonate and iron oxide are very inexpensive, due to abundant natural resources at world market prices below $1 per kg, allowing for economical use as a redox material. The concept is first validated in laboratory-scale experiments performed in an IR furnace, followed by scaling up the process to a 20 kW solar reactor.

Experimental

Synthesis of $SrFeO_{3-\delta}$ Particles

$SrFeO_{3-\delta}$ was synthesized by a solid state reaction between the raw materials, strontium carbonate, and iron oxide. The powders were mixed in a ball mill, then annealed at 1100 °C for 20 h in an alumina crucible. The desired phase composition was verified via XRD. For the experiments, 4 mm spherical particles were prepared by mixing-spheronization using microcrystalline cellulose as binder.

Laboratory-Scale Proof of Concept

50 g of the synthesized beads were packed in a fixed bed reactor with an inner diameter of 20 mm that allowed for rapid heating and cooling via infrared radiation.

The bed was kept horizontal and heated in gas flows of synthetic air and nitrogen while the oxygen content of the outlet gas was monitored continuously. For the reduction process, the sample was heated under a flow of nitrogen of approximately 500 sccm to 800 °C. Reoxidation was performed under either synthetic air or 1% oxygen and 99% nitrogen flow at 350 °C.

Results

Laboratory-Scale Proof of Concept

The goal of the experiments was to prove that very low oxygen partial pressures can be reached during the reoxidation step performed in synthetic air. After the high temperature reduction step, when oxidation begins, the outlet oxygen concentration drops to a value of approximately 3×10^{-6} bar for some time, giving an outflow of purified nitrogen with low oxygen impurities, from the gas entering the system with 20% oxygen. As the oxidation progresses, the oxygen vacancies in the material become filled and it slowly loses capability to remove oxygen at low concentrations, so that the outlet concentration increases over time. The approximate residence time for the gas flow in the bed is calculated to be less than 5 s for the air flow through the bed. This highlights the very high kinetic activity of the oxidation reaction under plug flow conditions.

To investigate the cyclability of the material, five cycles were performed at identical thermal conditions of 800 °C reduction and 350 °C oxidation for air purification. Additional five cycles were performed at identical conditions but with an inlet gas of 1% oxygen in a 500 sccm nitrogen flow. This is of interest as the $SrFeO_3$ cycle alone would be a rather energy intensive route to producing nitrogen, but coupling it with an already established technology for producing low purity nitrogen, such as a pressure swing adsorption unit, could offer a very efficient route to removing the remaining trace oxygen and producing high purity nitrogen. The production versus purity relationships for each cycle is given in Fig. 1. The narrow grouping between curves of different cycles shows the excellent repeatability of the process and lack of degradation of the material, at least over the limited number of cycles performed.

When removing oxygen from air, the 48 grams of material produced 0.602–0.662 l of gas with less than 10 ppm of oxygen impurities. When removing oxygen from a semi-purified stream, the same amount of material produced 2.014–2.661 l of gas with the same purity.

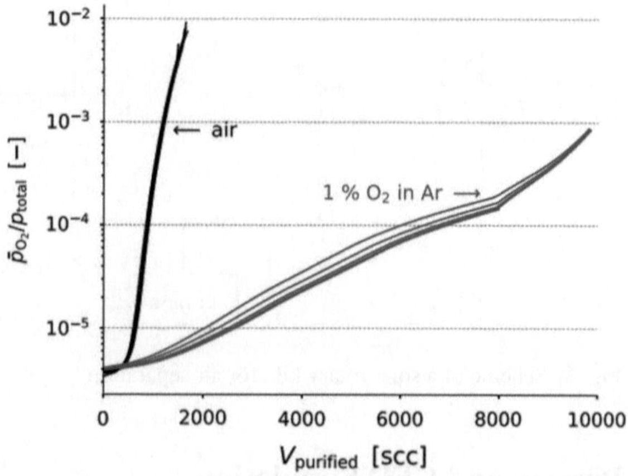

Fig. 1 Production curves for five identical purification trials from both synthetic air and from a mixture with 1% O_2

20 kW Solar Reactor Design

$SrFeO_{3-\delta}$ is suitable for use in the form of particles, therefore a number of reactor designs may be considered, such as a fixed bed reactor, rotary kiln, fluidized bed reactor or moving bed reactor. During the first step of the selection process, the latter two were rejected due to their relatively complex design and operation with concentrated sunlight, and only the fixed bed reactor and the rotary kiln were considered during the continuing evaluation.

The use of a fixed bed reactor is a simple solution that is similar to the laboratory-scale technology demonstration in the IR furnace, but it only allows batch mode operation, and since the whole reactor has to be cooled for the oxidation reaction step, the overall efficiency would be lower. A more promising option is to use a rotary kiln that would allow the separation of the reduction and oxidation step to reach higher energy efficiency. The reduction of the redox material would take place at a higher temperature in the kiln heated by concentrated sunlight, then it would be transferred to another reactor, where after cooling the air separation would take place. The rapid kinetics of the redox reaction by using perovskites means that the rotary kiln could be given a high mass flow and a high input power, improving the reactor's thermal efficiency. Solar rotary-driven kilns were already demonstrated to work well in previous studies, also with continuous particle flow [5].

The proposed reactor scheme is shown in Fig. 2. The solar radiation enters through a quartz window that is fixed to the stainless steel crucible with a flange that ensures a gas-tight connection. The gas inlet and outlet pipes are welded to the back of the crucible. The inlet pipe reaches inside the crucible close to the focal point of the radiation, which allows the gas to preheat before entering the reactor area. In addition, it serves to separate the inlet and outlet flows. The oxygen concentration of the outlet flow is monitored by a zirconium oxide-based lambda-sensor.

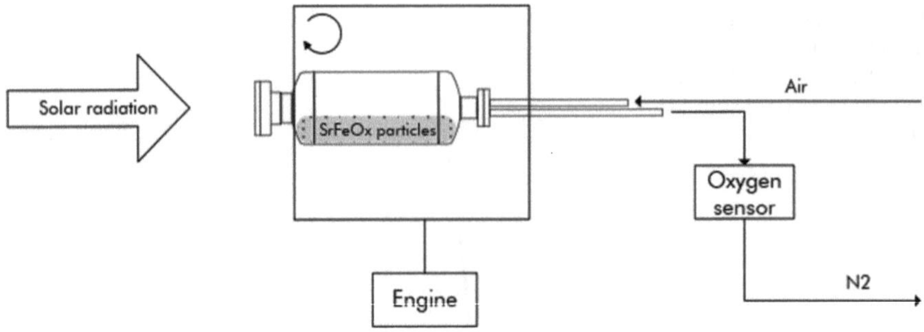

Fig. 2 Scheme of a solar rotary kiln for air separation

Process and CFD Simulation

Based on the thermochemical characterization of the redox material and the results from the laboratory-scale experiments, a CFD model of the solar rotary kiln (Fig. 2) is set up in order to investigate and understand the interaction of gaseous flow, chemistry of the redox processes and heat and mass transfer. Furthermore, the CFD model combines conventional CFD techniques with a discrete element modeling (DEM) approach, modeling the bed consisting of individual numerical model particles to account both for internal solid mixing of the bed as well as the fluid flow through the porous bed structures. The redox chemistry is included as heterogeneous surface reactions attached to the DEM particles. This model should also serve to optimize both the design and the operation of the solar rotary kiln and provide information about the reactor characteristics for a process model.

An overall process model is created to investigate the thermochemical feasibility of the process in general, as well as the related mass and energy balances in particular. It is based on process simulation software tools and components [6, 7] and the corresponding thermochemical substance data by GTT-Technologies, and can be used to investigate the effects of different operating conditions regarding the intended process design, thus also contributing to a comprehensive economic analysis.

Summary

Solar thermochemical air separation is a promising approach to provide nitrogen with the required purity to feed the Haber–Bosch process. Our current work focused on selecting a suitable redox material for the thermochemical cycle based on large-scale experimental screening aided by thermodynamic calculations. The viability of using the perovskite $SrFeO_3$ was verified by laboratory-scale experiments, where 50 g of redox material particles produces ca. 600 ml purified nitrogen per cycle. Based on these results, a 20 kW solar rotary kiln design was proposed for the scaled-up process that was optimized by process simulation software tools.

References

1. U.S. Geological Survey (2014) Mineral commodity summaries, pp 112–113
2. Säck J-P, Breuer S, Cotelli P, Houaijia A, Lange M, Wullenkord M, Spenke C, Roeb M, Sattler C (2016) High temperature hydrogen production: design of a 750 KW demonstration plant for a two step thermochemical cycle. Sol Energy 135:232–241
3. Goldschmidt VM (1923) Naturwissenschaften 477–485
4. Vieten J, Bulfin B, Call F, Lange M, Schmücker M, Francke A, Roeb M, Sattler C (2016) Perovskite oxides for application in thermochemical air separation and oxygen storage. J Mater Chem A 4(36):13652–13659
5. Tescari S, Agrafiotis C, Breuer S, de Oliveira L, Neises-von Puttkamer M, Roeb M, Sattler C (2014) Thermochemical solar energy storage via redox oxides: materials and reactor/heat exchanger concepts. Energy Procedia 49:1034–1043
6. Petersen S, Hack K (2007) The thermochemistry library ChemApp and its applications. Int J Mat Res 98(10):935–945. https://doi.org/10.3139/146.101551
7. Petersen S, Hack K, Monheim P, Pickartz U (2007) SimuSage—the component library for rapid process modeling and its applications. Int J Mat Res 98(10):946–953. https://doi.org/10.3139/146.101552

Part II
REWAS 2019: Education and Workforce Development

Sustainability as a Lens for Traditional Material Science Curriculums

Gabrielle Gaustad

Abstract The theoretical and methodological foundations of the sciences and technologies are essential to the removal of barriers to achieving sustainable systems. The teachings of these concepts still lie in traditional academic disciplines such as engineering, science, and mathematics. This structure can often manifest significant barriers to progress in tackling challenging sustainability issues due to an absence of a multifaceted, interdisciplinary, systems approach. This work will explore approaches for using current sustainability issues and problems to introduce both systems thinking and traditional material science discipline specific learning objectives to the classroom. Specific examples will be illustrated for a diverse set of courses and curriculum. Results show such an approach can improve recruitment and retention results in addition to improved teaching outcomes.

Keywords Pedagogy · Engineering · Education · ABET

The theoretical and methodological foundations of the sciences and engineering are essential to the removal of barriers to achieving sustainable systems. The teachings of these concepts still lie in traditional academic disciplines such as engineering, science, and mathematics. This structure can often manifest significant barriers to progress in tackling challenging sustainability issues due to the absence of a multifaceted, interdisciplinary, systems approach.

Material science has a particularly relevant set of foundational courses that lend themselves to interesting sustainability integration. Material selection approaches and software have begun to incorporate both economic and environmental "properties" in the decision analysis, highlighting the tradeoffs made in real applications [1]. Thermodynamics and kinetics principles can be illustrated in interesting energy conversion examples for next-generation renewable energy storage and production technologies. Other important contributions exist in fundamentals of mining, processing, alloying, phase equilibria, material flow analysis, etc. A variety of recent research is available with additional innovative suggestions [4, 6, 8].

G. Gaustad (✉)
Inamori School of Engineering, Alfred University, 1 Saxon Drive, Alfred, NY, USA
e-mail: gabrielle.gaustad@rit.edu

G. Gaustad et al. (eds.), *REWAS 2019*, The Minerals, Metals & Materials Series,
https://doi.org/10.1007/978-3-030-10386-6_6

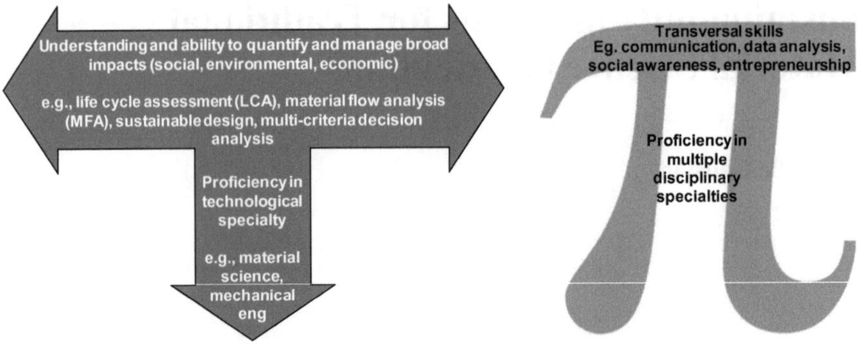

Fig. 1 Schematic representation of T-shaped and Pi-shaped competencies

Integration of sustainability issues into material science curriculum promotes nexus thinking. In many engineering curriculums, it can be challenging to promote interdisciplinary thinking when the curricular approach is inherently siloed. This can often lead to a reductionist spiral where "solutions" produce unintended consequences or additional problems. The famous inventor Thomas Midgeley is often used to illustrate such unintended consequences. While his introduction of MTBE to replace lead in automotive fuels was revolutionary and life-saving as it dramatically reduced emissions of lead to the air, the fuel additives he introduced have now been linked to endocrine disruption and ecotoxicity issues [11]. Another frequent example I give in class is the New York state rebate on new appliances. This rebate program was introduced to incentivize New Yorkers to replace older, less energy efficient appliances with newer ones in the hopes of reducing electricity consumption in the state. While the program was popular, it had the unintended consequence of a majority of participants buying a new refrigerator and keeping their old ones which actually increased appliance electricity consumption overall. Nexus thinking is a systems-based educational approach that ensures broader critical thinking skills; it promotes thinking through unintended consequences [7, 10].

Integration of sustainability into traditional disciplinary curriculums also promotes T-shaped or Pi-shaped student competencies (Fig. 1). Broad, transversal skills are being emphasized more by employers; it is imperative that today's student leave with not just a disciplinary degree but communication, organization, analysis, and critical thinking skillsets [2, 3]. These approaches also help to improve the translation of theory to practice, another key gap cited by employers. Students often struggle to take academic learning and use it directly for on the job skillsets; ABET has emphasized this need in its accreditation processes [5, 9].

Work presented at REWAS 2019 will explore approaches for using current sustainability issues and problems to introduce both systems thinking *and* traditional material science discipline-specific learning objectives to the classroom. Specific examples will be illustrated for a diverse set of courses and curriculum. Results show such an approach can contribute to improved recruitment and retention number and preliminary results appear to also enhance student learning outcomes measured via traditional assessment methods.

References

1. Ashby MF, Shercliff H, Cebon D (2013) Materials: engineering, science, processing and design. Butterworth-Heinemann
2. Connor A, Sosa R, Jackson AG, Marks S (2017) Problem solving at the edge of disciplines. In: Handbook of research on creative problem-solving skill development in higher education. IGI Global, pp 212–234
3. Faris J, Kolker E, Szalay A, Bradlow L, Deelman E, Feng W, Qiu J, Russell D, Stewart E, Kolker E (2011) Communication and data-intensive science in the beginning of the 21st century. OMICS J Integr Biol 15(4):213–215
4. Gipson KG, Prins RJ (2015) Materials and mechanics: a multidisciplinary course incorporating. In: Handbook of research on recent developments in materials science and corrosion engineering education, vol 230
5. Glasgow RE, Emmons KM (2007) How can we increase translation of research into practice? Types of evidence needed. Annu Rev Public Health 28:413–433
6. Gunister E, Ozturk F, Simmons RJ, Deveci T (2015) Innovative instructional strategies for teaching materials science in engineering. In: Handbook of research on recent developments in materials science and corrosion engineering education, vol 100
7. Hussey K, Pittock J, Dovers S (2015) Justifying, extending and applying "nexus" thinking in the quest for sustainable development. In: Climate, energy and water
8. Mainali B, Petrolito J, Russell J, Ionescu D, Al Abadi H (2015) Integrating sustainable engineering principles in material science engineering education. In: Handbook of research on recent developments in materials science and corrosion engineering education, vol 273
9. Passow HJ (2012) Which ABET competencies do engineering graduates find most important in their work? J Eng Educ 101(1):95–118
10. Stringer L, Quinn C, Berman R, Le H, Msuya F, Orchard S, Pezzuti J (2014) Combining nexus and resilience thinking in a novel framework to enable more equitable and just outcomes
11. Von Krauss MK, Harremoës P (2001) 11. MTBE in petrol as a substitute for lead. In: Late lessons from early warnings: the precautionary principle 1896–2000, vol 110

Corrosion Education for Materials Life Extension: Pathway to Improvement in Resource Productivity

Brajendra Mishra

Abstract Materials are nonrenewal resources that are created through an "unnatural" process. In addition to resource recovery and recycling of valuable materials to achieve sustainability, simple methods to extend the life of materials enhance the resource productivity. The natural process of corrosion tries to reverse the process of material extraction causing enormous loss of energy and impacts the environment. Corrosion costs the U.S. over $300 billion per year and also produces significant safety hazards. Corrosion control is, therefore, important to enhance the life of engineering metals and materials, which is also the focus of many government regulatory agencies such as the EPA, DOT, and OPS. Corrosion protection technology utilizes metallurgy, material chemistry, and physics as well as electricity to prevent or control corrosion degradation and therefore, the education of corrosion science and engineering is directly linked to improving material life. Education in corrosion control applies these sciences to control the chemical and mechanical aspects that are involved in the deterioration of properties. This paper will address the educational aspects of corrosion technology that allow resource productivity improvement of materials.

Keywords Corrosion · Sustainability · Education

Definitions

Resource productivity is the quantity of good that is obtained through the expenditure of unit resource and is expressed as monetary yield per unit resource. For example, agricultural output may be defined as "crop per drop".

B. Mishra (✉)
Mechanical Engineering Department, Worcester Polytechnic Institute, 100 Institute Road, Worcester, MA 01609, USA
e-mail: bmishra@wpi.edu

G. Gaustad et al. (eds.), *REWAS 2019*, The Minerals, Metals & Materials Series,
https://doi.org/10.1007/978-3-030-10386-6_7

Resource intensity is a measure of the resources (water, energy and materials) needed for the production, processing, and disposal of a unit of good.

Resource efficiency is maximizing the supply of materials to function effectively, with minimum wasted natural resources. Alternatively, it may be understood as using Earth's limited resources in a sustainable manner while minimizing environmental impact.

Corrosion can be considered a natural result of energy stored in the metal when it was refined and fabricated. Thus, when a metal corrodes, it is essentially the wastage of this stored energy. Corrosion is the natural process that reverses extraction and production of particularly metallic materials.

Dominant effects of corrosion that limit resource productivity are (a) material loss and replacement cost; (b) production loss: plant downtime; (c) pollution; (d) over design; (e) protection (over) costs; and (f) inspection, repair, maintenance costs.

Although corrosion is a natural process, it is controllable to a large extent. Reported data in 2015 demonstrates that for a GDP of 17.7 trillion, direct and indirect cost of corrosion total 6.2% of GDP, or a total corrosion cost in US of approx. 1.1 trillion. The cost on global scale exceeds 2.5 trillion. It is believed that approx. 35% can be saved by adopting best practices.

Corrosion Education

Improvement in resource productivity can be achieved through a proper understanding and knowledge of corrosion and needs to be included in sustainability curricula. The main aspects of this education for scientists and practicing engineers can be divided into three parts:

(a) Realizing: science, mechanisms
(b) Recognizing: identify and measure
(c) Responding: protection

Based on an NRC Report by Scully and Harris, the following figure describes the level of corrosion education needed for persons engaged in the manufacturing business.

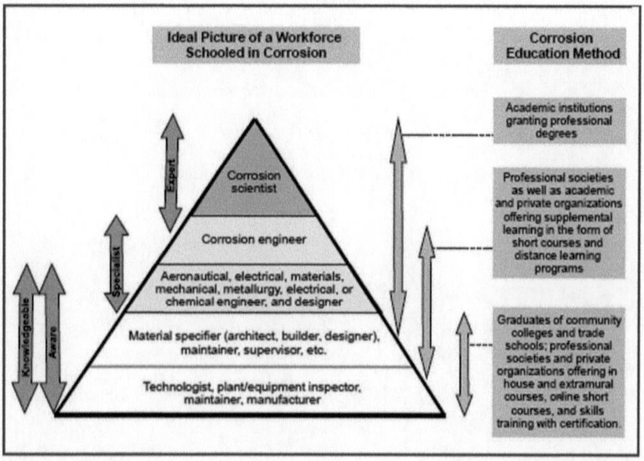

Summary

- Corrosion impacts materials sustainability extensively
- Cost of corrosion significantly impacts economies
- Protection against corrosion can improve resource productivity
- Knowledge of corrosion is essential for practicing engineers.

Material-Oriented Product Development by QFD4Mat Material Selection Strategy Approach

Fabrizio D'Errico

Abstract Beyond the already published SpringerBrief of the author on Material Selection strategy titled "Material Selections by a Hybrid Multi-Criteria Approach", US Springer, 2015, the QFD4Mat method is presented as it has been teaching to engineering students as material selection strategy learning tool, as well as it has been adopting in big automotive sector industry that employs training and knowledge transfer. Structured on the classic quality function deployment approach, one key of success of QFD4Mat method is its open-source customizable platform by which any multidisciplinary teamwork with varying specialties and languages can actively participate to specific product development decision-making process, together with material specialists. The QFD4Mat method goes beyond usual Quality Function Deployment (shortly, QFD) technique by creation of two useful graphic analysis tools: (a) the material value curves for product and the Performance–Cost–Receptiveness (PCR) bubble maps, namely, two immediate infographic snapshots of the best product value material solution for the specific challenge.

Keywords Material selection strategy · Manufacturing · New product development · Product competitiveness · Innovation

Introduction

Material specialists love to claim that materials largely contribute to product improvement and development processes. They like to consider the choice of material as a "quarterback" in several challenges, since such decisions could considerably impact the entire product value chain. Selection of materials is strictly linked to product manufacturing phases and its choice is a key issue in the process, which leads to a rise in the complexity of a product value chain.

F. D'Errico (✉)
Department of Mechanical Engineering, Politecnico di Milano, Via La Masa 34, 20156 Milan, Italy
e-mail: fabrizio.derrico@polimi.it

G. Gaustad et al. (eds.), *REWAS 2019*, The Minerals, Metals & Materials Series,
https://doi.org/10.1007/978-3-030-10386-6_8

After the product is on-the-shelf, namely, sold to customers or users (sometimes, these can be different subjects), the usage stage starts out: some factors connected with materials can substantially influence consumers' evaluation on usability, quality, and durability. Product development follows therefore a two-tier approach: one is externally projected to recognize customer needs, the other is internally focused mostly on the domain of engineering in that it aims to define such key features in a product that can efficiently and powerfully translate customer needs into a success story, possibly surpassing the customers' expectations. It is not rare to find that the choice of materials constitutes the added value of a product. For example, to the touch of hand, the aluminum used in smartphone cases is perceived as more appealing than polymers. Superior raw material and manufacturing costs for such choices have been powerfully deployed by engineers. Aluminum, they bet, would offer more resistance per gram, the cover case can be thinner, wider but lighter, thus suitable for a larger and more colorful screen. Apple produces a smartphone 40% of whose weight is made of metals (aluminum accounts for 18%, stainless steel for 16%) against a very low 1.8% weight of plastics, carefully shaping publicity in terms of saying they have created an environmentally friendly life cycle for electronic goods mass production. After the usage phase, the choices made on the material to utilize decisions which were taken in the design phase, and also determine end-of-life management since the product can be disposed of, reused or, in the best cases recycled. During both the usage and end-of-life phases, there are some negative externalities: the choice of higher carbon footprint materials could impact economically on revenues if Governments choose to reduce such externalities. As opposed to that, choice of recyclable materials would be a strategic decision progressively to reduce the higher cost of new metal alloys which are expensive today, but have high recycling potential. Recycling reduces the impact of the cost of virgin materials.

On the above premises, it seems there are several ways to create value in same product type. It could perform better and/or it could be easier to use; it could be fabricated at more convenient costs, but it could be less durable. Thus, by introducing the term *strategy*, the problem of material selection is necessarily enlarged to define not just a method, but plans and actions that will on a long-term basis have a substantial impact on the success of a product on the market, and consequently on the success of firms against competitive forces [1].

The selection of materials for market strategy is therefore a complex and multifaceted challenge that requires teams with multidisciplinary skills, ranging from people who remind engineers that something which is the best is not always something that can be sold if the customer does not perceive the difference to others who insist that the cost of materials can be less impacting on the projected long-life product costs if the cheaper solution in materials creates nonconformities that will need to be managed and to people that can explain technology and added value techniques in easy words.

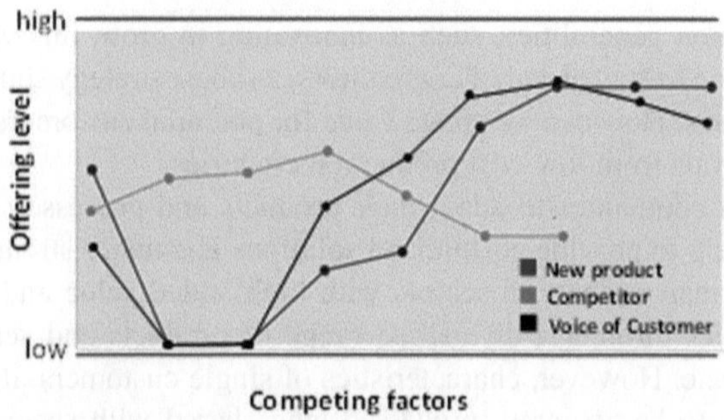

Fig. 1 A diagnostic framework depicting the value curve of service, the product, and the material. The horizontal axis contains the main factors on which industry interest is more focused, while the vertical axis qualitatively measures the offering level that buyers receive from all of these key factors

To that scope, effective strategy analysis of product performance is usually achieved by creating the product *value curve* which aims to identify how customers might assess product features that would reply to customer needs [2]. In Fig. 1, it is shown a simplified example of a value curve for a product. The curves inside the diagram allow to correlate the product key features identified with qualitative response attributed by customers. The black line in this example represents the voice of targeted customers, namely, how sample population of target customers are ideally more or less responsive to key features. The other colored lines represent the qualitative response of customers to your and competitors' product features, namely, considering how these features have been technically embedded in product and how they are appraised by customers.

The *product value curve* is therefore considered one basic component of product market strategy [2]. Reaching a more efficient value curve means the product has been developed with some features that are distinguishable and capable to impact on customers' preferences in better and efficient way than competing products. On the other hand, if the introduction of some feature in a product makes product complexity greatly increases as well product cost, the product value curve would allow developers to focus mainly on such features capable to propose value to customer. For example, application of recycled or biodegradable material for the making of capsules for domestic coffee machines has been increasing, and the analysis of product value curve depicts customers' environmental consciousness and need for an easier waste management.

Nowadays, the attention is slowly shifting from production volumes, process repeatability, standardization of operations, and modular components, to the concept of personalized products, the "one of a kind" attitude.

Beyond common generalities, such as innovating to grow, innovating to create value or innovating to stay ahead of competitors, a robust strategy should answer the following questions: How can we create value for potential customers? How can we defend against rivals from low-cost production countries?

The ability of companies to adapt their products and processes to the specific needs of customers to provide customized solutions is a major strength that allows you to maintain market share in sectors with high added value and compete with emerging countries through a diversified range of products and services that are difficult to replicate. However, characteristics of single customers, their needs, and preferences have to be acquired, processed, and shared with specialists from the product development teams working in different business units. The key for product development is therefore a *multidisciplinary team* that can take part to the process of the product designing; the team is made of engineers as well as procurement analysts or market analysts with a variety of personal background and different point of views on the features of the product. The decision to face engineering problems, such as modeling of the product, the production process, and the design of the manufacturing system, in an organic and structured way, often requires the definition of a collaborative and multidisciplinary network. In these cases, the main challenge is the integration and harmonization of the multidisciplinary knowledge, of the different working methods and of the mental approaches to problem-solving present in company.

To overcome such heterogeneity of the information, a comfort and easy language method and relative tool, QFD4Mat tool—acronym of Quality Function Deployment for Materials—has been developed by author to manage and elaborate data in order to identify management strategies for product improvement. The tool is available on qfd4mat.com website as freeware student edition that has been developed on an xls platform, slightly customizable as described in the Student Manual Guide, also available at website link. A commercial edition is based on Java language, available on web-server mode, or in resident full-licensed mode, by uploading the software on a company server. Advantages of commercial edition are unlimited number of material key features and product voice of customer key features to manage, access to material database, automatic and user-friendly guided step for filling in main data required for drawing the infographic maps and material value curves, and customizable reporting.

Material Selection Strategy for Product Improvement: An Overview

When we go to the bike shop and look around to buy our new bike, first we are conscious about our budget. Our budget is obviously related to the price, but not exclusively. We elaborate several aspects at the same time in order to judge whether our main needs could be satisfied by the bikes we are "screening" in front of our eyes. For example, whether they give us the right kind of feeling that they will be light and robust enough for the main type of riding we do. Weight and frame rigidity is often an issue for many bikers. A bike manufacturer knows that the targeted market segments are sensitive to this aspect, but customers are sufficiently specialized to understand that lightness must not exclude robustness and durability. All these external requirements are input requirements collected from the market (often referred to as the "*Voice of Customer*" or *VOC*) but they need to be translated into technical functions. The translation process from the external or customer needs to the internal or designer requirements is fundamental process for elaboration of matrix of correlation between product features and material technical features, according to common quality function deployment approach [3]. For example, while the customer thinks about a lightweight bike, the engineer thinks of the density of construction material. Furthermore, when a biker is willing to pay $700 for a lightweight bike but he wants to put it under pressure by riding it off track, he will probably expect the bike to be robust enough and not to break all of a sudden. The engineers therefore think about the toughness, resistance, or resilience capability of the frame material. This process is what we call in material selection strategy the translation of needs (or external) requirements into technical (or internal) requirements. If the reader has gone beyond, he or she will already have identified the following crucial key point: once engineers have translated, the customer needs into measurable parameters, namely, once they have defined the *metrics*, how can they set the optimal values to target? Actually, this is the crucial question that most literature in the selection of materials try to answer by introducing rigorous analytic methodologies [4]. Anyway, the introduction of internal metrics and targets allows engineers to *rank and screen* various candidates in order then to select the one that matches the targets and respects technical and nontechnical constraints. One of the major nontechnical constraints that bike engineers fix is a target price that is not lower than marginal cost, as they are not indifferent to the budget constraints of the buyer. Other constraints are directly expressed in the technical language of engineers, for example, how they translate the safe behavior required and expected of the bike frame when users ride down the hill. Engineers fix a minimum value of toughness resistance for the materials to screen, since they acknowledge that materials with lower values could possibly fail, leading to sudden and dangerous breakages at high speed.

What happens in every material strategy we plan is that several requirements that aim to answer customer needs will compete for fixed resources, such as the cost budgeted, the manufacturing and assembly time, and weight reduction, which contends against the strength of materials.

A material selection strategy is thus performed proceeding with three main consequential tasks as follows:

(1) The translation of customer or user needs (i.e., external analysis) for the product, influenced by material features into technical and nontechnical requirements (i.e., internal analysis);

(2) On the basis of a technical and nontechnical set of requirements, namely, the material key features inventory, as developed in (1) the formulation of performance metrics to measure how well a material matches a set of requirements;
(3) a search procedure, namely, a structured material selection method, in order to (a) explore a solution space, (b) identify materials that meet the constraints, and (c) rank them by their ability to meet the requirements.

A selection strategy works by defining how it is possible to convert a set of inputs, which we call the requirements of the product, into a set of outputs. Thus, material selection strategy deals with the selection of materials and processes to shape the external requirements and transfer the correct perception to customers and users so to succeed because their expectations have been satisfied (or, hopefully, more than satisfied).

It is a typical multi-objective optimization problem to be solved, with varying approaches that could be employed. Methods can be divided into nonderivative, or implicit methods, and derivative, or explicit, methods. *Derivative methods* for material selection describes (i.e., "make it explicit") a material "optimization function"—e.g., the weight-saving function of a component that is bend loaded—in terms of dependent material variables—e.g., density, elastic modulus, etc. Generalizing, such methods consists in deriving a function in those variables that can represent the objective function to maximize—or minimize—by acting on material variables and with respect to constraints. Most known is the Ashby's method [2] that derives for specific component functions—e.g., strength-limited design at minimum mass—the performance function dependent on group of material properties, namely, the material performance indexes; familiar examples are the specific stiffness, E/ρ, and the specific strength, $\sigma y/\rho$ (where σy is the yield strength or elastic limit, and ρ is the density). Such a method implies to write the optimization function for the specific case. It is actually recognized to be an inherently stable method, since it is insensitive to alternatives: the derivation of performance equation of the problem makes engineers keep distance from self-judgment on material properties.

Nonderivative methods, or implicit method, instead does not derive the objective functions in order to calculate the optimum. They are also known as "black box methods." Most known is the *weight-sum method*. It consists in assigning weight factors to each key feature that is thought to have direct or indirect impact on single or multiple objectives. Objective is not described by any explicit functions throughout key features (or variables), as it happens in explicit method; thus experts are asked to evaluate (a) design, product, or material key features that can causally impact on performance optimization, considering they can work contemporarily and concurrently, and (b) level of importance of certain features among others.

Each explicit or implicit method contains some advantages and drawbacks, especially in the case of industrial product development process that is key for competing on marketplace. However, if product development is recognized to be generally a step-by-step structured process that concerns insights into roles played by product functionality, product performance, easiness of assembly, maintenance and durability, capability for recycling, and interaction of all these features with cost estimation and manufacturability [5, 6], such a process typically requires both engineering, marketing and procurement experts, financial officers and others that can build such a cross-functional team capable to evaluate in deep several features that can impact on the product at a whole. Invaluable is therefore possibility to merge such a variety of personal background and different point of views for defining features of the product.

The Quality Function Deployment Framework Applied to Material Selection Strategy

The method consists of identifying and gathering information into three categories, namely, performance, cost, and receptiveness, in a nondimensional scale:

- The *Voice of Customer* (VOC) product features in common language of user/customer that are recognized as important/impacting on customer choice and needs; they will be classified into three categories;
- The *material key features*, namely, the technical translation of the VOC product features into inner material properties;

An example of this phase is shown in Table 1 for a didactic case of a bike saddle product [7].

Once the (outer) customer requirements and the (inner) engineering characteristics have been outlines, relationships between customer requirements and engineering characteristics are created throughout the relationship matrix, as shown in Fig. 2.

The last calculation step according to QFD approach consists in assignment of scoring our product and those of our competitors (Fig. 3).

After completing the assessment box (see again Fig. 3), two infographic powerful tools are realized: (a) the material *Value Curves* (Fig. 4) and the Performance—Cost–Receptiveness (or PCR) bubble maps (Fig. 5). They are graphs that are immediate snapshots of results obtained by QFD4Mat analysis that translates the information gathered by correlation matrix into an assessment box for various solutions.

The value curve tries to resolve the trade-off between differentiation and low cost and to create a new value curve:

- Which key features that a particular sector takes for granted should be eliminated? This question forces decision-makers to consider those features that reply to the requirements which companies are competing for, but which actually have a low or even no impact at all on product value;
- Which key features should be reduced well below the industry's standard? Some product features do not add value to the product, or they are overdesigned;
- Which key features should be raised well above the industry's standard? Quality appearance, image, and impact resistance by use of light metal alloys in electronics, for example, laptop covers, are today made of machined aluminum instead of the cheaper injection molding polymers, which give greater formability;

Table 1 Main key engineering issues obtained by translation of major bikers' requirements for bike saddles [7]

Voice of customer			Technical translation of *VOC*		
Primary	Secondary	Tertiary	Functional requirements, FR	Engineering characteristics (key factors influencing FR)	Metrics
Performance	It should help to reduce energy lost by pedaling	It should be light	Increase stiffness of the part	Maximum displacement produced by vertical force F	mm
		It should be rigid	Use thick sections	Cross-sectional thickness	mm
			Use light materials, reduce volume	Weight mass	kg
	It should look "high performance" and aggressive	It should appear aerodynamic in shape	Use thin section	Cross-sectional maximum thickness	mm
			Capability to vary profile sections (look for aerodynamic shapes)	Manufacturing process for very free shape design	–
	It should not cause pain after prolonged sitting	It should be comfortable	Dumping capacity/deform under cyclic load	Maximum width of material hysteresis under load cycle	mm/mm
		It should feel soft	Have a soft seat	Local pressure absorbed, p	MPa
			Respect the anatomical conformation	Central channel width, w	mm
			Reduce the pressure on the perineum	Height of back seat h_{sup}/depression height of the nose, h_{inf}/Central channel width w	mm
			Adjust for any seats	Length of the bar frame, l	mm
			Limit too much soft and thick padding	Local pressure absorbed, p	MPa
Cost	It should have low operating costs	It should not require any maintenance by me	Achieve high durability	Salt spray corrosion test duration	hours
		It should be durable		Prolonged duration test in saline environment without intermediate cleaning	days
		It should be easy to clean	Can clean the materials using water	Wash test in water	None

- Which key features should be created, responding to unexplored product requirements for the sector, which the industry has never offered? This forces us to seek out solutions that break out of the industry's sector boundaries, to explore new offer contexts.

The PCR bubbles maps diagram (see Fig. 5) is instead based on action made to separate the side of demand that pertains to choices and preferences (needs) of consumers from the side of supply, pertaining to how are evaluated solutions [7]. The data by which bubbles are created are taken by the assessment box, the "foot" of the matrix—see Fig. 3—namely, the results of the assessment of material key features for each candidate material. Data are simply reorganized by grouping into the P, C, or R category.

The bubble map outlines on the same diagram for the supply-side and the demand-side two circles with their centers of coordinates, respectively, P and C, and with their diameters R. The ideal situation is obviously when the PCR bubble, representing the supply-side, completely matches the demand-side PCR bubble. When the VOC bubble is partially covered by the bubble of the material key features, it means the product value that the suppliers propose, partially satisfies the consumers' expectations.

Customer Requirements: Primary	Secondary	Tertiary	Customer Degree of Importance	Maximum displacement produced by	Cross-section thickness	Weight mass	Prolonged duration test in saline environment	Cycles before failures	Maximum hysteresis under load cycle	Local pressure absorbed, p	Central channel width, w	Heigh of back seat h_{sup}	Depression heigth of the nose, h_{inf}	Length of the bar frame, l	Salt spray corrosion test duration	Capability to clean materials by water
Performance	It reduces energy lost by pedaling	It is light	5		●	●			▽		▽					
		It is rigid	5	●	●	○	●		○		○			○		
	It looks high performance and aggressive	It appears aerodynamic in shape	4		●						○		○			
	It does not cause pain after prolonged sitting	It is comfortable	3							●	●	●	●			
		It feels soft	3							●						
		It is adjustable for any seats	4											●		
Cost	It has low operating costs	Avoid any maintenance by user	3					●							●	○
		Be durable	4					●							●	○
		Be easy to clean	3												▽	●
	It has low purchase cost	Be cheaper	3													
Safety	It is compliant with safety requirements	Respect the international specs	5					●								
		Be safe when it has a failure crash	5		○											

Functional Requirements

Fig. 2 Example of correlation matrix linking material key with product key features in QFD4Mat analysis [7]

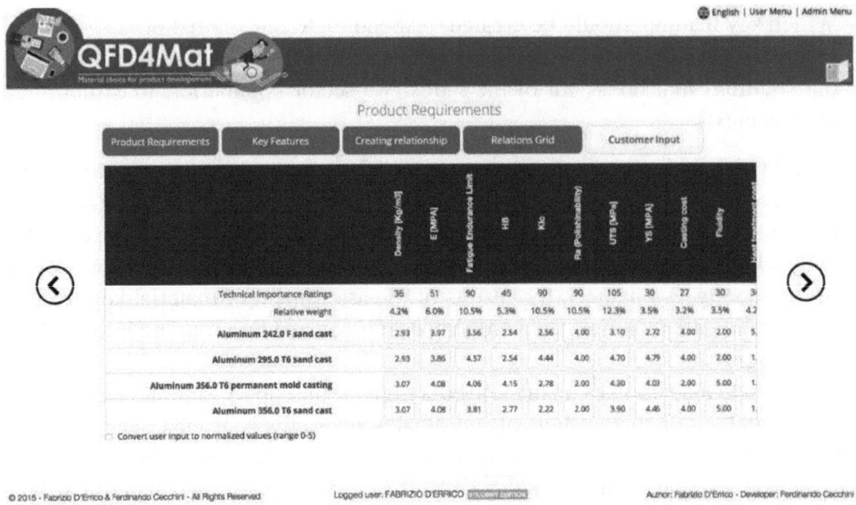

Fig. 3 Example of calculation box for the assessment of key features and the relative weighted score for competing products in a classroom case studied regarding with new lightweight automotive suspension arm. The screenshot is provided by the commercial edition of the tool

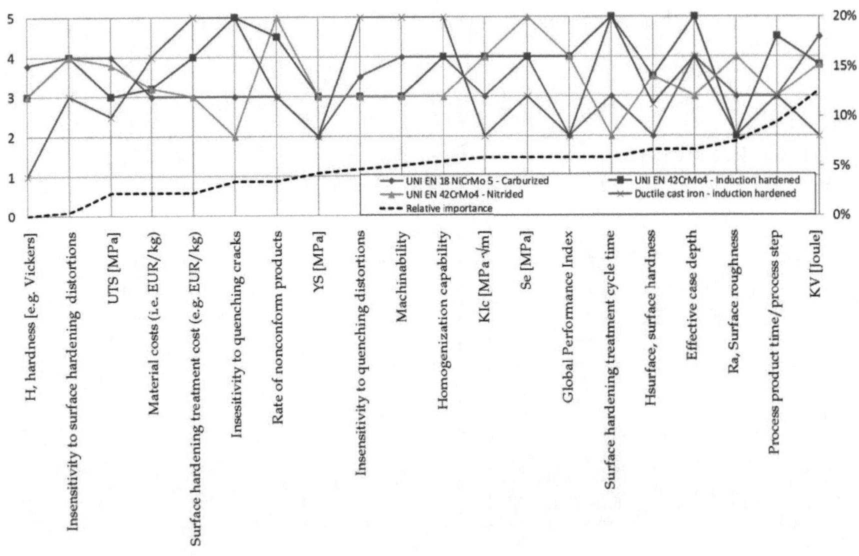

Fig. 4 The value curve visualizing the final results of QFD4Mat in the crankshaft industrial case study [7]

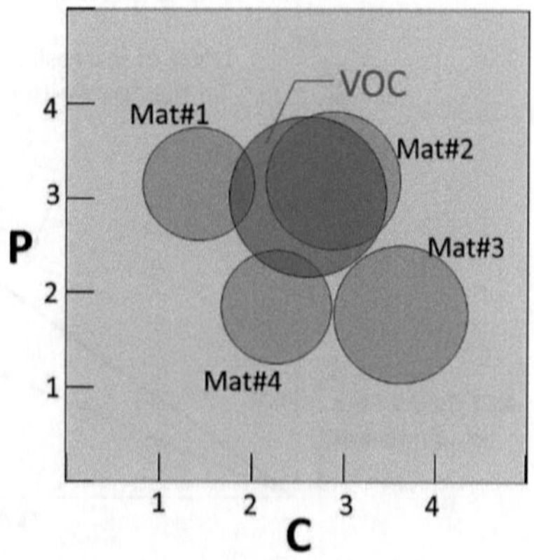

Fig. 5 Mapping of material solutions by QFD4Mat bubble map graphic output: the customer-based analysis for checking best-for product value for customer solution

Embedding the LCA Analysis: The QFD4Mat Approach for Growing Environmental Consciousness in Engineers

Frequently, material selection problems are of greater complexity, when several aspects must be contemporarily taken into account for a successful marketplace positioning of products. The challenge might sometimes be to create an enlarged point of view, sometimes embedding further aspects product engineers do not familiarize, as well. For example, which class of material is the best candidate to give the most powerful and efficient contribution to weight saving in cars but in environmentally friendly way? As usual, material specialists love to claim that materials largely contribute to product improvement and development processes. We like to consider the choice of material as a "quarterback" in several challenges, since such decisions could considerably impact the entire product value chain.

Thus, today it is common knowledge; the choice of lighter materials to put onboard a vehicle fueled by gasoline can contribute to reducing emissions, namely, the carbon footprint of the vehicle over its life. Otherwise, if we point out the linear function emissions over travel distance as it is depicted in Fig. 6, we can just consider two scenarios: the baseline scenario, namely, the conventional heavy solution has always a higher slope curve for the emissions vs travel in the usage phase of vehicle than a light metal solution adopted. But with surprise, lighter components made of energy-intensive materials, when they are not employed by recycling routes, such as aluminum, magnesium, and titanium, show "dirtier" behavior; even though the direct emissions from their usage are largely lower (i.e., gentle slope of emission vs onboard travel curve) than that of heavier materials. This fact is due to all emissions "stored" in manufacturing process: greenability of material must be accounted for in the net balance between the cleaner and dirtier phase when, in order to make a decision, we compare a lighter scenario with that of the baseline [7].

Such a practical example for automotive components shows clearly efforts young engineers should put into concept design for assuring sustainable but competitive products reach the marketplace. Selection of materials may give control of the quan-

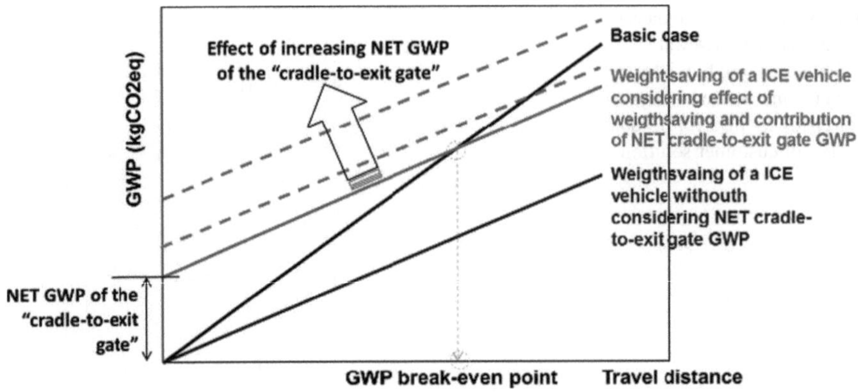

Fig. 6 Scheme for assessment of positive or negative global warming potential (GWP) of lighter automotive component: seeking the GWP break-even point [7]

tity and types of materials to be used in the manufacturing phase. Main technical, economic, and environmental issues, once evaluated and investigated, should be adequately accounted for and finally "merged".

Due to several features to be considered, furthermore each of them having its own unit dimension (e.g., kg for mass part; euro per kilogram for raw material costs; kg CO_2 eq per kilogram for unitary pollutant equivalent emissions for raw material used, etc.), an explicit method is quite unpracticable to select best candidate material when all those features must be monitored. On the other hand, implicit methods, such as the QFD, give possibility to treat the assessed key features of product in nondimensional way, thus allowing users to easily merge results.

Specifically, the environmental aspect of product is typically included in a QFD4Mat analysis in the receptiveness category, as today the environmental performance of products is considered an added value by multinational enterprises, as it would increase the attractiveness of products for several users.

By this way, the QFD4Mat acts as a common language among all the diversities of a multidiscipline team that wants to take the product design as a whole. Using it puts people from each company department on equal footing, all able to stimulate the decision-making process, no one being sidetracked by partial views, just like a roundtable, which, in a most effective metaphor, has no head, and at which everyone who has decided to sit does so with equal status and power for discussion.

Conclusions

The master idea of this multidisciplinary approach came out by education program included in academic course student held in past 10 years in on advanced metallurgy and process topics. On the other hand, the QFD4Mat method has been employing in industry just being introduced by former student; invaluable is the fact young engineers realize that no point of view ever has minor importance in multidisciplinary teams, as technical aspect is just a single facet of the multi-sided problem. There can only be an insufficient way of sharing disciplines, because, in such cases, there are so many incomprehensible languages. As in a board game with common rules for all the players, even when they have different strategies or scopes, the approach here shortly described would aim to communicate possible common rules, so aiding the single player, though he or she may have a genius for approaching a problem from his/her point of view, to put his or her real talent to use for the team as a whole.

References

1. Porter ME (2008) The five competitive forces that shape strategy. Harvard Bus Rev 86(1):57–71
2. Kim WC, Mauborgne R (2005) Blue ocean strategy: from theory to practice. Calif Manag Rev 47(3):105–121
3. Lai-Kow C, Ming-Lu W (2002) Quality function deployment: a literature review. Eur J Oper Res 143:463–497
4. Ashby MF (2000) Multi-objective optimization in material design and selection. Acta Mater 48:359–369
5. Ulrich, Karl T, Eppinger, Steven D (2004) Product design and development, 3rd edn. McGraw-Hill, New York
6. Mital, Anil et al (2008) Product development. Elsevier Science & Technology, Burlington
7. D'Errico F (2015) Material selections by a hybrid multi-criteria approach. Springer, NY, USA

Part III
REWAS 2019: Rethinking Production

Recycling Steel Manufacturing Wastewater Treatment Solid Wastes via In-process Separation with Dynamic Separators

Naiyang Ma

Abstract In steel manufacturing, various solid wastes are generated in wastewater treatment. Iron, carbon, and fluxes (CaO and MgO) are the main beneficial components in these solid wastes for recycling in the ironmaking and steelmaking process. However, the wastewater treatment solid wastes often also contain some undesirable components. Separation of those unwanted components from the wastewater treatment solid wastes is a prerequisite to recycle the solid wastes safely, economically, and environmentally. In this contribution, producing clean wastewater treatment solid wastes via dynamic separation at ArcelorMittal is reviewed and discussed, and some case studies are presented.

Keywords Steel manufacturing · Wastewater treatment · Solid wastes · Recycling · Separation · Dynamic separation

Introduction

In steel manufacturing, wastewater is constantly generated in many locations, such as sinter plant wet scrubbers, blast furnace wet scrubbers, BOF spark boxes, BOF wet scrubbers, casters, and hot rolling and cold rolling mills [1, 2]. The wastewater is routinely treated for reuse of the water in the steel manufacturing process and for blowdown of the water to the environment. The wastewater treatment generally includes removal of suspended solids, control of heavy metals, control of oil and grease, breakpoint chlorination, thermal treatment, sewage treatment, and biological treatment [2]. In the various treatments, chemical compounds are added into the wastewater to destroy undesirable components, to convert soluble components into precipitated particles and to have minute particles coalesce into large particles. Sedimentation, centrifugal separation, and filtration are commonly used for removal of the suspended solid particles from the wastewater. Eventually, along with the

N. Ma (✉)
ArcelorMittal Global R&D - East Chicago Laboratories, 3001 E Columbus Dr., East Chicago, IN 46312, USA
e-mail: Naiyang.Ma@arcelormittal.com

G. Gaustad et al. (eds.), *REWAS 2019*, The Minerals, Metals & Materials Series,
https://doi.org/10.1007/978-3-030-10386-6_9

regeneration of clean water, various solid wastes are simultaneously generated in the wastewater treatment process.

The wastewater treatment solid wastes are generally rich in iron and/or carbon and/or fluxes (CaO and MgO). Iron, carbon, and fluxes are beneficial components needed by the ironmaking and steelmaking process. However, these solid wastes also contain some undesirable components like zinc, alkalis, oil, sulfur, chlorine, etc. [2–4]. If these undesirable components return to the ironmaking and steelmaking process in some large quantities with the wastewater treatment solid wastes, they may damage ironmaking and steelmaking equipment, destabilize ironmaking and steelmaking operations, degrade steel qualities, and cause environmental violations [2, 5]. Therefore, how to produce clean solid wastes with minimal undesirable components is a great challenge for sustainable recycling of the wastewater treatment solid wastes.

The undesirable components in the steel manufacturing wastewater treatment solid wastes may come from several sources. Raw materials like iron ore, coke, coal, fluxes, and steel scrap can bring zinc, lead, alkalis, sulfur, and chlorine into the steel manufacturing process. Oil and grease are widely applied in steel production as lubricant and coolant for vehicles, machines, bearings, hydraulic lifters, and rolling rolls and they can find ways into wastewater. Alkalis, sulfur, and chlorine can also come from chemical compounds added to the wastewater treatment process.

Separation plays a key role in producing recyclable steel manufacturing wastewater treatment solid wastes. There are three general separation strategies [6–11]. One is to separate the undesirable components from raw materials and to prevent the undesirable components from entering the steel manufacturing process. This strategy will significantly increase raw material cost for further stricter raw material specifications. In addition, this strategy is unable to prevent oil/grease and alkali-based, sulfur-based, and chlorine-based wastewater treatment chemicals from entering the wastewater. The second strategy is to treat the dirty end-of-pipe solid wastes which are already formed and stabilized. Separation of the multiple undesirable components from the end-of-pipe wastewater treatment sludge is generally not economical. The third strategy is in-process separation to selectively collect solid particles in the wastewater treatment process so that undesirable components can be separated from the beneficial components. In-process separation has the advantages of lower separation cost, lower energy consumption, and less environmental concerns than the other two separation strategies.

One of the in-process separation scenarios is to selectively collect suspended solid particles from the turbulent wastewater in the early stage of the wastewater treatment process with dynamic separators. In the rest of this paper, production of clean wastewater treatment solid wastes with dynamic separation will be reviewed and discussed, and some case studies at ArcelorMittal will be presented.

Overview of Separating Suspended Solids in the Existing Wastewater Treatment Process

In the existing wastewater treatment process of steel manufacturing, the main objective is to produce clean water for recycling of the water in the steel manufacturing process and for blowdown of the water to the environment. Sedimentation followed by filtration is the most common practice for removal of the suspended solids and for regeneration of clean water. The sedimentation consists of settling by gravity and is carried out in various settling devices, such as scale pits, settling tanks, thickeners, clarifiers, inclined plate separator, and so on.

The sedimentation can take place in a single settling device or a series of settling devices. Scale pits and settling tanks are used for pretreatment of the wastewater to remove coarse suspended solid particles. Thickeners, clarifiers, and inclined plate separators are utilized for removal of fine suspended solid particles and for precipitation, coagulation, and flocculation of soluble undesirable components. For producing recyclable water in the steel manufacturing process, sedimentation alone is generally sufficient. However, for blowdown of the water to the environment, further treatment and removal of the very fine suspended solids by filtration are required to meet environmental regulations.

Before the wastewater enters the settling devices, chemical compounds are sometimes added into the wastewater for destruction, precipitation, coagulation, and flocculation of undesirable components. The wastewater in the settling devices is moving very slowly and leaves sufficiently long retention time for the suspended solids to settle. The settled solids are reclaimed from the top or pumped from the bottom of the settling devices and delivered to dewatering facilities, where scale or grit or sludge is generated. Unsettled suspended solids leave the settling devices with the overflow water. They either return to the steel manufacturing process with recycled water or are captured by filters and sent back to the settling devices by back-flush water. Clearly, the existing practice of removing suspended solids from the wastewater does not separate beneficial components from undesirable components.

In an ArcelorMittal USA blast furnace ironmaking plant, wastewater from the wet scrubber is treated in thickeners first to remove most of the suspended solids. Polymers are added into the thickeners to accelerate sedimentation of the suspended solid particles. Part of the overflow wastewater is recycled to the scrubber and rest of the wastewater is further treated in a clarifier to remove fine suspended solids. The overflow water from the clarifier is recycled for granulation of blast furnace slag. The slag granulation wastewater is treated in a settling pond. Part of the wastewater is recycled back to the slag granulation process. Part of the wastewater for blowdown undergoes various treatments for destruction, precipitation, coagulation, and flocculation of undesirable components. Suspended solids in the blowdown water are separated by settling in a clarifier and by being filtered in a sand filtration device followed by an activated carbon filter. Blast furnace sludge from the thickeners and the clarifier before the slag granulation is generated in the process, and the sludge is unrecyclable in general due to high levels of undesirable components, especially zinc.

In an ArcelorMittal USA BOF steelmaking plant, suspended solids in BOF spark box wastewater settle in settling tanks. Overflow of the spark box wastewater is combined with BOF wet scrubber wastewater. The combined wastewater is treated in thickeners first. Part of the overflow wastewater from the thickeners is recycled back to the wet scrubbers, and rest of the overflow wastewater is filtered by multimedia filters. BOF sludge is generated in the process, and the sludge cannot be recycled to the sintering–ironmaking process due to high concentrations of undesirable components, especially zinc.

In an ArcelorMittal USA hot strip mill, wastewater takes mill scale into scale pits. Part of the wastewater in the pits is pumped back into the hot strip mill for reuse. Rest of the wastewater is further treated in clarifiers, various reactors, and multimedia filters. Issues with the current practice include high oil concentration in the reclaimed pit mill scale and high generation rate of oily mill scale sludge from the clarifiers.

In summary, in the existing wastewater treatment process of steel manufacturing, gravity separation of the suspended solids from wastewater is the mainstream separation method, and the existing process is not deemed to produce clean recyclable wastewater treatment solid wastes, instead the solid wastes become paradises of various undesirable components. The pretreatment of the wastewater for removal of scale and grit could produce recyclable solid wastes, but it has issues of high undesirable components in the collected solid wastes, high moisture content in the solids, and low efficiency of the scale pits and the settling tanks.

Dynamic Separation for Production of Clean Wastewater Treatment Solid Wastes

There are two kinds of suspended solids in the settling devices. One is pre-existing suspended solids in the incoming wastewater and the other is newly formed suspended solids in the settling devices. The newly formed suspended solid particles are products of precipitation reactions of soluble undesirable components with added chemical compounds. These suspended solids of undesirable components cannot be recycled into the ironmaking and steelmaking process. When the formation of undesirable particles takes place in the settling devices, the undesirable components can contaminate the pre-existing suspended solids by coating, bonding, and mixing.

The pre-existing suspended solid particles in the incoming wastewater can be divided into three categories. The first category of the solid particles is mainly composed of beneficial components, the second mainly undesirable components, and the third beneficial components coated by or bonded with undesirable components. The suspended solid particles of beneficial components are more recyclable if containing less undesirable components. Therefore, clean recyclable wastewater treatment solid wastes can only be produced prior to the settling devices and before the addition of various chemical compounds.

Separation of suspended solids by gravity in settling devices is regarded as static separation which is characterized by slow motion of the wastewater, minimal disturbing to the wastewater, and long retention time of the suspended solid particles in the settling devices. The disadvantages of the static separation include contamination of undesirable components to beneficial components, large space required, and great hold-up of the wastewater. For removing soluble undesirable components, the static settling devices might be necessary. However, for producing recyclable wastewater treatment solid wastes, the static separation is far from satisfaction.

Different from static separation, dynamic separation will make use of strong turbulence and fast motion of the wastewater to separate clean solid particles from the wastewater and from the undesirable components. The dynamic separation is characterized by fast motion and strong turbulence of the wastewater. Objective of dynamic separation is to produce clean recyclable wastewater treatment solid wastes in addition to making clean water.

There are a few reasons why dynamic separation can separate beneficial components from undesirable components. First of all, in dynamic separation, large particles and/or heavy particles will be separated, but small particles and/or light particles will stay with the wastewater. It has been validated that concentrations of undesirable components in solid wastes are reversely proportional to particle size and density [12–15]. As a result, large and/or heavy particles will contain less undesirable components than small and/or light particles. Hence, dynamic separation can separate beneficial components from undesirable components.

Second, in strongly turbulent wastewater, liquid undesirable components, such as oil and grease, will disperse into small droplets. Compared to large patches of the liquid undesirable components formed in the static separation, the small droplets have less possibility to contaminate the suspended solid particles due to the strong drag forces on the solid particles and on the oil droplets [16–18].

Third, the strong turbulence will accelerate dissolution of soluble undesirable components, such as alkali chlorides [19]. The dissolved undesirable components will not be collected by dynamic separators and will stay in the wastewater to be further treated.

As to dynamic separators, any separating devices can work as long as they can separate suspended solid particles from the wastewater and from the undesirable components in the wastewater by making use of strong turbulence and fast motion of the wastewater. Such separators may include hydrocyclones, magnetic liquid traps, deflective separators, and various swirl concentrators (also called as vortex separator). These dynamic separators are already widely available on market. However, it has not been well understood how to produce clean recyclable wastewater treatment solid wastes with these dynamic separators.

Dynamic Separation of Grit from BOF Spark Box Wastewater with a Swirl Concentrator

One of ArcelorMittal USA BOF steelmaking plants is equipped with spark boxes in its BOF off-gas cleaning system. A series of water spray nozzles are installed in the spark boxes. While the BOF vessels are running, the spray nozzles inject atomized water to cool the hot off-gas. Part of the water evaporates and becomes steam going with the BOF off-gas. Rest of the water washes the off-gas and takes dust particles to the wastewater treatment system.

A block diagram of the spark box wastewater treatment system is shown in Fig. 1. Make-up water is pumped into the second mixing tank so that wastewater with sufficient flowrate can be pumped and delivered into the settling trailer box without causing clogging issues in the pipelines. The spark box wastewater is top charged into the trailer box. Grit particles settle onto the bottom of the trailer box, and overflow wastewater takes unsettled solids and flows to the clarifier for further treatment with other wastewater streams.

Problems with the existing spark box wastewater treatment shown in Fig. 1 include high water/grit ratio (about 9–1 by weight) in the trailer box, low separation efficiency of the trailer box, and high zinc in the separated solids.

A swirl concentrator coupled with a screw classifier was selected for the study of separation of grit from the spark box slurry. Integration of the swirl concentrator and the classifier is schematically shown in Fig. 2. The swirl concentrator was connected to the pipeline right before the trailer box shown in Fig. 1. The spark box wastewater tangentially entered the concentrator. Suspended solids in the wastewater settled and fell onto the bottom of the concentrator and cleaned water floated up and exited as overflow. Underflow water took the settled solids and tangentially flew into the screw classifier chamber. The screw classifier moved the settled solids up, dewatered the solids, and discharged the solids.

The study was a full-scale industrial trial. The swirl concentrator received all spark box wastewater which had been directed into the trailer box. The influent wastewater and the effluent wastewater were sampled. Concentrations of total suspended

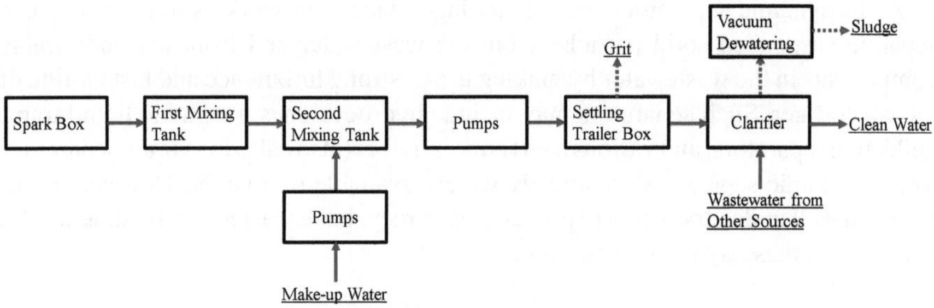

Fig. 1 Block diagram of spark box wastewater treatment at an ArcelorMittal USA BOF steelmaking plant. Solid arrowed lines represent water flows and dash arrowed lines denote flows of solids

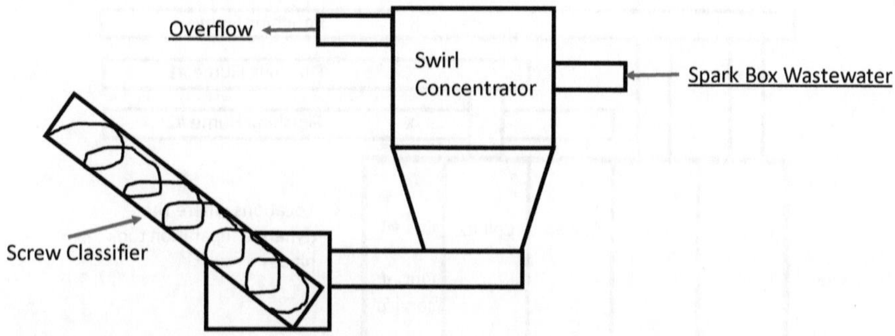

Fig. 2 Integration of a swirl concentrator with a screw classifier

solids in the influent and the effluent samples were analyzed. The separated solids and the solids recovered from the influent samples, and the effluent samples were analyzed for size distributions and chemical compositions. The separation efficiency was calculated by the following formula:

$$\eta = \frac{(S_{in} - S_e)}{S_{in}} \times 100\%, \quad (1)$$

where η is the separation efficiency in %wt, and S_{in} and S_e are the suspended solid concentrations in the influent and effluent water samples in g/cc (gram of solids per cubic centimeter of water sample).

The trial lasted for 4 days, and spark box wastewater generated in 22 heats was treated with the swirl concentrator. On average, 587 kg grit was collected for every heat of steel made. Average moisture level in the grit was about 17% by weight. The separation efficiency was between 95 and 98%. Average zinc concentrations in the samples of influent suspended solids, effluent suspended solids, and the separated grit were 0.17%, 1.63%, and 0.08%, respectively. As a result, the dynamic separation of the suspended solids from the BOF spark box wastewater with the swirl concentrator resulted in rather dry recovered solids, high separation efficiency, and significant reduction of zinc in the reclaimed solids.

Dynamic Separation of Mill Scale from Hot Rolling Mill Wastewater with a Magnet

An ArcelorMittal USA hot strip mill had two scale pits receiving wastewater from the hot rolling process. The #1 pit received wastewater from the reheat furnace discharge ends, the first descaling table, and a few roughing mill stands. The #2 pit received wastewater from last a few roughing mill stands, the second descaling table, and the

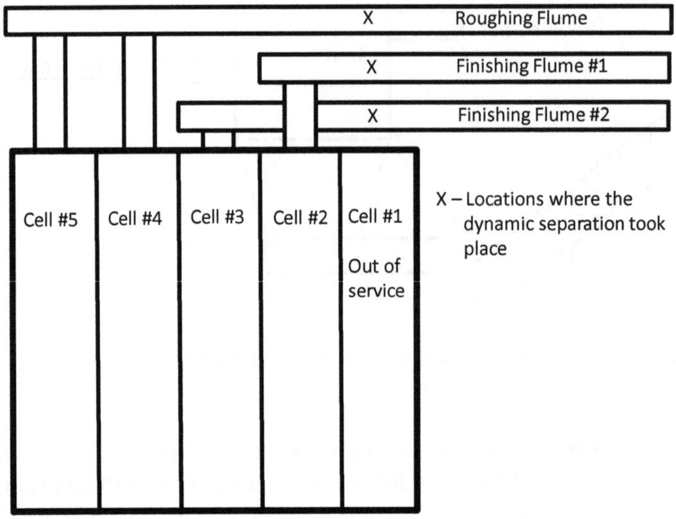

Fig. 3 The #2 scale pit and the flumes connected to the pit at an ArcelorMittal USA hot strip mill

finishing mill stands. The wastewater from both pits was combined and pumped to multimedia filtration tanks.

Mill scale settled in the scale pits was top reclaimed by a clamshell bucket at the #1 pit and by chain scrapers at the #2 pit. The #2 pit had constantly generated oily mill scale which could not be recycled at the sinter plant since the high concentration of oil in the mill scale could cause high volatile organic compounds (VOC) emissions at the sinter plant. It had been a great challenge to reduce the oil concentration in the mill scale.

The #2 pit is schematically shown in Fig. 3. The pit consisted of 5 cells. The first cell had been out of service. The second and third cells received wastewater from finishing flumes #1 and #2, respectively. The roughing flume wastewater split right before the scale pit and flew into cells #4 and #5, respectively. The wastewater in the flumes was moving at high speed in order to prevent any sedimentation and clogging of mill scale particles in the flumes. There was one chain scraper for each active cell. While the plant was running, the chain scrapers continuously removed mill scale from the scale pit. Trailer boxes were positioned at the discharge ends of the chain scrapers and the chain scrapers discharged mill scale into the trailer boxes.

A 1.1T permanent magnet was chosen for this study. The magnet was mounted on a stainless steel pole and positioned in the flumes. The magnet attracted mill scale particles. The mill scale was recovered and analyzed. In the same time, samples from the trailer boxes were taken and analyzed. Figure 4 demonstrates the mill scale attracted by the 1.1T magnet mounted on a stainless steel pole.

The trial lasted for 3 days. Average oil concentrations in the samples recovered by the magnet and by the chain scrapers are presented in Table 1. It is clear that the mill

Fig. 4 Mill scale recovered by the 1.1T magnet separator mounted on a stainless steel pole

Table 1 Average oil concentrations in mill scale samples, %wt in dry basis

Separation method	Roughing	Finishing #1	Finishing #2
Static separation	5.15	0.27	1.30
Dynamic separation	0.27	0.04	0.07

scale samples recovered by dynamic separation with the strong magnet contained significantly less oil than the mill scale samples from the existing static separation with settling in the pit and reclaiming by the chain scrapers.

Conclusions

In the existing wastewater treatment process of steel manufacturing, static separation is the mainstream of separation of suspended solids from the wastewater. Existing wastewater treatment is clean water oriented, but solid wastes are notorious nuisances contaminated with various undesirable components. Dynamic separation, making use of strong turbulence and fast motion of the wastewater, can produce significantly cleaner wastewater treatment solid wastes. Effective locations of dynamic separation are prior to static settling devices and before adding chemical compounds for destruction, precipitation, coagulation, and flocculation of undesirable components. In the trials of separating grit from BOF spark box wastewater with a swirl concentrator and separating mill scale from hot strip mill wastewater with a strong magnet at ArcelorMittal, clean wastewater treatment solid wastes were produced with significantly low zinc and oil contents, respectively. It thus has been validated that dynamic separation can produce clean recyclable wastewater treatment solid wastes in the wastewater treatment process of steel manufacturing.

References

1. U.S. Environmental Protection Agency Office of Water (2002) Development document for final effluent limitations guidelines and standards for the iron and steel manufacturing point source category. United States Environmental Protection Agency report, EPA-821-R-02-004
2. Remus R, Aguado Monsonet MA, Roudier S, Sancho LD (2013) Best available techniques (BAT) reference document for iron and steel production. Publications office of the European Union, Luxembourg. http://eippcb.jrc.ec.europa.eu/reference/BREF/IS_Adopted_03_2012.pdf. Accessed 22 Aug 2018
3. International Iron and Steel Institute (1994) The management of steel plant ferruginous by-products. International Iron and Steel Institute, Brussels, Belgium
4. American Iron and Steel Institute (2001) Steel technology roadmap—iron unit recycling. US DOE. http://energy.gov/sites/prod/files/2013/11/f4/roadmap_chap3.pdf. Accessed 22 Aug 2018
5. Geerdes M, Chaigneau R, Kurunov I, Lingiardi O, Ricketts J (2015) Modern blast furnace ironmaking, 3rd edn. IOS Press BV, The Netherlands
6. Ma NY (2014) Sustainable recycling of solid wastes via in-process separation. In: Yurko J et al (eds) EPD congress. Wiley Publishing Company, pp 529–536
7. Ma NY (2008) On the separation of zinc from dust in ironmaking and steelmaking off-gas cleaning systems. In: Howard SM (ed) EPD congress 2008. The Minerals, Metals & Materials Society, pp 547–552
8. Ma NY, Andrade MW (2018) Integrated treatment of offgas, wastewater and solid wastes for minimization of environmental footprints. In: Proceedings of AISTech annual conference. AISTech
9. Ma NY, Andrade MW (2015) In-process separation of zinc from blast furnace offgas solid wastes. Iron Steel Technol 3:84–95
10. Ma NY, Atkinson M, Neale K (2012) In-process separation of zinc from BOF offgas cleaning system solid wastes. Iron Steel Technol 4:77–86
11. Ma NY (2016) Recycling of basic oxygen furnace steelmaking dust by in-process separation of zinc from the dust. J Clean Prod 112:4497–4504
12. Uno S et al (1979) Dezincing equipment and operation based on wet classification of wet-cleaned BF dust. Nippon Steel Tech Rep 13:80–84
13. Heijwegen CP, Kat W (1983) Beneficiation of blast furnace sludge. World Steel Metalwork 5:35–39
14. Butterworth P, Linsley K, Aumonier J (1996) Hydrocyclone treatment of blast furnace slurry within British steel. La Revue de Metallurgie-CIT 6:807–815
15. Toda H et al (1979) Separation of nonferrous metals from blast furnace flue dust by hydrocyclone. Nippon Steel Techn Rep 13:73–79
16. Ma NY (2012) In-process separation of mill scale from oil at steel hot rolling mills. In: Zhang L et al (eds) EPD congress 2012. Wiley, pp 323–329

17. Ma NY, Houser JB, Wood LA (2018) Production of cleaner mill scale by dynamic separation of the mill scale from the fast moving flume water at a hot rolling mill. J Clean Prod 176:889–894
18. Ma, NY (2017) A method for separating mill scale from wastewater. International Patent WO2017037540 A1, 9 March 2017
19. Ma NY (2009) Production of high-grade potassium chloride from a sinter plant baghouse dust. In: Howard S et al (eds) EPD congress 2009. The Minerals, Metals & Materials Society, pp 927–929

Tannic Acid—A Novel Intumescent Agent for Epoxy Systems

Matthew Korey, Alexander Johnson, William Webb and John A. Howarter

Abstract Tannic acid (TA) is a bio-based high molecular weight organic (HMWO), aromatic molecule. Although biologically sourced, TA currently pollutes industrial wastewater streams, and there is an industrial desire to find applications to downcycle this molecule. Epoxy thermosets have revolutionized many industries, but are too flammable to be used in many applications without additives which augment its flame retardancy (FR). Many flame retardants used in epoxy thermosets are synthesized from petroleum-based monomers leading to significant environmental impacts at the industrial scale. Various bio-based modifiers have been developed to improve the FR of the epoxy resin; however, increasing FR of the system without trade-offs with other properties has proven to be challenging. In this work, TA was incorporated into the thermoset. The molecule was found to increase the intumescence of the system without significant changes to the Tg and strength paving the way for biological intumescent agents to be used industrially.

Keywords Epoxy · Tannic acid · Thermoset chemistry and properties

Introduction

For the past several decades, halogenated flame retardants (HFRs) have been used as additives in a variety of different polymer systems in order for these materials to meet the stringent flame retardant (FR) standards required in the US and UK. HFRs are prevalent throughout the world in a variety of materials including pentabromodiphenyl ether (PentaBDE) in furniture, hexabromocyclododecane (HBCD) in building insulation, and 1,2,3-trichloro-4-(2,3,4-trichlorophenyl)benzene (Aracor) and tetrabromobisphenol A (TBBPA) in the printed circuit boards of electronics. However, more recent research has shown that bioaccumulation of HFRs in humans can lead to lower birth weight and length of children[5], impair neurological development, and even cause cancer. Furthermore, the health impacts of halogenated flame

M. Korey (✉) · A. Johnson · W. Webb · J. A. Howarter
West Lafayette, IN, USA
e-mail: mkorey@purdue.edu

G. Gaustad et al. (eds.), *REWAS 2019*, The Minerals, Metals & Materials Series,
https://doi.org/10.1007/978-3-030-10386-6_10

retardants are felt more strongly in less-wealthy regions of the world where individuals are found to have elevated concentrations of HFRs in breast milk and water. Due to these concerns, there is currently a high demand in industry for alternative chemicals to HFRs in a variety of applications such as automotive, electronics, and construction.

Researchers have developed synthetic, non-brominated flame retardant alternative compounds like triphenyl phosphate (TPP) for electronics and *tris*(1-chloro-2-propyl) phosphate for building insulation, but these compounds still bioaccumulate and have high persistence and toxicity. There has also been a significant amount of research on biologically based flame retardant additives, such as cellulose, deoxyribonucleic acid, lignins, condensed tannins, and tannic acid but these compounds are often limited in application by a solubility mismatch between compounds and hydrophobic polymer systems, such as epoxy. There are several methods that have been used to prevent this problem. Surfactant can be used to assist dispersion and increase the overall dispersion of the additive, but excess surfactant can affect the resulting thermomechanical properties—which are important for epoxy. Using solvent has been found to greatly increase the dispersion of these additives in polymers but will also lead to the environmental emission of volatile organic compounds. Chemically modifying the additive to better match the system into which it is being inserted can increase the interfacial compatibility of the additive and the polymer matrix, but can potentially make the product more difficult to commercialize due to the added chemical steps in manufacturing. Impregnation of the additive is another method for increasing dispersion of these particles, but it requires a large-scale and optimized lyophilization process to preserve shape integrity. Due to these concerns, the commercializability of biologically sourced additives into polymer matrices is dependent on the minimization of the complexity of the polymer dispersion processing mechanism.

There is a need for a new, simpler processing mechanism for the dispersion of biologically sourced HFR alternatives in epoxy. However, there is very limited research on this topic for several bio-based chemicals above, especially tannic acid (TA). TA is a natural polyphenolic compound found in nuts, galls, seeds, and tree bark. TA is a hydrolyzable tannin that—when dispersed into polymer matrices—has been found to enhance the thermal properties of plastics. When thermally activated in the presence of a fire, TA was found to crosslink into the polymer network forming a char barrier between the surface of the material and the propagating flame, therefore retarding the spread of the fire. Because of this, TA has been explored as an FR additive in a variety of polymer matrices including poly(lactic acid), nylon 6, formaldehyde-based polymers (P), polyesters, and urethane foams. TA shows favorability for use as an HFR alternative compound, but its limited dispersibility in epoxy is the main barrier to its use in this application. Uniform dispersion of high amounts of TA in epoxy has only recently been achieved due to the distinct compatibility differences between TA and epoxy resin, but there is no research on the flame retardancy of tannic acid in epoxy.

In this study, a novel dispersion method of tannic acid in epoxy resin is explored. It is observed that increasing the temperature of the TA/epoxy resin mixture can increase overall compatibility allowing for significantly increased dispersion. Samples were heated at 105 °C in N_2 gas for 1 h and then heated at 150 °C in N_2 for 2 h. Through optical microscopy and visual inspection, it was observed that a significant increase in dispersion was achieved in all samples up to 8 wt% TA, but all samples remained a viscous liquid rather than a hard crosslinked thermoset at this loading level of TA. Samples above 8 wt% TA became too viscous to mix during the first stage of the heating procedure and were not explored for this study.

Samples were characterized using thermal and thermomechanical analyses, chemical characterization, and flame testing using mass loss calorimetry and cone calorimetry. Thermogravimetric analysis was run on the TA–DGEBA composites to further understand this stark change in compatibility through heating. Thermograms indicated significant increases in thermal stability as compared to the control samples alone. The results from TGA showed that all samples containing TA and DGEBA had elevated temperatures of thermal degradation (T_d) as compared to either control. No significant differences were found between the temperature of degradation in any composite system, but the magnitude of the maximum value of the derivative weight curve decreased as the concentration of TA in the composite system increased. A second degradation was observed in all TADGEBA composites at 425 °C and the magnitude of the maximum value of the derivative weight curve increased as the concentration of TA in the composite increased. The seven thermal degradations of the DGEBA control sample at 280 °C have been found by this group in previous research to be associated with degradation of the epoxy rings within the molecule. Similarly, the degradation of TA at 260 °C has been found by previous research to be associated with the phenol groups on the molecule. No TA–DGEBA composite samples had thermal degradation peaks at either of these values and the peak temperatures of thermal degradation associated with these composites were significantly different from the controls, although the curves were quite broad in nature.

Additionally, samples were characterized using differential scanning calorimetry and Fourier transform infrared (FTIR) spectroscopy. After crosslinking, all samples showed a glass transition temperature at −20 °C regardless of TA loading into solution. However, control and samples containing 1 wt% TA were shown on first heat to have a melting temperature at 34 °C, and no significant changes in melting temperature were observed, whereas samples containing 3.2 wt% or more TA were found to have no melting peak. There were no observed crystallization peaks on the cooling curve and no observed melting points on the second heating curve, although the glass transition was observed in all samples regardless of TA loading. FTIR was run on samples after heating to help understand the behavior observed in both TGA and DSC for all TA–DGEBA systems. Absorbance values for the epoxy peak at 915 cm^{-1} were found to significantly decrease as the samples were heated and with increased TA loading.

The initial explanation for this behavior is that the observed change in compatibility in the solutions was not a product of TA simply dissolving into the DGEBA system but was instead a product of TA reacting with it (Scheme 1). The proposed mechanism for this phenol group on the surface of the TA molecule reacts with the epoxy group on the DGEBA causing the oxirane ring to open and results in a nonreversible chemical linkage between the two molecules. FTIR absorbance values corroborate this claim, as the increased presence of oxirane rings resulted in a decreased absorbance as TA was loaded in the sample. Considering that TA has 25 potential reactive hydrophilic phenol groups on its surface, the reaction between TA and DGEBA results in a more hydrophobic DGEBA-functionalized surface on TA molecules which is far more compatible with the DGEBA matrix. The DGEBA-functionalized TA molecules are then more likely to dissolve into the matrix and become further functionalized. These results are consistent with previous literature.

The TA–DGEBA solutions show the initiation of T_d within the range of both TA and DGEBA control samples, but a peak that is quite broad indicating a broad range of chemical surface functionalization and little uniformity between surface-functionalized TA systems in the resulting prepolymer system. Thermograms showed decreased molecular mobility and a resulting delay in the initiation of T_d due to the presence of large DGEBA-functionalized TA molecules, and this initiation was consistent between all TA–DGEBA solutions. As the magnitude of the thermal degradation at 325–340 °C in all DGEBA–TA samples decreased as TA wt% loading in the samples was increased, the magnitude of the thermal degradation at 425 °C increased. Previous researchers have found that the thermal degradation at 425 °C was due to chemical crosslinks in solutions containing high amounts of TA. The thermograms for this work indicate that the ability for TA–DGEBA solutions to crosslink increased as TA loading was increased. These results indicate that at low loading levels, DGEBA reacts with TA to functionalize its surface, but the odds of DGEBA-functionalized TA molecules finding additional phenol groups in solution are rare—due to the extreme excess of epoxy groups in the solution. However, at loading levels at or above 1% TA, the molecules in the solution are present enough to find each other and react forming a crosslinked network. DSC results corroborate this claim, as surface-modified TA molecules act as nucleation sites in the solution resulting in higher crystallinity and a larger magnitude to the crystallization peak in the first heat. This crosslinked network is more thermally stable than the DGEBA-functionalized TA molecules in solution and degrades at higher temperature—425 °C. However, due to the broad nature of the degradation peak at 425 °C, there is a distribution of crosslink densities throughout the system, with some crosslinks being less dense and degrading below 400 °C, while others are more dense and degrade near 450 °C.

After hardening the TA–DGEBA prepolymers using GP2074, the epoxy samples were characterized to determine their mechanical properties. At low TA loading levels ($\leq$1 wt% TA), an observable decrease in ultimate strength, stiffness, and fracture toughness were seen. However, at higher TA loading levels ($\geq$3 wt% TA), samples were not found to be significantly different from the control sample. The preliminary explanation for this behavior is that surface-functionalized DGEBA–TA molecules in small concentrations decreased the overall crosslinking density of the sample resulting in a less strong, stiff, and tough thermoset. These results are consistent with the results obtained from DMA for the prepolymer systems.

To understand this behavior, epoxy samples were analyzed using mass loss calorimetry (MLC) and limited oxygen index (LOI) analysis. Results indicate that at low TA loading levels in DGEBA that have been hardened with GP2074, there was no significant change in the peak heat release rate. Surprisingly, in samples containing TA in 3.2 wt% or higher, the peak heat release rate actually increased and did not correlate with greater TA loading. The time to ignition did not significantly change from sample to sample, and in some cases, decreased with TA loading. It is important to note that air flow in the hood was observed parallel to the surface of the sample while collecting this data and may have had an impact on the heat release rate data for all samples. The mass loss rate behavior was the same regardless of TA loading. Past a 1% TA loading level, the peak of the mass loss rate curve increased significantly as compared to the control but not relative to the amount of TA present in the sample. These results indicate that TA which is surface functionalized with DGEBA is unable to function as a flame retardant once hardened with GP2074. The total mass loss was consistent between all samples; the TA did not increase the amount of the sample remaining after MLC analysis. Samples containing TA showed slightly decreased percent mass loss during the duration of the test which indicated that in the fire itself, TA can form small amounts of char, but the percent mass loss was most decreased in the highly TA-loaded samples.

Although the quantifiable data obtained by MLC shows increases in the peak heat release rate and mass loss rate of highly loaded TA samples, visually, the samples ignited quite differently. With increased TA loading, the fire propagated slowly across the surface. The samples that had decreased mass loss and delayed fire propagation across the surface were also the samples that showed phase separation and clumping when analyzed with optical microscopy. The initial explanation is that unmodified TA was available in the system to delay the propagation of the fire. It is also of note to mention that samples containing TA did not show significantly increased char formation by mass but did show an increase in intumescent behavior. Limited oxygen index was also performed on GP2074-hardened samples of TADGEBA, and all samples with TA loading in any amount did not appear to have significantly increased or decreased limited oxygen index (data is available in the Supplementary Materials and Data section). This result indicates that the presence of TA does not noticeably increase or decrease the ignitability of the samples, which matches the data for the time to ignition measured by MLC and cone calorimetry.

These results indicated that the ability for TA to retard the spread of fire is directly proportional to the presence of the phenol groups on the surface of the molecule. Removing the flame retardant –OH groups from the surface of the molecule did not allow for the TA molecules to retard the spread of the fire significantly, something not seen before in literature for this molecule. These results also indicated that it was not the phenol groups that affect the molecule's ability to increase a polymer's intumescent behavior, but instead it is the internal structure of the molecule that results in intumescent behavior. Using this knowledge, a more flame retardant and intumescent TA-based additive for epoxy systems can be developed. This result is consistent with what was expected after analyzing the unhardened TA–DGEBA composites where two temperatures of thermal degradation were observed. In samples containing TA, there are regions of high crosslink density and low crosslink density, so regions of low crosslink density will degrade first, pushing up on the resultant char causing an increased expansion of the apparent volume of char. Cone calorimetry was also performed and preliminary results corroborated these conclusions (data is available in the Supplementary Materials and Data section).

This work shows that in order to increase the compatibility of TA in DGEBA, the system must be heated. However, upon heating the TA and DGEBA, a surface functionalization of the molecule takes place in which the galloyl groups are reacted and disappear, as evidenced by chemical characterization. Once this takes place, the molecule no longer retains its flame retardant ability in the resulting thermoset, although it does serve as an intumescent agent resulting in a significant increase in the volume of the resulting char. Although the apparent volume of the char does increase with TA loading, the mass of the resulting char does not significantly change between samples. These results indicate that the phenol groups on the surface of the molecule are directly responsible for its ability to serve as a flame retardant in this application, and the inner structure of TA is responsible for its intumescent behavior. Further, it was discovered that TA loading impacted its resulting behavior in the prepolymer system. Samples with less than 3 wt% TA were found to surface functionalize the TA molecule without crosslinking, whereas samples with 3 wt% TA or more crosslinked resulting in increased thermal stability within the system and increased apparent crosslinking density. The results from this work answer many questions previously unanswered by the scientific literature, and it is believed that this work can be used by future researchers to determine the most effective way to incorporate TA into an epoxy network.

Effect of CO Partial Pressure on Extraction of Alumina from Coal Fly Ash During Carbothermal Reduction Process

Yang Xue, Wenzhou Yu, Zhixiong You and Xuewei Lv

Abstract Comprehensive utilization of coal fly ash is a promising industry with great environmental protection and resource recycling. Alumina extraction from coal fly ash usually produces a large amount of silicon-containing waste, which would have adverse effects on the environment. In order to reduce the production of silicon-containing slag, a new process of carbothermal reduction is proposed. Owing to the generation of CO in solid phase reaction, it is necessary to study the effect of CO partial pressure on the reaction. By changing argon gas flow rates, the effect of CO partial pressure on carbothermal reduction process is analyzed. Moreover, the experiments in vacuum furnace have been studied. The results show that the mullite phase in coal fly ash can be more easily decomposed with low CO partial pressure at the same reduction temperature. This paper provides a new sight for the industrial application of alumina extraction from coal fly ash.

Keywords Coal fly ash · Carbothermal reduction · Alumina · Ferrosilicon alloys · CO partial pressure

Introduction

Coal fly ash (CFA) is an industrial solid waste which is derived from coal combustion in thermal power plants [1, 2]. The alumina content of CFA, produced by the coal-fired power plants in the southern Inner Mongolia of China, is about 40–50% [3, 4]. Therefore, CFA is a kind of potential raw material for extracting aluminum industrial. Profitable utilization of CFA in alumina extraction has attracted extensive attention recently.

Y. Xue · W. Yu (✉) · Z. You · X. Lv
College of Materials Science and Engineering, Chongqing University, Chongqing 400045, People's Republic of China
e-mail: yuwenzhoucd@cqu.edu.cn

Y. Xue · W. Yu · Z. You · X. Lv
Chongqing Key Laboratory of Vanadium-Titanium Metallurgy and Advanced Materials, Chongqing University, Chongqing 400044, People's Republic of China

G. Gaustad et al. (eds.), *REWAS 2019*, The Minerals, Metals & Materials Series,
https://doi.org/10.1007/978-3-030-10386-6_11

Techniques of the aluminum extraction from coal fly ash can be classified into acidic [5, 6] and alkali [7–10] method. The acidic method requires acid-resistant processing equipment, which makes it costly. The alkali method mainly comprises limestone sinter process and lime-soda sinter process. Usually, a large amount of silicon slag is produced simultaneously by those treatment processes.

Recently, a new approach for alumina recovery and ferrosilicon alloys preparation from CFA using Fe_2O_3 as an activator and charcoal as a reducing agent was proposed. The approach involves the following procedures. First, we need to calcine the mixture of CFA, charcoal, and Fe_2O_3 to produce aluminum and ferrosilicon alloys. Second, the calcined samples need to be grinded into fine powder. After that, aluminum and ferrosilicon alloys were separated by magnetic separation [11]. Owing to the high separation efficiency of alumina and ferrosilicon alloys, this approach could be used for both large-scale utilizations of CFA and production of ferrosilicon alloys.

Mullite is a main phase in CFA, and aluminum in CFA is mainly found in mullite phase. With addition of Fe_2O_3, mullite reduction by carbon is presented by the following reaction:

$$Al_6Si_2O_{13} + 3Fe_2O_3 + 13C = 3Al_2O_3 + 2Fe_3Si + 13CO$$

The overall reduction reaction includes the following steps: (1) carbon-oxygen chemical reaction, and (2) CO mass transfer from the reaction interface to the gas phase. The resistance due to CO mass transfer in gas phase can hinder the reduction kinetics. Therefore, it is necessary to study the effects of CO partial pressure on the reduction of mullite phase.

Experimental

Materials

CFA was obtained from a coal-fired power plant located in Inner Mongolia, China; its sizes were smaller than 0.15 mm. The chemical compositions analyzed by X-ray fluorescence spectrometry (XRF, Shimadzu XRF-1800) are as follows: 35.37 mass% Al_2O_3, 49.83 mass% SiO_2, 2.45 mass% Fe_2O_3, 6.20 mass% CaO, and 0.66 mass% MgO. The mineralogical compositions of the samples were determined by X-Ray Powder Diffraction (XRD) analysis using a Rigaku D/max 2500 PC X-ray diffractometer with Cu Kα radiation ($\lambda = 0.154$ nm, 40 kV, 150 mA) at a scan rate of 0.3°/s. Their XRD patterns are presented in Fig. 1. The main phases in the sample are mullite ($Al_6Si_2O_{13}$) and quartz (SiO_2).

Analytically pure reagent of iron oxide powder was used in experiments. Its composition was more than 99.99 mass% Fe_2O_3. Moreover, charcoal used in experiments was purchased from market, and its sizes were smaller than 0.5 mm. Its composition

Fig. 1 XRD pattern of coal fly ash raw material used in this study

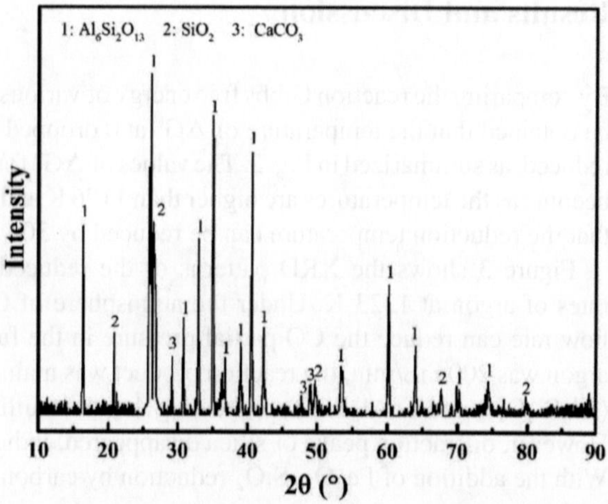

was 1.57 mass% ash content, 4.45 mass% volatile content, 6.14 mass% moisture content, and 87.84 mass% solid carbon content.

Methods

Samples were prepared from CFA, charcoal, and Fe_2O_3 with 5:2:3 mass ratios. Then they were pressed into pellets and were dried at 120 °C for 12 h before use. After that, pellets were roasted in a $MoSi_2$-heating furnace for 2 h at 1523 K. A flow of argon was maintained through the furnace to prevent oxidation of the samples. The flow rates of argon were 200, 400, and 600 mL/min.

Moreover, pellets were roasted in a graphite-heating vacuum furnace for 2 h at temperatures of 1223, 1323, and 1423 K. The gas pressure in furnace was about 0.001 atm.

Thermodynamic Analysis

The main objective of the thermodynamic analysis was to predict the effect of the reduction temperature and partial pressure of CO on the chemical composition of the reduction products. The standard free energy of the reactions was calculated using the reaction module of FactSage 6.2 software.

Results and Discussion

By comparing the reaction Gibbs free energy of various partial pressures of CO, it can be obtained that the temperature of ΔG^{θ} at 0 dropped when the CO partial pressure reduced, as summarized in Fig. 2. The values of ΔG^{θ} for the P_{CO}/P^{θ} of 1 and 0.001 atm become as the temperatures are higher than 1176 K and 872 K, respectively, implying that the reduction temperature can be reduced by 304 K.

Figure 3 shows the XRD patterns of the reduced samples with different flow rates of argon at 1523 K. Under the atmosphere of CO and Ar, increasing the Ar flow rate can reduce the CO partial pressure in the furnace. When the flow rate of argon was 200 mL/min, the reaction product was mainly composed of Fe_3Si, mullite ($Al_6Si_2O_{13}$) and $CaAl_2Si_2O_8$, indicating that the mullite in CFA did not decompose. However, diffraction peaks of silica disappeared, indicating that silica was reduced. With the addition of Fe_2O_3, SiO_2 reduction by carbon is presented by the following reactions:

$$2SiO_2 + 3Fe_2O_3 + 13C = 2Fe_3Si + 13CO$$

When the Ar flow rate was 400 mL/min, the diffraction peaks of Al_2O_3 appeared, indicating that the mullite in CFA began to decompose. When the Ar flow rate increased to 600 mL/min, the diffraction peaks of SiC appeared and those of Fe_3Si strengthened. Moreover, the diffraction peaks of $Al_6Si_2O_{13}$ disappeared. These changes indicate that low CO partial pressure could promote the reduction of mullite phase. It was because low CO partial pressure was beneficial to the escape of CO, which favored this reaction. The SEM image of the reduced sample is shown in Fig. 4. The grain size of ferrosilicon alloys is about 35 μm.

Figure 5 shows the effect of the CO partial pressure on the possible reactions during the reduction process, which exhibits that, for the same reaction, the reaction

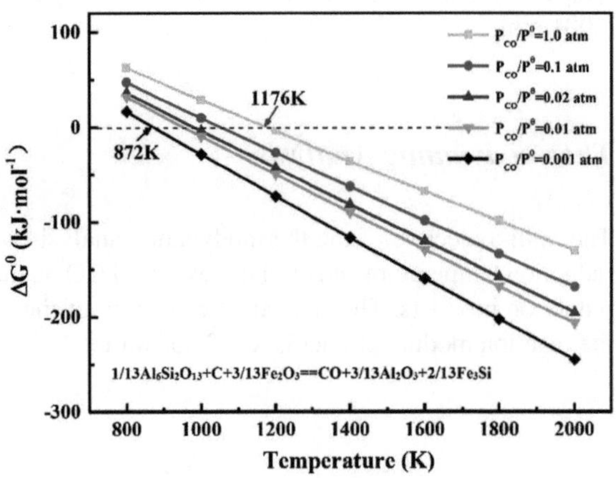

Fig. 2 Effect of the CO partial pressure on the standard free energy variation for the reaction

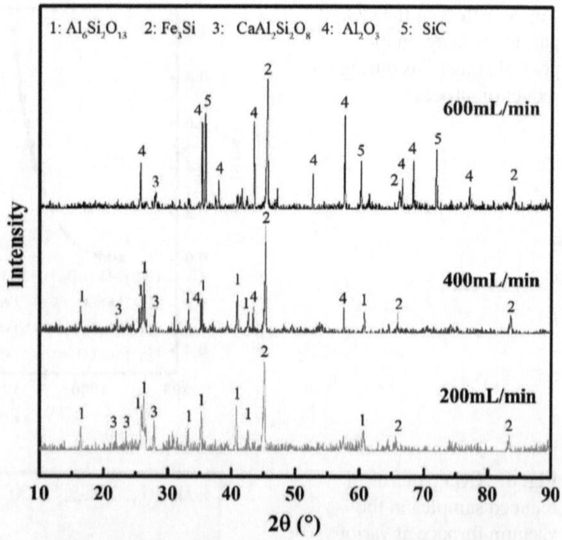

Fig. 3 XRD patterns of the reduced samples with different flow rates of argon at 1523 K

Fig. 4 SEM image of the reduced sample

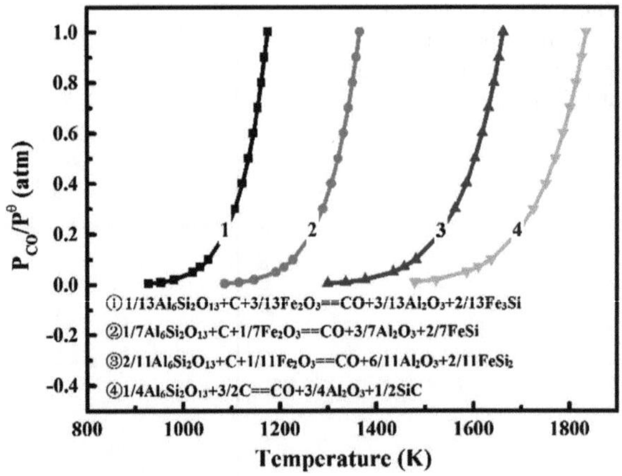

Fig. 5 Effect of the CO partial pressure on the possible reactions during the reduction process

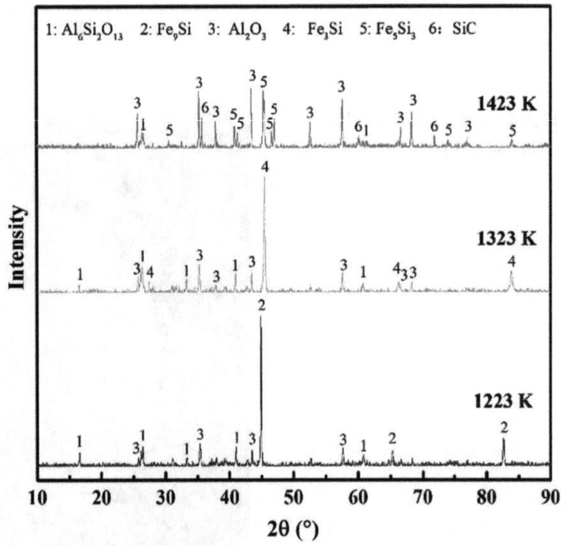

Fig. 6 XRD patterns of reduced samples in the vacuum furnace at various temperatures

temperature can be reduced by decreasing the partial pressure of carbon monoxide. Besides, Si content in ferrosilicon alloys advanced as the reduction temperature increased under the same CO partial pressure. Figure 6 shows the XRD patterns of reduced samples in the vacuum furnace at various temperatures. The gas pressure in the vacuum furnace was about 0.001 atm. The relationship between gas pressure in the vacuum furnace and CO pressure is

$$P_{VF} = P_{CO}$$

where P_{VF} is the gas pressure in the vacuum furnace and P_{CO} is the CO pressure.

When the reduction temperature was 1223 K, the reaction product was mainly composed of Fe_9Si, mullite, and Al_2O_3, indicating that a small part of mullite was reduced. At temperature of 1323 K, the diffraction peaks of Fe_3Si replaced those of Fe_9Si. When the temperature increased to 1423 K, the diffraction peaks of $Al_6Si_2O_{13}$ weakened while those of Al_2O_3 strengthened, and those of SiC appeared. Moreover, the diffraction peaks of Fe_5Si_3 replaced those of Fe_3Si. These changes indicate that Si content in ferrosilicon alloys advanced as the reduction temperature increased, which coincided with the theory. The reason for the existence of mullite at 1423 K was that the reaction time was only 2 h.

Conclusions

(1) In the $MoSi_2$ heating furnace, increasing the Ar flow rate can reduce the CO partial pressure in the furnace, which was beneficial to the reactions.
(2) The reaction temperature in vacuum furnace with gas pressure of 0.001 atm is about 100 K lower than that in $MoSi_2$-heating furnace with Ar flow rate of 600 mL/min.
(3) Increasing the reduction temperature is beneficial to the increase of silicon content in ferrosilicon alloys.

References

1. Ahmaruzzaman M (2010) A review on the utilization of fly ash. Prog Energy Combust Sci 36(3):327–363
2. Ding J, Ma S, Shen S et al (2017) Research and industrialization progress of recovering alumina from fly ash: a concise review. Waste Manag 60
3. Blissett RS, Rowson NA (2012) A review of the multi-component utilisation of coal fly ash. Fuel 97(7):1–23
4. Cao DZ, Selic E, Herbell JD (2008) Utilization of fly ash from coal-fired power plants in China. J Zhejiang Univ-Sci A (Appl Phys Eng) 9(5):681–687
5. Nayak N, Panda CR (2010) Aluminium extraction and leaching characteristics of Talcher Thermal Power Station fly ash with sulphuric acid. Fuel 89(1):53–58
6. Wu CY, Yu HF, Zhang HF (2012) Extraction of aluminum by pressure acid-leaching method from coal fly ash. Trans Nonferrous Metals Soc China 22(9):2282–2288
7. Bai GH, Teng W, Wang XG et al (2010) Alkali desilicated coal fly ash as substitute of bauxite in lime-soda sintering process for aluminum production. Trans Nonferrous Metals Soc China 20(s1):s169–s175
8. Padilla R, Sohn HY (1985) Sodium aluminate leaching and desilication in lime-soda sinter process for alumina from coal wastes. Metall Trans B 16(4):707–713
9. Jiang JC, Zhao YC (2008) Current research situation of Al extraction from fly ash. Nonferrous Metals Eng Res (in Chinese)
10. Jiang X (2017) Technique situation and development trend of extracting alumina from coal fly ash. Nonferrous Metals Eng (in Chinese)
11. Wang CL, Yang HF, Jiang BP et al (2014) Recovery of iron from lead slag with coal-based direct reduction followed by magnetic separation. Adv Mater Res 878(878):254–263

Removal of Sulfur from Copper Smelting Slag by CO_2

Wang Yun, Zhu Rong, Hu Shaoyan, Wang Hongyang and Guo Yaguang

Abstract In the process of extracting iron from copper slag, the iron obtained by reduction is difficult to use because of the high content of copper and sulfur. In this paper, the thermodynamics of CO_2 desulphurization of slag is simulated using FactSage 7.1. In tube furnace, the desulfurization of copper slag was investigated by changing the temperature and flow rate of CO_2. Thermodynamic simulation using FactSage 7.1 shows that increasing temperature and flow rate of CO_2 can promote the removal of sulfur, and the desulfurization rate can reach up to 54.97%. The experiment shows that the removal trend of sulfur in the slag is the same as that of the thermodynamic calculation, and the maximum removal rate is 56.64%. The utilization rate of CO_2 increased with the increase of temperature, but decreased with the increase of CO_2 flow rate. Compared with air, CO_2 is not easy to cause slag peroxidation.

Keywords Copper slag · Residue sulfur · Desulfurization rate · Utilization rate of CO_2

Introduction

Copper slag contains a large number of iron elements [1–3]. By deep reduction and adding tempering agent, not only valuable iron can be extracted from the slag, but also gangue can be fully utilized in the production of mineral wool, glass-ceramics, stone, and so on [4–6]. However, copper slag contains a lot of Cu, S, and other elements; the molten iron obtained is of poor quality, difficult to desulfurize, and difficult to be used in large quantities as raw materials for steelmaking [7–10].

W. Yun (✉) · Z. Rong · H. Shaoyan · W. Hongyang
School of Metallurgical and Ecological Engineering, University of Science and Technology Beijing, Beijing 100083, China
e-mail: wangyun2109@163.com

G. Yaguang
Chinese Enfei Engineering Technology Co. Ltd, Beijing 100038, China

G. Gaustad et al. (eds.), *REWAS 2019*, The Minerals, Metals & Materials Series,
https://doi.org/10.1007/978-3-030-10386-6_12

The removal of sulfur from copper slag can produce copper slag with low sulfur content. The use of air, oxygen, and other oxidants because of the strong oxidation can achieve the rapid removal of sulfur from slag [11, 12]. Despite the good desulfurization effect, it will also lead to the problem of slag overoxidation [13]. However, the removal effect of copper slag by weaker oxidant has not been studied, and the study of thermodynamic equilibrium critical oxygen partial pressure of melt desulfurization has not been reported.

Compared with air and oxygen, CO_2 has a weak oxidizing property at high temperature [14, 15], and it also has oxidizing property to sulfur in slag, which can remove sulfur from slag. However, there is a lack of research report on desulfurization of CO_2.

In this study, the molten copper slag was desulphurized using CO_2 with weaker oxidizing property. The desulphurizing effect of CO_2 on slag and the oxidizing property of slag were studied. The thermodynamic equilibrium process of copper slag oxidation was simulated using FactSage 7.1 thermodynamic software. The effects of temperature and flow rate of CO_2 on the residual sulfur content in slag, the oxidation of slag, the desulphurization rate, and the utilization rate of CO_2 were explored.

Experimental Device and Method

Materials

As shown in Table 1, the copper slag is the slag from bottom-blown oxygen-enriched bath. The slag without diluted is cooled slowly in the slag yard and then taken out for the experiment. Part of copper slag was crushed below 100 mesh for experiment and chemical analysis. The slag was sent to chemical analysis center of University of Science and Technology Beijing for component analysis. Fe_2O_3 cannot be detected directly, but it can be calculated from total Fe and FeO in slag.

Experimental Process and Results

The experiments were conducted in an airtight tubular furnace (with a maximum internal furnace temperature of 1600 °C) using argon as a protective gas at a rate of 100 mL/min. The sample was contained in an alumina crucible (φ40 mm × 120 mm),

Table 1 Chemical analysis of copper slag (wt%)

TFe	Cu	S	Al_2O_3	SiO_2	FeO	Fe_2O_3	Fe_3O_4
46.41	4.28	2.97	2.85	19.50	46.80	14.30	20.74

held in an alumina safety crucible (φ60 mm × 150 mm) to protect the furnace tube from inner crucible failure. The furnace temperature was increased to target temperature in 150 min, and then the temperature remains 30 min.

Basic conditions: 100 g copper slag was put into tube furnace, then 100 mL/min Ar protective gas into the furnace, heating up to 1350 °C in 150 min, and then holding for 30 min. The top blowing flow rate of CO_2 is 200 mL/min. After top blowing of 30 min, the sample was quenched in water.

The quenched samples were sent to chemical analysis to detect S, FeO, and Fe_2O_3. The main factors to be investigated are temperature, CO_2 flow, and time, as shown in Table 2. Inspection indexes are residual S, (Fe^{3+}/Fe^{2+}), desulphurization rate, and utilization rate of CO_2.

The input data to FactSage 7.1 for calculation were 46.80 g FeO, 14.30 g Fe_2O_3, 19.50 g SiO_2, 2.85 g Al_2O_3, 4.28 g Cu, and 2.97 g S. The gas pressure in Equilib module was 1.0 atm. The conditions and output data are shown in Table 3.

Experimental Results and Discussion

Residual Sulfur Content in Slag

As shown in Fig. 1, the sulfur content in slag decreases significantly with the increase of temperature and CO_2 flow rate, which is consistent with the FactSage 7.1 simulation results. FactSage 7.1 calculation is the final reaction equilibrium of sulfur removal, but the experimental results show that the sulfur removal is lower than FactSage 7.1 calculation. This phenomenon may be caused by the erosion of cru-

Table 2 Experimental conditions and data

No.	Temperature (°C)	Flow rate of CO_2 (mL min^{-1})	Mass of final slag (g)	Composition of slag (wt%)		
				FeO	Fe_2O_3	S
1	1300	200	97.9	50.2	10.34	1.73
2	1325	200	97.8	51.55	9.13	1.66
3	1350	200	97.7	52.85	7.95	1.6
4	1375	200	97.6	53.66	7.19	1.55
5	1400	200	97.6	53.87	6.99	1.5
6	1350	0	98.0	51.26	9.14	1.99
7	1350	100	97.8	52.59	8.02	1.77
8	1350	200	97.7	52.84	7.95	1.61
9	1350	400	97.6	52.86	8.23	1.32

Table 3 The conditions and output data from FactSage 7.1

No.	Temperature (°C)	Volume of CO_2 (L)	Mass of gas (g)	Sulfur in gas (g)	Composition of slag (wt%)	
					FeO	Fe_2O_3
1	1300	6.00	13.96	1.27	50.2	10.34
2	1325	6.00	14.01	1.31	51.55	9.13
3	1350	6.00	14.06	1.36	52.85	7.95
4	1375	6.00	14.11	1.41	53.66	7.19
5	1400	6.00	14.15	1.45	53.87	6.99
6	1350	0	1.96	0.99	51.26	9.14
7	1350	3.00	8.07	1.20	52.59	8.02
8	1350	6.00	14.05	1.36	52.84	7.95
9	1350	12.00	20.00	1.63	52.86	8.23

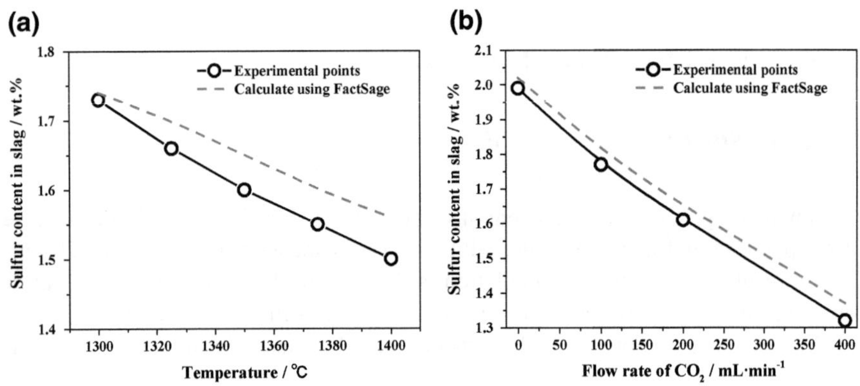

Fig. 1 Relationship between residual sulfur content in slag and **a** temperature and **b** flow rate of CO_2

cible in slag, which dilutes the elements in the slag and reduces the sulfur content in the slag.

Thermodynamic analysis of slag desulfurization shows that self-desulfurization occurs during slag heating, and the sulfide in slag reacts mainly with Fe_2O_3.

$$6Fe^{3+} + S^{2-} + 2O^{2-} = 6Fe^{2+} + SO_2 \quad (1)$$

When CO_2 is injected into the bath, CO_2 will react with sulfide in the slag to produce CO and SO_2.

$$S^{2-} + 3CO_2 = O^{2-} + 3CO + SO_2 \quad (2)$$

Equation (2) reaches equilibrium

$$\Delta G = A_1T + B_1 + RT\ln\left[\frac{p_{SO2}}{p^{\ominus}} \cdot \left(\frac{p_{CO}}{p_{CO2}}\right)^3 \cdot \frac{a_{O2-}}{\gamma_{S2-} \cdot w_S}\right] = 0 \quad (3)$$

where T is the temperature, K; A_1 and B_1 are the constants, and $\Delta G^0 = A_1T + B_1$ is the standard Gibbs free energy of Eq. (2); R is the constant of 8.314 J mol^{-1} K^{-1}; p_{SO2}, p_{CO}, and p_{CO2} are the partial pressures of SO_2, CO, and CO_2, respectively, atm; $p^{\ominus}$ is the standard atmospheric pressure, 1.0 atm; a_{O2-} is the activity of O^{2-} in molten slag; γ_{S2-} is the activity coefficient of S^{2-} in molten slag; and w_S is the mass fraction of sulfur in slag (sulfur in slag almost all exists in the form of S^{2-}), wt%.

The activity of oxygen ions in slag can be shown as follows:

$$2Fe^{2+} + CO_2 = 2Fe^{3+} + CO + O^{2-} \quad (4)$$

$$\Delta G = A_2T + B_2 + RT\ln\left[a_{O2-} \cdot \frac{p_{CO}}{p_{CO2}} \cdot \left(\frac{\gamma_{Fe3+}}{\gamma_{Fe2+}} \cdot \frac{w_{Fe3+}}{w_{Fe2+}}\right)^2\right] = 0 \quad (5)$$

where A_2 and B_2 are the constants, and $\Delta G^0 = A_2T + B_2$ is the standard Gibbs free energy of Eq. (4); γ_{Fe3+} and γ_{Fe2+} are the activity coefficients of Fe^{3+} and Fe^{2+}, respectively; and w_{Fe3+} and w_{Fe2+} are the mass fractions of Fe^{3+} and Fe^{2+} in slag, respectively, wt%.

According to Eqs. (3) and (5),

$$\ln(w_{S2-}) = \frac{AT + B}{RT} + \ln\left(\frac{p_{SO2}}{p^{\ominus}}\right) + 2\ln\frac{p_{CO}}{p_{CO2}} - 2\ln\frac{w_{Fe3+}}{w_{Fe2+}} - 2\ln\frac{\gamma_{Fe3+}}{\gamma_{Fe2+}} - \ln(\gamma_{S2-}) \quad (6)$$

where A and B are the constants, and $A = A_1 - A_2$, $B = B_1 - B_2$.

According to Eq. (6), sulfur content in slag is affected by temperature, p_{SO2}, p_{CO}/p_{CO2}, and w_{Fe3+}/w_{Fe2+}. The sulfur content in slag can be reduced by increasing temperature, which is consistent with the curve of Fig. 1a.

Slag Oxidation (Fe^{3+}/Fe^{2+})

The oxidation of slag can be measured by the ratio of w_{Fe3+}/w_{Fe2+}($Fe^{3+}/Fe^{2+} = w_{Fe3+}/w_{Fe2+}$). The larger the ratio of Fe^{3+} to Fe^{2+}, the stronger the oxidation of slag, that is, the higher the content of Fe_2O_3 in slag.

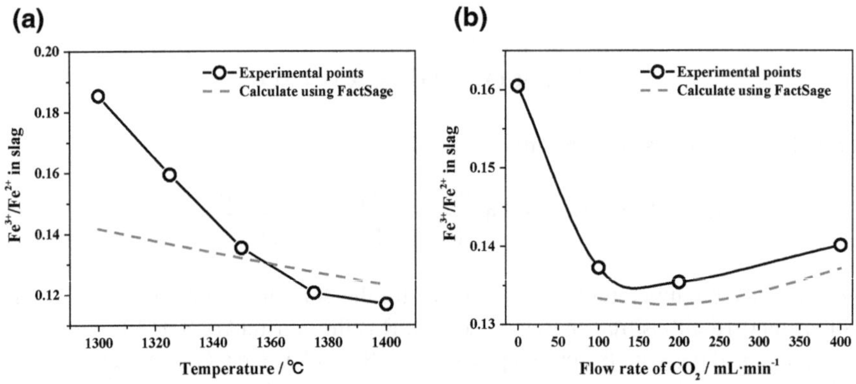

Fig. 2 Relationship between Fe^{3+}/Fe^{2+} in slag and **a** temperature and **b** flow rate of CO_2

Figure 2a shows that the oxidation of slag decreases with increasing temperature, which is consistent with the results of the calculation using FactSage 7.1. The change of Fe^{3+}/Fe^{2+} with temperature is more significant at 1300–1350 °C, but the decrease is not significant when the temperature is 1375–1400 °C. However, the calculation results using FactSage 7.1 show a linear relationship between temperature and Fe^{3+}/Fe^{2+}.

As shown in Fig. 2b, with the increasing CO_2 flow rate, the Fe^{3+}/Fe^{2+} in the slag decreases first and then increases slightly. When CO_2 is not blown, the slag is not agitated and the reaction is slow and the initial Fe_2O_3 content in the slag is higher. However, when CO_2 is blown into slag, the Fe_2O_3 will be consumed greatly. When the CO_2 continues to increase, the oxidative effect of CO_2 on slag will appear, making the slag oxidizability increase again, but the increase is small, which is consistent with the results of calculation using FactSage 7.1. Compared with air and oxygen, CO_2 is not easy to cause slag overoxidation, which can keep slag oxidation in a stable range.

Desulfurization Rate of Slag

In order to study the effect of CO_2 on the desulphurization of slag, the desulphurization rate of slag is defined as η.

$$\eta = (1 - \frac{m \cdot w_S}{m_0 \cdot w_{0S}}) \times 100\% \quad \text{(Experimental)} \tag{7}$$

$$\eta = \frac{m_S}{m_{0S}} \times 100\% \quad \text{(FactSage 7.1)} \tag{8}$$

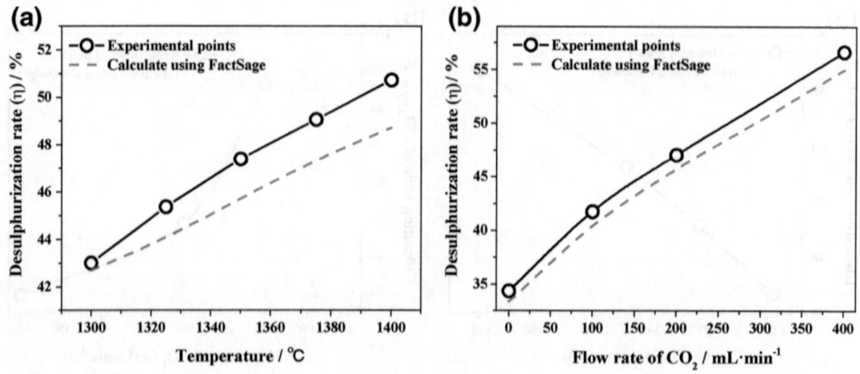

Fig. 3 Relationship between desulphurization rate and **a** temperature and **b** flow rate of CO_2

where m is the mass of slag after experiments, g; w_S is the sulfur content in slag after experiments, wt%; m_0 is the mass of initial slag, g; w_{0S} is the sulfur content in initial slag, wt%; m_S is the mass of sulfur in gas obtained from FactSage 7.1, g; and m_{0S} is the mass of sulfur input to FactSage 7.1, g.

With the increase of temperature, the desulfurization rate increases gradually. When the temperature rises from 1300 to 1400 °C, the desulfurization rate increases from 43.00 to 50.73%, while the desulfurization rate calculated using FactSage 7.1 rises from 42.69 to 48.72%. The increase curve of desulfurization rate obtained from calculation is almost straight line, but the increase of desulfurization rate in experiment is not. The slope of curve decreases when the temperature is higher, that is, the increase of desulfurization rate becomes slower.

With the increase of CO_2 flow rate, the desulfurization rate increased significantly. The desulfurization rate increased from 34.35 to 56.64% with the CO_2 flow rate from 0 to 400 mL/min in experiments, and the desulfurization rate increased from 33.32 to 54.97% according to FactSage 7.1 calculation. The experimental results are in good agreement with the calculation results in Fig. 3b, which is quite different from Fig. 3a.

Utilization Rate of CO_2

According to Eq. (2), 3 mol CO_2 can remove 1 mol sulfur in slag. However, not all the CO_2 blown in slag can take part in the reaction. In order to study the ratio of CO_2 participating in desulfurization reaction to the total amount of CO_2 blown in slag, the utilization ratio of CO_2 was defined as φ

$$\varphi = \frac{(m_0 \cdot w_{0S} - m \cdot w_S)/32 \times 3 \times V_{m,0}}{Q \cdot t/1000} \times 100\% \text{ (Experimental)}$$

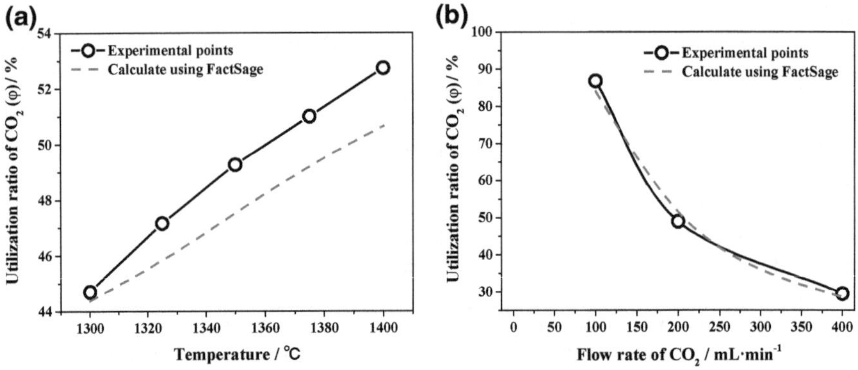

Fig. 4 Relationship between utilization rate of CO_2 and **a** temperature and **b** flow rate of CO_2

$$\varphi = \frac{m_S/32 \times 3 \times V_{m,0}}{V_{CO2}} \times 100\% \quad \text{(FactSage 7.1)}$$

where $V_{m,0}$ is the constant, 22.4 L/mol; Q is the flow rate of CO_2, mL/min; t is the blowing time, 30 min; and $V_{m,0}$ is the volume of CO_2 input to FactSage 7.1 at 1.0 atm.

With the increase of temperature, the utilization ratio of CO_2 increases gradually, which is consistent with the trend of FactSage 7.1 calculation. When the temperature rises from 1300 to 1400°C, the utilization rate of CO_2 rises from 44.70 to 52.73%, while the utilization rate of CO_2 calculated using FactSage 7.1 rises from 44.37 to 50.64%.

With the increase of CO_2 flow rate, CO_2 utilization rate decreases significantly in Fig. 4a. The experimental results almost coincide with the curve calculated using FactSage 7.1. The CO_2 flow rate rises from 100 mL/min to 400 mL/min, the CO_2 utilization rate decreases from 86.77 to 29.44%, and the corresponding calculated CO_2 utilization rate decreases from 83.96 to 28.57%. Compared with air, CO_2 has a weak ability to remove sulfur [13], but it also can avoid slag overoxidation, that is, excessive CO_2 blown will not cause a large number of slag to be oxidized, CO_2 can be used as the end-point gas of oxidative desulfurization, which has a certain oxidizing effect while playing the role of gas stirring, and it is not easy to cause slag overoxidation when excessive blowing.

Conclusion

(1) Blow of CO_2 can promote the removal of sulfur from copper slag, mainly because (1) CO_2 has a weak oxidizability, which can oxidize sulfide in slag into SO_2; (2) enhanced gas stirring accelerates the self-desulfurization process of slag at high temperature.

(2) The increase of temperature can promote the removal of sulfur from slag. The experimental results show that the effect is consistent with the theoretical analysis and FactSage 7.1 calculation. However, under the experimental conditions, the effect of temperature on residual sulfur content and desulfurization rate in slag is not as significant as the effect of CO_2 flow rate.
(3) The oxidation of slag decreases with the increase of temperature, i.e., Fe^{3+}/Fe^{2+} decreases; while Fe^{3+}/Fe^{2+} in slag decreases first and then increases slightly with the increase of CO_2 flow rate. Compared with air, CO_2 has a weaker oxidation, and the utilization rate will decrease significantly with the increase of gas flow rate, causing that CO_2 is not easy to cause slag peroxidation.

Acknowledgements This study was supported by the National Natural Science Foundation in China [grant numbers: 51474024, 51674021, 51574021].

References

1. Lei LI, Jian-Hang HU, Hua W (2011) Study on smelting reduction ironmaking of copper slag. Chinese J Process Eng 11(1):65–71
2. Zhang J, Yuan-Hong QI, Yan DL et al (2015) A new technology for copper slag reduction to get molten iron and copper matter. J Iron Steel Res (International) 22(5):396–401
3. Wang H (2012) Experimental research of deep reduction and magnetic separation process of a high-iron copper slag. Metal Mine 41(11):141–144
4. Liu Y, Caibin W U, Lei C et al (2014) Comprehensive iron recovery technology from copper tailings of copper smelting slag. Nonferrous Metals Sci Eng 5(5):141–144
5. Havanagi VG, Sinha AK, Prasad PS et al (2009) Copper slag as an alternative material for road construction. In: Proceeding of, international conference on solid waste technology and management, Philadelphia, USA, pp 133
6. Ambrish E, Dhavamani DS, Shanmuganathan N et al (2017) Partial replacement of copper slag as fine aggregate. Int J Civil Eng 4(3):18–23
7. Xian-Lin Z, De-Qing Z, Jian P et al (2015) Utilization of waste copper slag to produce directly reduced iron for weathering resistant steel. Trans Iron Steel Inst Jpn 55:1347–1352
8. Zheng P (2016) Study on direct reduction of copper and iron from copper slag. Non-Ferrous Mining Metal 32(1):36–38
9. Zhang BJ, Zhang TA, Niu LP et al (2018) Desulfurization of copper-iron reduced from copper slag. In: TMS Meeting & Exhibition. Springer, Cham, pp 15–23
10. Yang Z, Ma Z (2017) The effect of basicity and calcium fluoride on glass-ceramic production and iron recovery from copper slag. Steel Res Int 88(3)
11. Gyurov S, Rabadjieva D, Kovacheva D et al (2014) Kinetics of copper slag oxidation under nonisothermal conditions. J Therm Anal Calorim 116(2):945–953
12. Naganuma Y, Kobayashi S, Tsuru T et al (1991) Improvement in the properties of copper smelting slag through oxidation at high temperature. Shigen-to-Sozai, 107(12):919–923
13. Wang Y, Zhu R, Chen Q (2018) Experimental study on oxidative desulfurization and selective reduction of molten copper slag. In: TMS Meeting & Exhibition. Springer, Cham, pp 233–243
14. Shibata K, Kitamura T, Tokumitsu N (2009) Kinetic model for the reaction between iron oxide in molten slag and carbon in molten iron via CO–CO_2 bubble. Tetsu-to-Hagane 76(11):2011–2018
15. Tran T, Sun S, Jahanshaht S (1997) Interfacial reactions of low iron-containing slag with CO_2-CO mixtures. In: International conference on molten slags, fluxes and salts, pp 115–123

Sustainable Use of Precious and Rare Metals Through Biotechnological Recycling

Norizoh Saitoh, Toshiyuki Nomura and Yasuhiro Konishi

Abstract We proposed using new biotechnologies to recycle platinum group metals (PGMs) and gold from the end of life wastes, which will lead to the sustainable use of precious and rare metals. When targeting leachate of spent automotive catalysts, the metal ion-reducing bacterium *Shewanella algae* was found to reduce and deposit aqueous PGMs ions (Pd(II), Pt(IV) and Rh(III)) as metallic particles within the bacterial cells at room temperature and pH 6 within 60 min, using formate as the electron donor. We also found that the baker's yeast *Saccharomyces cerevisiae* can be applied as a biomaterial for adsorbing Au(III) ions from aqueous acidic solutions. When processing leachate of spent electronic components, *S. cerevisiae* cells were able to rapidly and selectively collect Au(III) ions from strongly acidic solutions. Unlike conventional hydrometallurgical methods, our proposed microbial methods enable the attractive and eco-friendly recovery of PGMs and gold from secondary sources.

Keywords Recycling · Platinum group metals · Gold · Biotechnology · Biomineralization · Biosorption

Introduction

End of life wastes, such as electronic goods and catalytic converters in cars, are important sources of precious and rare metals (PRMs), such as platinum group metals (PGMs) and gold. Conventional thermal and chemical recycling techniques remain the best methods for recycling PRMs, so these domestic resources of PRMs have yet to be fully utilized. Therefore, more research and development is needed to recycle PRMs from secondary sources.

We believe that biological technologies now provide an attractive and eco-friendly alternative strategy. We have recently developed new recovery technologies that are

N. Saitoh · T. Nomura · Y. Konishi (✉)
Department of Chemical Engineering, Osaka Prefecture University, 1-1 Gakuen-Cho, Sakai, Osaka 599-8531, Japan
e-mail: yasuhiro@chemeng.osakafu-u.ac.jp

G. Gaustad et al. (eds.), *REWAS 2019*, The Minerals, Metals & Materials Series,
https://doi.org/10.1007/978-3-030-10386-6_13

designed to reduce and deposit aqueous ions of Au(III) and PGMs (Pt(IV), Pd(II), and Rh(III)) as metallic nanoparticles, a process known as "biomineralization" [1–9]. We have also proposed the separation of rare metal In(III) ions from very dilute solutions and its concentration within bacterial cells, a process known as "biosorption" [10, 11]. This paper describes our research results from using new biotechnologies to recover PGMs and gold from leachates of spent automotive catalysts and electronic waste (Central Processing Units, CPU). We also propose recycling flows of PGMs and gold from secondary sources using new biotechnologies, which will lead to the sustainable use of PRMs.

Microbial Recovery of Platinum Group Metals from Aqua Regia Leachate of Automotive Catalysts

Experimental Procedures

We used the metal ion-reducing bacterium *S. algae* (ATCC 51181) to recover soluble PGMs. *S. algae* was grown anaerobically in ATCC medium 2 containing sodium lactate and iron (III) citrate at 25 °C and pH 7.0 [1–4]. After 24 h of batch inoculation, *S. algae* cells were harvested by centrifugation, resuspended in KH_2PO_4/NaOH buffer (pH 7.0), and pelleted again by centrifugation. The washed *S. algae* cells were subsequently resuspended in KH_2PO_4/NaOH buffer (pH 7.0). The cell suspension was bubbled with N_2 for 10 min and immediately used for microbial recovery of PGMs.

Ceramic car catalyst substrate blocks were crushed and ground up, so that the materials had a particle size range of less than 30 μm. The metal contents in the particulate materials were 0.24 wt% PGMs (platinum, palladium, and rhodium), 20.9 wt% aluminum, 12.1 wt% silicon, 4.7 wt% magnesium, 4.8 wt% rare earths (cerium and lanthanum), and 1.9 wt% iron, respectively. An aqueous 50% aqua regia solution was used to extract the PGMs from the ground catalytic materials into aqueous solution. At 60 °C and atmospheric pressure, above 95% of the PGMs in the sample were leached out into the solution within 24 h. However, aluminum and other heavy metals also leached out.

It was necessary to adjust the aqua regia leachate of catalyst substrate to an optimal pH range for *S. algae* activity between pH 6.0 and 7.0. To do so, sodium phosphate was added into the acidic aqua regia leachate, to prevent coprecipitation of the PGMs ions during the pH adjustment. Phosphate was a useful precipitation agent for aqueous heavy metal ions, like aluminum, ferric iron, cerium, and lanthanum. This treatment completely removed the heavy metal ions as insoluble phosphates under acidic pH conditions. By adding phosphate and sodium hydroxide, we were successfully able to adjust the pH of the aqua regia leachate to pH 6.

A typical recovery experiment was carried out at 25 °C using a single continuous-flow stirred tank reactor (CSTR), in which the aqua regia leachate, sodium formate, and *S. algae* cells were continuously added to and removed from the reactor under anaerobic conditions. The feed PGMs concentration was 1.4 mol/m^3, with 200 mol/m^3 sodium lactate as the electron donor. The cell concentration in the reactor was 5.0×10^{15} cells/m^3, and the solution was buffered at pH 6 with KH_2PO_4/NaOH. The mean residence time (reactor volume/effluent suspension flow rate) varied from 0.3 to 10 min. After the continuous PGMs recovery run was started at a fixed mean residence time, an aliquot of aqueous solution was periodically withdrawn from the reactor. The concentration of PGMs in the liquid samples was determined by ICP spectroscopy, and the number of *S. algae* cells in the solution was counted in a Petroff-Hausser counting chamber with an optical microscope.

Microbial Reduction and Deposition of PGMs Using the Metal Ion-Reducing Bacterium S. algae

A batch recovery run was initiated by inoculating *S. algae* cells into the aqua regia leachate of catalyst substrate at 25 °C and pH 6. When *S. algae* cells were combined with formate as an electron donor under anaerobic conditions, the aqueous PGMs concentration rapidly decreased from its initial value of 1.4 mol/m^3 to almost zero within 30 min. In this case, the appearance of the bacterial cell suspension visibly changed from yellow to black due to the formation of PGMs nanoparticles. However, without formate, a slight decrease in aqueous PGMs concentration was likely due to the adsorption of PGMs ions by *S. algae* cells. In this case, the color of the suspension did not change. Furthermore, no change in the PGMs concentration was observed in the absence of *S. algae* cells, indicating that formate did not chemically reduce the aqueous PGMs ions. Therefore, we concluded that the marked decrease in the PGMs concentration reflected the reduction of PGMs ions to metallic nanoparticles by *S. algae* cells.

To process larger amounts of the catalyst leaching solution, we carried out microbial recovery using a stirred tank reactor under continuous operation (Fig. 1). The catalyst leaching solution and *S. algae* suspension were continuously supplied to the reactor, and the resulting products were discharged continuously. At PGMs solution feed concentration of 1.4 mol/m^3, 95% recovery of the PGMs (platinum, palladium, and rhodium) was achieved under steady-state operation, even when the mean residence time was decreased to 5 min. Under these operating conditions, we were able to achieve a maximum recovery rate of 3.5 kg/h/m^3. Our microbial procedure has the capacity for rapid and highly efficient recovery of PGMs at room temperature.

Figure 2 shows a flow diagram of our method for concentrating PGMs in the bacterial cells. When wet, metal-containing bacterial cells were dried at 50 °C for 20 h, and the PGMs content of the dried cells was 20%. This represented an approximately 100-fold increase in PGMs concentration compared with that of the catalyst precursor materials, which was 0.24%. Furthermore, we were able to produce a mixture of PGMs by firing the dried cells in an electric furnace.

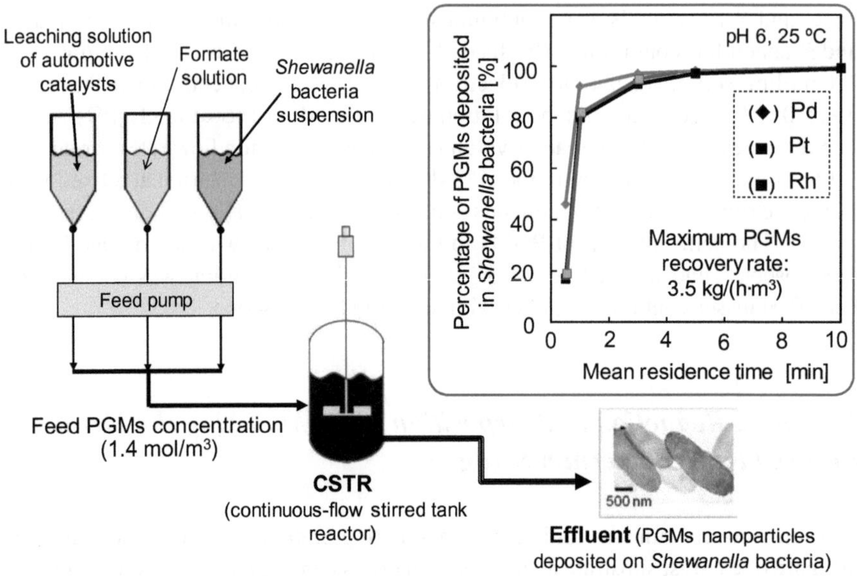

Fig. 1 Microbial recovery of soluble PGMs (platinum, palladium, and rhodium) in a continuous-flow stirred tank reactor

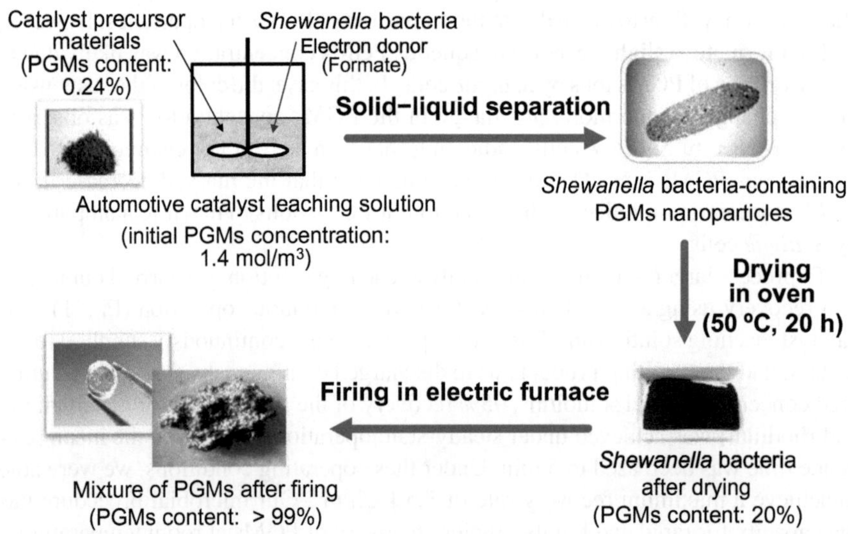

Fig. 2 Flow diagram of method for concentrating and recovering PGMs (Pd, Pt, Rh) collected in bacterial cells

Microbial Recovery of Gold from Aqua Regia Leachate of CPU

Experimental Procedures

We used the baker's yeast *S. cerevisiae* as an inexpensive biomaterial for adsorbing soluble Au(III) from aqueous acidic solutions. *S. cerevisiae* (NBRC 2044) was obtained from the Biological Resource Center, NITE (National Institute of Technology and Evaluation) and grown anaerobically in glucose-yeast-peptone medium at 33 °C and pH 7.0. After 24 h of batch inoculation, *S. cerevisiae* cells were harvested by centrifugation, resuspended in KH_2PO4/NaOH buffer (pH 7.0), and pelleted again by centrifugation. This washing was repeated twice.

We obtained central processing units (CPU) that are sourced from waste electrical and electronic equipment (WEEE), and then the CPU were cut and crushed into small sizes of 2–4 mm. The aqua regia leachate containing Au(III) ions was prepared by dissolving a known weight of CPU in a 50% aqua regia which was diluted with deionized water in the ratio 1:1. Above 99% of gold in the CPU was leached out in 2 h at 70 °C and atmospheric pressure, and the aqua regia leachate also contained copper, nickel, and iron. The aqua regia leachate of CPU was adjusted to an optimal pH 1.0 for *S. cerevisiae* activity by adding 10.0 kmol/m^3 NaOH solution. The aqua regia leachate with pH 1.0 had the following metal concentration (in mol/m^3): Au, 1.88; Cu, 7.40; Ni, 33.8; and Fe, 53.4.

For a typical adsorption experiment at 33 °C, 5 cm^3 of *S. cerevisiae* cell suspension was mixed with 10 cm^3 of the aqua regia leachate with pH 1.0 under air atmosphere. The *S. cerevisiae* cells used in the adsorption runs were grown in glucose-yeast-peptone medium at pH 7.0 and were not preadapted to strongly acidic solution at pH 1.0. The concentrations of *S. cerevisiae* cells ranged between 0.5×10^{14} and 5.0×10^{14} cells/m^3. An aliquot of this mixture was periodically withdrawn and analyzed for gold. The concentration of gold in the liquid samples was determined by ICP spectroscopy.

Microbial Adsorption of Soluble Gold Using the Baker's Yeast S. cerevisiae

It was evident that resting cells of *S. cerevisiae* are applicable to the recovery of Au(III) ions from the aqua regia leachate with pH 1.2 at 33 °C, even when the leachate with pH 1.2 has a higher salt concentration due to pH adjustment of aqua regia solution. The marked decrease in aqueous Au(III) concentration presumably reflects the adsorption of the Au(III) ions by the *S. cerevisiae* cells, and the microbial recovery rate of Au(III) ions was markedly enhanced as the cell concentration was increased from 0.5×10^{14} to 5.0×10^{14} cells/m^3. It is likely that *S. cerevisiae* has a

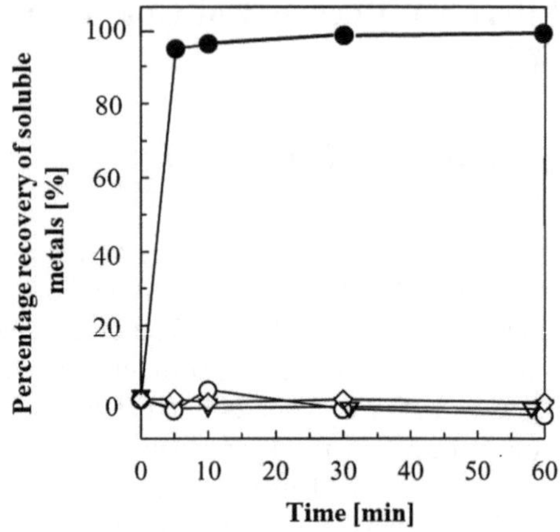

Fig. 3 Selective recovery of Au(III) ions in an aqua regia leachate of CPU with pH 1.2 at 33 °C and 5.0 × 10^{14} cells/m^3 *S. cerevisiae* cells. (●) gold; (○) copper; (▽) nickel; (◇) iron

cell surface consisting of biological materials containing functional groups, which are responsible for the adsorption of aqueous Au(III) ions at pH 1.2. Since the quantitative relation between cell number and dry weight of *S. cerevisiae* cells was measured to be 3.1 × 10^{10} cells/g-dry cells, the amount of Au(III) recovered by *S. cerevisiae* cells was determined as 16.0–23.6 mg-Au/g-dry cells. The Au(III) uptake ability of *S. cerevisiae* was comparable with that of a core–shell-type anion exchange resin for aqua regia leachate of electronic parts, 19.0–19.7 mg-Au(III)/g-dry cells [12].

Figure 3 indicates that 98% recovery of 1.25 mol/m^3 Au(III) ions at pH 1.2 was achieved at 5.0 × 10^{14} cells/m^3 within 10 min. Importantly, the *S. cerevisiae* cells did not react with other metal ions, such as Cu(II), Ni(II), and Fe(III). Since the point of zero charge (PZC) is approximately pH 4.0 for *S. cerevisiae* cells [13], the cell surface of *S. cerevisiae* carries positive charges at pH 1.2. The main chemical species of gold in aqua regia leachates is $AuCl_4^-$, which enhances the electrostatic force of attraction to the yeast surface. As a result, microbial adsorption occurs more readily when the pH is around 1. However, the main chemical species of Cu(II), Ni(II), and Fe(III) in acidic solutions becomes positive charges. This is one possible reason for the selective adsorption of Au(III) ions in the aqua regia leachate. The microbial ability to selectively and rapidly recover only gold under strongly acidic conditions demonstrates the potential for commercialization of this microbial recovery process, comprising a new system for microbial recycling of gold from WEEE.

Sustainable Use of Platinum Group Metals and Gold Through Biotechnological Recycling

Figure 4 shows flow diagrams comparing the conventional chemical process and the new bioprocess for the recycling of precious metals. In contrast to the conventional process, the new bioprocess is integrated, unifying a multi-step method into a single-

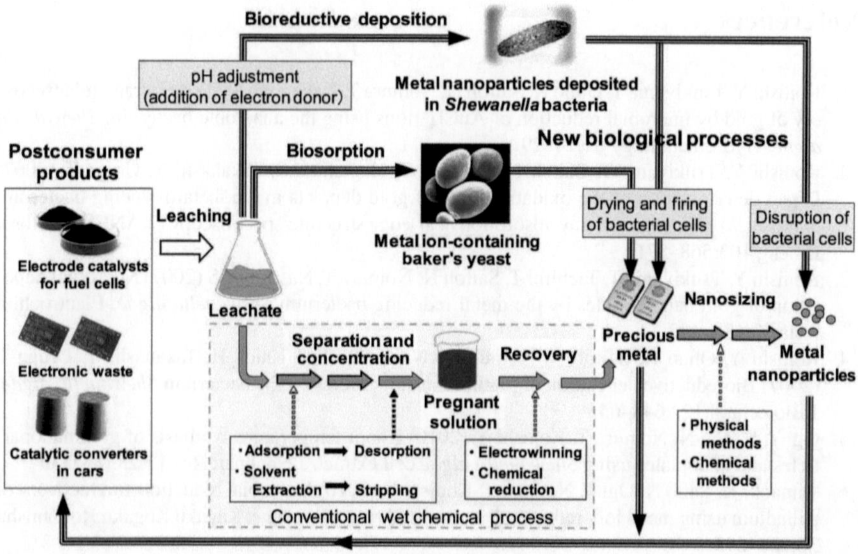

Fig. 4 Sustainable use of platinum group metals (PGMs) and gold through biotechnological recycling

step procedure that separates and concentrates precious metals from a dilute solution, enabling the formation of metal nanoparticles. However, the addition of an electron donor and pH adjustment is required for the bioprocess to operate.

Our proposed biotechnologies are good examples of using room temperature microbial reactions for low-energy and low-cost recycling processes, resulting in decreased CO_2 emissions. Our highly efficient process fits into a small unit and could be introduced at local collection points for end of life wastes and operate as a regionally distributed technology for recycling PRMs.

Conclusions

We successfully developed new biotechnologies to extract platinum group metals and gold from post-consumer products. Unlike conventional hydrometallurgical methods, our proposed microbial methods enable the attractive and eco-friendly recovery of PGMs and gold from secondary sources.

Acknowledgements This work was supported by JSPS KAKENHI Grant Numbers JP18H03846, JP20360411, JP23360406.

References

1. Konishi Y, Tsukiyama T, Ohno K, Saitoh N, Nomura T, Nagamine S (2006) Intracellular recovery of gold by microbial reduction of $AuCl_4^-$ ions using the anaerobic bacterium *Shewanella algae*. Hydrometallurgy 81:24–29
2. Konishi Y, Tsukiyama T, Saitoh N, Nomura T, Nagamine S, Takahashi Y, Uruga T (2007) Direct determination of the oxidation state of gold deposits in the metal-reducing bacterium *Shewanella algae* using X-ray absorption near-edge structure spectroscopy (XANES). J Biosci Bioeng 103:568–571
3. Konishi Y, Tsukiyama T, Tachimi T, Saitoh N, Nomura T, Nagamine S (2007) Microbial deposition of gold nanoparticles by the metal-reducing bacterium *Shewanella algae*. Electrochim Acta 53:186–192
4. Konishi Y, Ohno K, Saitoh N, Nomura T, Nagamine S, Hishida H, Takahashi Y, Uruga T (2007) Bioreductive deposition of platinum nanoparticles on the bacterium *Shewanella algae*. J Biotechnol 128:648–653
5. Ogi T, Saitoh N, Nomura T, Konishi Y (2010) Room-temperature synthesis of gold nanoparticles and nanoplates using *Shewanella algae* cell extract. J Nanopart Res 12:2531–2539
6. Tamaoki K, Saito N, Ogi T, Nomura T, Konishi Y (2010) Microbial reduction and recovery of palladium using metal ion–reducing bacterium *Shewanella algae*. Kagaku Kogaku Ronbunshu 36:288–292
7. Ogi T, Makita K, Tamaoki K, Saitoh N, Konishi Y (2011) Synthesis of gold nanoparticles by the metal ion-reducing bacterium *Shewanella oneidensis*. J Soc Powder Technol 48:160–166
8. Ogi T, Honda R, Tamaoki K, Saitoh N, Konishi Y (2011) Direct room-temperature synthesis of a highly dispersed Pd nanoparticle catalyst and its electrical properties in a fuel cell. Powder Technol 205:143–148
9. Tamaoki K, Saitoh N, Nomura T, Konishi Y (2013) Microbial recovery of rhodium from dilute solutions by the metal ion-reducing bacterium *Shewanella algae*. Hydrometallurgy 139:26–29
10. Higashi A, Saitoh N, Ogi T, Konishi Y (2011) Recovery of indium by biosorption and its application to recycling of waste liquid crystal display panel. J Jpn Inst Metals 75:620–625
11. Ogi T, Tamaoki K, Saitoh N, Higashi A, Konishi Y (2012) Recovery of indium from aqueous solutions by the Gram-negative bacterium Shewanella algae. Biochem Eng J 63:129–133
12. Cyganowski P, Garbera K, Lesniewicz A, Wolska J, Pohl P, Jermakowicz-Bartkowiak D (2017) The aqua regia leachate of electronic parts using a core-shell type anion exchange resin. J Saudi Chem Soc 21:741–750
13. Dengis PB, Nelissen LR, Rouxhet PG (1995) Mechanisms of yeast flocculation: comparison of top- and bottom-fermenting strains. Appl Environ Microbiol 61:718–728

Control of Leachate Contamination from Mine Wastes Through an Appropriate Operating Practice

Kenneth Sichone

Abstract Mine wastes pose a significant environmental risk. Natural waters in areas that have a long history of mining tend to be polluted by release of metals resulting in environmental issues such as acid mine drainage. Interventions to mitigate such risks can be part of the operating practice and need not be assigned to the post-mining era. In this study, the leachability potential of slag at a smelter slag dump of a former mine site was assessed using the modified Ontario leachate extraction procedure (MOLEP) and other regulatory leachate quality criteria. Results showed that the release of contaminants (Cu, Co, Pb, Zn, As, Fe and Cd) can be maintained within industry-regulated standard limits mainly by adopting a smelter operating practice that produces a slag composition with adequate self-neutralising capacity.

Keywords Leachate · Discard slag · MOLEP · Acid mine drainage (AMD) · Self-neutralising

Introduction

Almost every mining activity introduces some potential environmental risk that requires careful consideration and management. The unintended risk arising from mining and processing of minerals usually receives attention only after the mine has ceased their operation. The high environmental and economic costs of hazardous waste management require mining companies to research into pre-emptive measures to mitigate the problem during the life of the mine or mineral processing operation. It is often more cost-effective to control the amount of pollution generated in the first place than attempting to find engineering fixes that deal with it after it has been created [4]. Mineral processing plants have an important role of waste reduction and

K. Sichone (✉)
Formerly UR-CST, School of Mining and Geology, Kigali, Rwanda
e-mail: bmkspure@gmail.com

G. Gaustad et al. (eds.), *REWAS 2019*, The Minerals, Metals & Materials Series,
https://doi.org/10.1007/978-3-030-10386-6_14

mitigating pollution risk by adopting operating practices that take into account the life cycle of generated hazardous wastes. The problem of hazardous mine wastes relates to their potentiality of contaminating surface and groundwater with heavy metals that tend to be mobilised by vectors such as rainwater or wind. For example, smelting operations generate slags as waste products. The slags are usually stored on the surface where they become susceptible to atmospheric exposure and subsequent weathering of inherent minerals. Sulphide minerals get oxidised by chemical and microbial processes to sulphuric acid and heavy metal ions that are deported into surface and groundwaters where they cause environmental issues such as acid mine drainage (AMD) or acid rock drainage (ARD). The chemical Eq. (1) is a typical AMD reaction.

$$4FeS_{2(s)} + 15O_{2(g)} + 14H_2O_{(l)} = 4Fe(OH)_{3(s)} + 8SO_4{}^{2-}_{(aq)} + 16H^+_{(aq)} \quad (1)$$

This paper shows one of the options a copper or base metal smelter can use to mitigate heavy metal loads and leachate contamination from mine wastes through an appropriate fluxing or operating practice in the primary smelting furnace.

Copper Smelting Process Description

The major elements in copper-bearing materials which come to a copper smelter are iron, sulphur and copper associated with which are some quantities of refractory oxides such as SiO_2, Al_2O_3, CaO, MgO, etc. Copper smelting is usually carried out in two stages, the primary smelting stage and the secondary smelting or converting stage (Fig. 1). In the primary stage, concentrates are mixed with fluxes and heated to remove impurities (mainly oxides of Fe, Si, Al, Ca and Mg). Fluxes such as CaO are added for the purpose of reducing the melting temperature of impurity oxides. Upon melting, impurities get oxidised and collect in a slag layer that floats above the copper-rich phase. The copper-rich phase referred to as matte, also collects significant quantities of other contained valuable metals such as cobalt. The matte is processed into metallic copper in the secondary smelting stage (Fig. 1). Wastes from the second stage are recycled to the primary smelting stage. In the primary smelting stage, the separation of slag from matte, which is mainly by gravity, is not perfect, and hence some metal sulphide valuables report to the slag. The primary stage slag is usually discarded on sites referred to as slag dumps. The slag at the dumps is susceptible to leaching. Once the slag comes into contact with rain whose pH is acidic, it can mobilise heavy metals and result in increasing heavy metal loads and leachate contamination of surface and subsurface water systems of the environment.

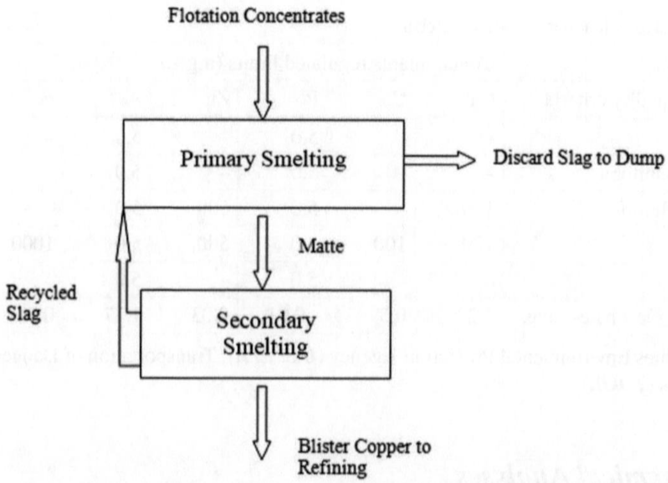

Fig. 1 Simplified block diagram of a typical smelting process for sulphidic copper ores

Experimental Methods and Materials

Sampling

In order to achieve a uniform coverage of the slag dump, the stratified random sampling pattern, which entailed division of the dump into five cells of equal size and siting at least five sampling points chosen at random within each cell, was adopted. Each sample, about 2.5 kg was taken from a depth of 200 mm.

Leachability Test Procedure

Slag leachability test procedure was based on the modified Ontario leachate extraction procedure (MOLEP) [3]. The slag sample was crushed to less than 9.5 mm and a 50 g portion was placed in a cylindrical bottle with a 60/40 weight percent mixture of sulphuric acid and nitric acid in a ratio of liquid (mls): solids (g) of 20:1 at pH 4.5. The bottle was tumbled for 24 h with its contents using a rotary shaker before removing the solids by filtration. The filtrate was analysed for Cu, Co, Pb, Zn, SO_4^{2-}, Fe, As, Cd and its pH was measured. The concentrations of dissolved metals were compared against the various leachate quality criteria regulated limits shown in Table 1. Leachate extraction results falling below the leachate quality criteria were deemed non-hazardous and were not subject to hazardous waste disposal requirements.

Table 1 Various leachate quality criteria

Description	Contaminants regulated limits (mg/L)						
Leachate quality criteria	Cu	Co	Pb	Zn	As	Fe	Cd
U.S. EPA*	–	–	5.0	–	5.0	–	1.0
Ontario/Manitoba	–	–	5.0	–	5.0	–	0.5
British Columbia	100	–	5.0	500	5.0	–	0.5
Alberta	100	100	5.0	500	5.0	1000	1.0
TDGR*	–	–	5.0	–	5.0	–	0.5
Actual slag leachate—avg.	1.22	0.59	<0.01	0.03	1.77	0.73	<0.01

*United States Environmental Protection Agency (*U.S. EPA*); Transportation of Dangerous Goods Regulations (*TDGR*)

Slag Chemical Analyses

The composition of each test slag sample is given in Table 2. Except for one slag, Sample 19, that appeared to have excessive matte carry-over, as inferred from its high copper and sulphur contents 12.51 and 3.80, respectively, all the slags were within the normal range of the smelter operating practice limits. The fluctuation in lime CaO content is probably due to variation in charge blends smelted with some being high in iron content and self-fluxing, and hence requiring less or no lime flux addition.

Theory

The problem of acid mine drainage is a well-known phenomenon that occurs when sulphides of various metals such as arsenic, lead, iron, copper and zinc are oxidised by either atmospheric oxygen or dissolved oxygen. The resulting products such as sulphuric acid and metal ions enter the percolating groundwater and flow from the mine waste. The reduction of heavy metal loads in contaminated water by precipitation at various pH levels is well established. In their studies, Balintova and Petrilakova [1] demonstrated the effectiveness of precipitation in decreasing contents of Fe, Cu, Zn, Al and Mn in acid mine drainage from an abandoned Smolnik mine site. From Fig. 2, it is evident that at low pH, most metals tend to be soluble, hence, increasing metal loads of leachates and at higher pH values the metal ion concentration decreases as the insoluble hydroxides are formed. The presence of CaO in smelter slags facilitates formation of metal hydroxides and may, therefore, be beneficial in the control of AMD and leachate contamination from mine wastes.

Table 2 Chemical composition of test slags

Slag sample	Composition (wt%)						
	Cu	Co	Pb	Zn	MgO	CaO	S
01	1.62	0.89	0.03	0.05	1.70	7.30	0.80
02	1.09	0.64	0.01	0.03	1.60	12.40	0.80
03	0.88	0.58	0.01	0.03	1.60	12.90	0.80
04	0.67	0.68	<0.01	0.03	1.60	12.40	0.60
05	0.97	0.84	0.02	0.03	1.60	6.00	0.70
06	1.19	0.42	<0.01	0.04	1.70	7.20	0.90
07	1.40	0.84	0.03	0.06	1.70	6.60	0.70
08	1.03	0.76	0.01	0.04	1.70	7.80	0.60
09	0.50	0.99	0.02	0.03	1.60	8.60	0.90
10	0.39	0.59	0.02	0.04	1.60	5.80	0.80
11	1.31	0.96	0.03	0.04	1.70	9.10	1.10
12	0.95	0.90	0.02	0.04	1.70	7.50	0.90
13	1.24	0.60	0.01	0.04	1.60	12.60	0.80
14	2.35	0.65	0.02	0.04	1.60	9.10	1.00
15	2.04	0.87	0.04	0.03	1.60	8.60	1.20
16	0.42	0.85	0.03	0.03	1.60	10.5	0.70
17	1.85	0.53	0.02	0.04	1.70	7.70	0.70
18	1.95	0.53	0.02	0.04	1.60	8.20	0.90
19	12.51	0.56	0.01	0.03	1.70	10.30	3.80
20	1.06	0.70	0.01	0.03	1.60	11.70	0.90
21	0.70	0.65	<0.01	0.03	1.60	10.80	1.00
22	2.03	0.83	0.02	0.03	1.60	8.90	0.80
23	1.02	0.90	0.01	0.03	1.60	10.50	0.70
24	1.26	0.58	0.01	0.03	1.60	11.70	0.70
25	0.64	0.61	0.01	0.03	1.60	11.10	0.60
26	0.63	0.76	0.01	0.03	1.60	10.30	0.60

Discussion

The Alberta leachate quality criteria cover all the seven contaminants investigated and have been found useful in assessing leachate extraction results, especially in respect of copper and cobalt, which are not included in other criteria. Leachate extraction results were below the regulated limits; therefore, the slags are not subject to hazardous waste disposal by the standards stipulated in various leachate criteria. This is in agreement with results from another study which concluded that the copper

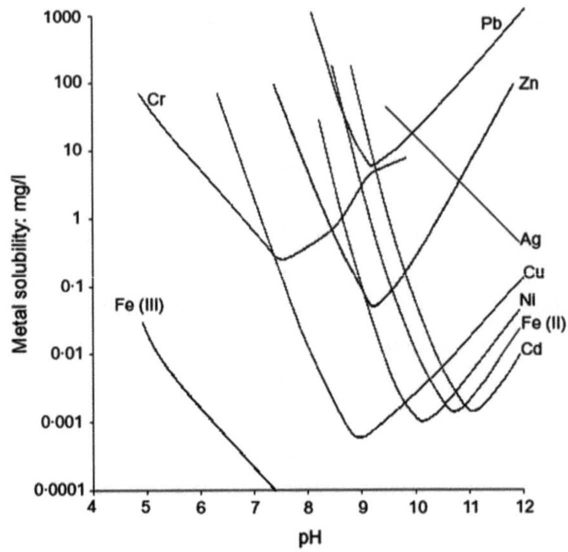

Fig. 2 Solubilities of metal hydroxides as a function of pH [2]

smelter slags are inert [5]. The presence of basic oxides in slags can mitigate leachate contamination. In Table 3, the self-neutralising effect of smelter slag in this study is evident from the increase in pH from an initial value of 4.5 to a maximum of 8.0 and average of 7.0.

The neutralising capacity of the slag results mainly from the basic nature of the slag and has been known to help reduce metal loadings of slag leachate.

Conclusion

The metal loads of leachates from all slags were below the regulatory leachate quality criteria. Therefore, the slags were deemed not to be hazardous waste. The presence of CaO in smelter discard slags enhances the neutralising effect of the slag, consequently reducing metal loads of leachates from mine wastes. Utilising CaO-containing slags in copper smelting practice is a viable option to reducing leachate contamination emanating from smelter slag dumps.

$FeO–CaO–SiO_2$ or CaO-containing slags have an additional benefit to smelting operations such as increasing the matte/slag separating efficiency due to lower viscosities and densities compared to $FeO–Fe_2O_3–SiO_2$ slags [6].

Most smelting furnaces generate flue dust that is usually captured in an adjoining dust-handling system. The dust-handling system is periodically cleaned to avoid clogging and pressurising the furnace. In some operations, the dust-handling system cleaning cycle coincides with the slag-skimming period. Cleaning of the dust-handling system or ashing of boilers during skimming operations as practiced in

Table 3 Leachate extraction results

Slag sample	Contaminant concentrations (mg/L)							pH
	Cu	Co	Pb	Zn	As	Fe	Cd	
R1PO1	1.51	0.26	<0.01	0.06	0.89	0.57	<0.01	6.9
R1P02	1.06	0.18	<0.01	0.03	0.61	0.29	<0.01	6.6
R1P03	0.60	0.51	<0.01	0.02	0.89	0.75	<0.01	7.5
R1P04	1.33	0.99	<0.01	0.03	0.59	0.57	<0.01	7.8
R1P05	1.63	1.01	<0.01	0.03	2.77	1.08	<0.01	7.1
R2P01	0.92	0.22	<0.01	0.04	3.01	0.58	<0.01	7.4
R2P02	1.65	0.41	<0.01	0.04	3.00	0.57	<0.01	7.2
R2P03	1.80	1.94	<0.01	0.05	1.98	1.16	<0.01	7.4
R2P04	1.06	0.26	<0.01	0.02	2.28	1.65	<0.01	7.5
R2P05	1.28	0.35	<0.01	0.02	2.24	0.57	<0.01	7.3
R3P01	1.73	0.61	<0.01	0.02	1.97	0.58	<0.01	7.5
R3P02	2.13	0.65	<0.01	0.04	0.59	0.76	<0.01	7.2
R3P03	1.72	0.44	<0.01	0.03	4.01	0.76	<0.01	7.7
R3P04	2.28	0.87	<0.01	0.03	1.94	0.39	<0.01	7.8
R3P05	1.29	1.82	<0.01	0.03	0.61	0.58	<0.01	7.5
R4P01	2.80	0.58	<0.01	0.01	1.78	0.67	<0.01	7.8
R4P02	1.19	0.71	<0.01	0.05	1.35	1.19	<0.01	7.9
R4P03	0.48	0.28	<0.01	0.02	1.95	0.48	<0.01	7.7
R4P04	1.15	0.35	<0.01	0.03	1.96	1.15	<0.01	7.8
R4P05	0.97	1.19	<0.01	0.03	1.74	0.97	<0.01	7.9
R4L611	0.75	0.32	<0.01	0.02	2.65	0.75	<0.01	7.5
R5P01	0.77	0.72	<0.01	0.01	1.97	0.77	<0.01	8.0
R5P02	0.68	0.27	<0.01	0.02	2.15	0.68	<0.01	7.8
R5P03	0.57	0.32	<0.01	0.02	1.94	0.57	<0.01	7.8
R5P04	0.75	0.41	<0.01	0.03	2.70	0.75	<0.01	7.4
R5P05	0.77	0.33	<0.01	0.02	2.27	0.77	<0.01	7.8
Average	1.22	0.59	<0.01	0.03	1.85	0.73	<0.01	7.0
Alberta leachate quality criteria	100	100	5.0	500	5.0	1000	1.0	–

some smelters should be discouraged. Smelter streams or effluents containing water-soluble compounds such as flue dust should not be mixed with slag. Therefore, cleaning of the dust-handling system or ashing of boilers should be carried out separately from skimming operations in order to avoid sending water-soluble compounds containing flue dusts to the slag dump. Boiler ashes contain significant amounts of water- and acid-soluble compounds, which might exaggerate contaminant loads of run-off water.

References

1. Balintova M, Petrilakova A (2011) Study of pH influence on selective precipitation of heavy metals from acid mine drainage
2. Boardman DI, Glendinning S et al (2004) The influences of iron (III) and lead (II) contaminants on lime-stabilised clay. Géotechnique 54(7):467–486
3. Koren DW, Wilson LJ et al (1997) Investigations of leach test protocols for slags. In: Finch JA, Rao SR, Huang L (eds) Processing of complex ores. Montreal, Canadian Institute of Mining, Metallurgy and, Petroleum, pp 339–354
4. Masters GM (2000) Introduction to environmental engineering and science. Prentice-Hall Inc, Englewood Cliffs, NJ
5. Shanmuganathan P, Lakshmipathiraj P, Srikanth S, Nachiappan AL, Sumathy A (2008) Toxicity characterization and long-term stability studies on copper slag from the ISASMELT process. Resour Conserv Recycl 52(4):601–611
6. Sohn HY (2014) Non-ferrous process principles and production technologies. Treatise Process Metallurgy Seetharaman S, vol 3. Amsterdam, Elsevier, p 1097

Part IV
REWAS 2019: Rethinking Production: Poster Session

Degradation of Ore Collector with Photooxidation UV/H_2O_2 and Photo-Fenton

Isabela F. B. Alves, Marcela P. Baltazar, Jorge A. S. Tenório and Denise C. Romano

Abstract Several types of surfactants have been used in mineral process flotation. Studies appoint an environmental danger potential for some surfactants due to its persistent pollutant characteristic. Oxidation techniques, such as photo-Fenton, are better for surfactants degradation than conventional treatments. The aim of this work is to study the influence of reagent concentrations in the photo-Fenton reaction to degrade a surfactant used on mineral flotation. Experiments were conducted in 3L UV reactor with approximately 260 ppm surfactant solution, 10, 15, 22, 33, and 45 mM H_2O_2 and 1.4, 1.9, 2.2, 28, 80.13, 83, and 84.4 Fe^{2+}/H_2O_2 molar rate. From this result, different pH ranges are available showing the 2.5–3 obtained best result. Samples were taken every 20 min in total of 2 h. The degradation was measured through analysis of total organic carbon. Preliminary results indicated that 1.4 molar rate is more efficient showing 70% of degradation when compared to another concentration.

Keyword Flotation · Surfactant · Photo-Fenton

Introduction

Flotation processes are widely used in iron mining to remove impurities or concentrate ores. In reverse iron flotation, silica and quartz are the principal impurities of interest to remove. Impurities can be aggregated through substances with collecting and depressant substances, which are cationic surfactants such ether amines and starch [1].

Cationic surfactant to flotation has so far been based on unsubstituted alkyl amines and quaternary ammonium compounds. Ether amines are a type of cationic surfactant where the dissociation in the aqueous medium depends on pH. When the pH of the solution is above 8, amines surfactants surfer protonation like demonstrated in Eq. 1, R is the alkyl chain attached [1]. Biodegradation of these substances can leave 1 month when in small concentrations and has high toxicity to the aquatic species

I. F. B. Alves (✉) · M. P. Baltazar · J. A. S. Tenório · D. C. Romano
São Paulo, Brazil
e-mail: isabela_falconi@usp.br

G. Gaustad et al. (eds.), *REWAS 2019*, The Minerals, Metals & Materials Series,
https://doi.org/10.1007/978-3-030-10386-6_15

(with $LC_{50} = 3.58$ ppm for *B. rerio* species) [2, 3]. LC_{50} is a lethal concentration when 50% of the exposed population dies.

$$RNH_2 + H_2O \rightarrow RNH_3^+ + OH^-. \tag{1}$$

Final products in the iron mining are sinter, pellet feeds, pebble or lump ore with high concentration of iron and low silica concentration (less than 2%).

Before flotation, reagents remain together and are referred to tailing sludges of the process. Cationic collectors are strongly bonded to suspended solids or activated sludge by electrostatic attraction. Under aerobic conditions, a largest part of cationic surfactants is readily biodegradable and mineralized. However, the same does not occur in anaerobic conditions with the presence of these substances in non-aerated sludge. In cold habitats, the biodegradation of surfactant occurs at very slow rates occurring bioaccumulation [2, 4]. The bioaccumulation it's due to complexes form between amine surfactants and metals, increasing the toxicity and implying that fish may accumulate these more readily [2].

Oxidation is alternative to degrade and mineralize pollutants like ether amines and other surfactants. Your benefits consist in loss or no generation of sludge, in addition, with easily available reagents. Use of UV radiation can grow pollutants in excited energy state transforming them in oxidation product. Combination between UV and H_2O_2 is of much use to degradation of organic matrices, because it can produce hydroxyl radicals [5].

Another oxidative process utilized is photo-Fenton (Eq. 2 when k_1 is velocity kinetic constant) that reduces much organic forms in water, carbon dioxide, and salts or other gases. A process known as mineralization is realized from oxidation with hydroxyl radicals generated with the oxide reduction of ferrous salt and hydrogen peroxide [6, 7]. Hydroxyl radical reacts with organic molecules generating oxidation products until there is complete mineralization [8].

$$Fe^{2+} + H_2O_2 \rightarrow Fe^{3+} + OH^{\cdot} + OH^- \qquad \mathrm{k_1 = 76\,M^{-1}\,s^{-1}}, \tag{2}$$

$$OH^{\cdot} + RH \rightarrow R^{\cdot} + H_2O. \tag{3}$$

Hydroxyl is a powerful oxidant (+2.8 V of standard potential reduction) being no more reactive only than fluorine (+3.06 V) and appropriated to degrade organic and recalcitrant substances. Fe^{3+} ions will continue to react with hydrogen peroxide in a slow reaction (Eq. 4) and result in hydroperoxyl radical less reactive than hydroxyl radical [9].

$$Fe^{3+} + H_2O_2 \rightarrow Fe^{2+} + HO_2^{\cdot} + H^+ \quad \mathrm{E_a = 126\,kJ\,mol^{-1}} \quad \mathrm{k_2 = 0.001 - 0.01\,M^{-1}\,s^{-1}}. \tag{4}$$

UV radiations catalyze the Fenton reaction through one effect known as ligand-to-metal charge transfer(LMCT). This effect reduces Fe^{3+} to Fe^{2+} giving continuity to the reaction Fenton (Eq. 2) increasing the hydroxyl produced and consequently the degradation, and this is represented in Eqs. 5 and 6 [6, 10].

$$Fe^{3+} + H_2O \rightarrow Fe(OH)^{2+} + H^+, \quad (5)$$

$$Fe(OH)^{2+} + hv \rightarrow Fe^{2+} + OH^{\cdot}. \quad (6)$$

Reaction support by UV radiation is known as photo-Fenton. Important variable in these processes is pH operation. The dependence of iron state in solution is increasing or decreasing the efficiency of hydroxyl radicals, and better operation for this occurs in 2.5–3 range. But the dissociation of collecting surfactants depends on the pH of as mentioned above. This work studied the ether monoamine collector used in reverse flotation of iron ore, degradation utilizing UV/H_2O_2, and homogenous photo-Fenton system and influence of reagents concentration and pH.

Materials and Methods

Photoreactor Utilized

Experiments are conducted on airlift photoreactor with lights coupled around vessel, 6 lamps UVB low-pressure mercury with 7.2 W OSHIO®, 306 nm and 6 lamps is UVC 7 W with 254 nm, OSRAM®, totaling about 12 lamps around the acrylic (polymethyl methacrylate) vessel. Aeration system coupled is adjusted in 5 vvm for to guarantee homogenous agitation in solution.

Solutions with 250 ppm ether monoamine mixtures (Flotigam EDA-C, Clariant) were used for conducting the degradation assays.

Photo-Fenton Assays

Peroxide hydrogen (35%), ferrous sulfate heptahydrate ($FeSO_4 \cdot 7H_2O$), and oxalic acid ($C_2H_2O_4 \cdot 2H_2O$) were used for degradation. First, the better hydrogen peroxide concentration is done varying the concentration in the ranges of 10, 15, 22, 33, and 45 mM (160, 270, 390, 590, and 810 ppm, respectively). Then, photo-Fenton reaction in different molar rates between ferrous sulfate and hydrogen peroxide is done in 1.4, 1.9, 2.2, 28, 80.13, 83.7, and 84.4. Such values were chosen through studies conducted for degradation of drugs and surfactants [11, 12]. These experiments are conditioned in pH 2.5–3 with solutions NaOH 2 M and H_2SO_4 0.1 M. Additional trials were done to evaluate the pH influence in ether amine degradation, and they are varied in the following bands: 1.5–2, 3–3.5, 4–4.5, 5–5.5, and 7.5–8. Each degradation experiment was carried out in 2 h.

Amounts Prepared

Aliquots were withdrawn each 20 minutes and adding a solution of sodium sulfate in excess (Na_2SO_4, 1 M) for interrupting the hydroxyl radical reactions [12]. Then, precipitates are separated with 0.4 μm membrane filters and the supernatant is diluted three times for analyses in total organic carbon (TOC).

TOC Shimadzu equipment is used for quantification of organic and inorganic carbon concentrations. Degradation was measured as in Eq. 6, through decreasing organic carbon concentration dissolved in solution. Complete mineralization of organic substances reduces molecule mainly in CO_2 [5]. Thus, TOC values should decrease over time in exposure to photooxidative processes.

$$(\%)degradation = 100 - \frac{[TOC]_t}{[TOC]_0} \times 100 \quad (6)$$

The total carbon (TC) concentration is adjusted in calibration curve in 100 and 10 ppm carbon utilizing potassium hydrogen phthalate. And, for inorganic carbon (IC) concentration calibration curves in 100 and 10 ppm are adjusted using sodium hydrogen carbonate and sodium carbonate. Quantification of total organic carbon is result of the difference between TC and IC concentrations.

Results and Discussion

Degradation in Presence of UV and Increasing H_2O_2

Solution used for assays has ether monoamine collector, hydrogen peroxide, and deionized water. In turn, these degradation calculations are based on organic carbons dissolved in solution. Table 1 shows carbon concentrations in deionized water utilized and ether monoamine solution prepared.

Deionized water has a low organic carbon concentration (0.495 ppm) in comparison with the ether monoamine solution (257.32 ppm). After water analysis realization, "zero shift" function was set to leave water interferences when TOC has low concentrations.

Table 1 Organic, inorganic, and total carbon concentrations

Solution	TOC (ppm)	TC (ppm)	IC (ppm)
Ether monoamine solution	257.9	275.32	17.67
Deionized water	0.483	0.495	0.011

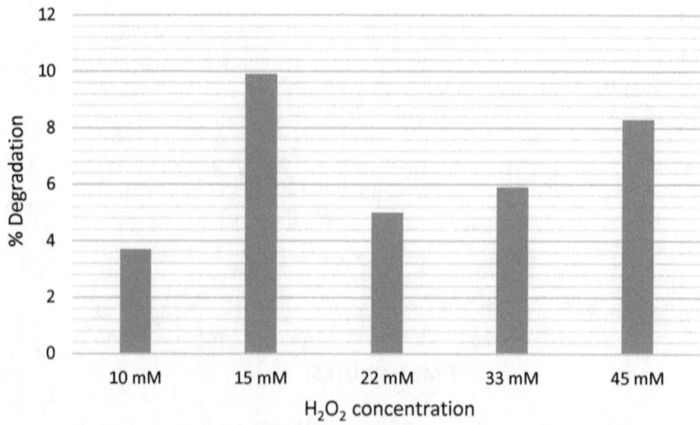

Fig. 1 Degradation percentage for different H_2O_2 concentrations in 2 h test

Assays to evaluate better concentration of hydrogen peroxide for degradation of the ether monoamines are realized in different concentrations. Figure 1 demonstrates the results in 2 h.

In Fig. 1, it is possible to note that the better degradation result was obtained in 120 min with 15 mM and 10% of degradation. Ether monoamine is a possible persistent substance in aqueous medium. Hydrogen peroxide under UV radiation follows Eq. 6, in specific 254 nm wavelength of your molar absorption coefficient is approximately 19.6 L mol^{-1} cm^{-1} and 0.5 quantum yield (φ) to the number reactant molecules. In 300 nm, this value is approximately 0.9 L mol^{-1} cm^{-1} and 0.3 quantum yield. While in 200–204 nm, this value is approximately 179.2 L mol^{-1} cm^{-1}. Molar absorption is an important propriety in photooxidation, because these correspond to photoproduct molecules formed per unit time divided by the number of photons absorbed during this period [5, 12]. When solution substrate has a molar absorption value close to hydrogen peroxide, there may be competition for photons emitted by UV radiation [13]. Therefore, perhaps the degradation obtained from the ether monoamines did not reach significant values. Mercury lamps used in photoreactor emit radiation in 254 and 306 nm.

In 10 mM, it has horse degradation value and for 15 mM it is better. For higher concentration values (22, 33 and 45 mM), degradation has less degradation result in 15 mM. Thus, collector degradation does not accompany the growth hydrogen peroxide. When hydrogen peroxide is present in excess it can behave as a scavenger of hydroxyl radicals, because its concentration is larger than hydroxyl radical in solution [13]. This fact is in common related to works using sonolysis Fenton, with it is appointed the excess of hydrogen which has a negative effect [14]. But, in other way it has been related that growing in hydrogen peroxide is behavior degradation growing, this fact is depending on the substrate to be degraded [15].

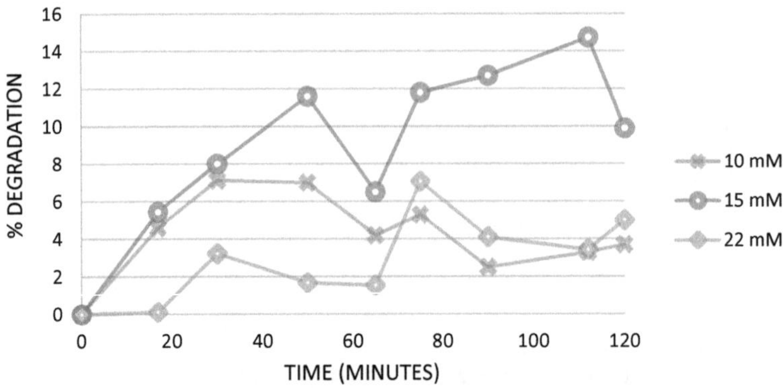

Fig. 2 Variation in degradation value in 10, 15, and 22 mM hydrogen peroxide

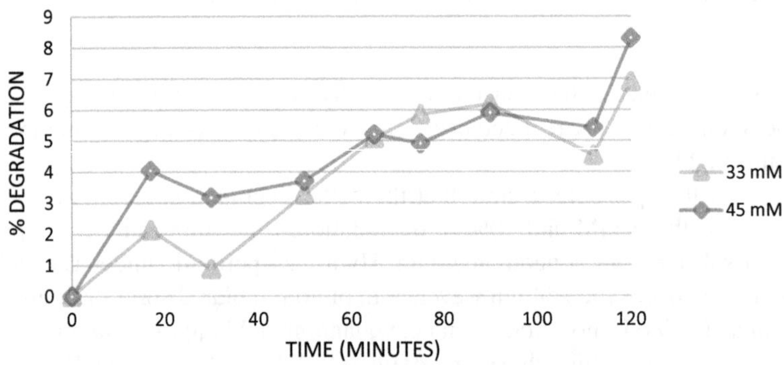

Fig. 3 Variation in degradation value in 33 and 45 mM hydrogen peroxide

Results for degradation in 10, 15, and 22 mM of hydrogen peroxide in function of TOC and time are shown in Fig. 2. For 33 and 45 mM the results are shown in Fig. 3.

Behavior for degradation (hence, TOC increases and decreases) oscillate over time with curves presenting nonlinear comportments for all peroxide concentrations. 15% of degradation in 110 min was reached for the concentration of 15 mM, while a better result appears for 22 mM concentration in 70 min with 7% degradation. For 33 and 45 mM hydrogen peroxide concentration, the highest value is reached in 120 min.

The organic substances need degradate in intermediate reactions before to achieve complete mineralization in CO_2, water and inert mineral salts. These intermediate reaction form intermediate substances, the acetone (C_3H_6O) for example before completed mineralization pass through 50 reactions [5]. Equations 7 and 8 describe that the CO_2 dissolved in solution can form carbonic acid and react again to another intermediate or ether monoamine (R) leaving to be inorganic carbon and returning in organic form [16].

$$CO_{2(aq)} + H_2O \rightarrow H_2CO_3, \quad (7)$$

$$H_2CO_3 + R \rightarrow HCO_3 - R + H^+. \quad (8)$$

Oppenländer [17] affirms that reaction 6 (Eq. 6) occurs in very small amounts in acid solutions forming 0.1% carbonic acid. But the aeration adjusted at 5 vvm in assays can interfere these processes increasing carbonic acid formation being oscillating the TOC in system.

Photo-Fenton Degradation with Increasing Molar Ratio (Fe^{2+}/H_2O_2)

Results for each concentration used in photo-Fenton system are shown in Table 2.

Degradation values vary with molar rate (H_2O_2/Fe^{2+}) and do not follow a linear behavior, in the same way as for the UV/H_2O_2 system. High degradation value was obtained at 1.4 molar rate with removal of 70% in 60 min. In UV/H_2O_2 system with 15 mM concentration it has the better result, but in photo-Fenton system this occurs in 10 mM. Increasing in molar rate does not exactly increase degradation, the higher value in molar rate (84.4) degradation presents low (16.15%). But, in closest value of 83.7 degradation increased for 46% and for 80.13 rate removal gone 2.1%. Another peak is present in 2.2 molar rate showing that the relation between degradation and molar rate has various peaks for maximize collector mineralization. Proximity in molar rates and larger variation in degradation demonstrate a great influence in reagents concentration in photo-Fenton reaction. When this value is closest to estequiometry 1:1, a large growth in degradation was obtained, except for 1.9 molar rate.

In 70% degradation, 1.4 molar rate in 60 min is achieved, a time lowest of the other results because the majority TOC removed is obtained in 120 min (exception

Table 2 Results for the homogeneous system in different molar rates in 2.5–3 pH range

Molar rate H_2O_2/Fe^{2+}	H_2O_2 (mM)	Better result	
		Time (min)	Degradação (%)
1.4	10	60	70
1.9	45	120	7.3
2.2	15	50	43.5
28	45	120	8.27
80.13	22	120	2.1
83.7	10	120	46
84.4	45	75	16.5

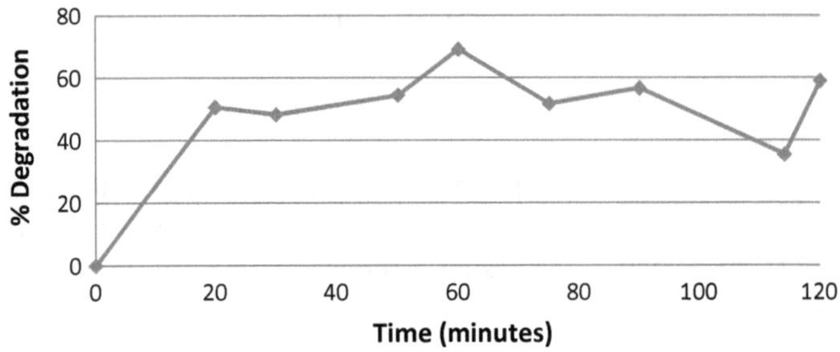

Fig. 4 Degradation with 1.4 molar rate (Fe^{2+}/H_2O_2)

for 2.2 and 84.4 molar rates). Results for degradation over time in 1.4 molar rate are represented in Fig. 4.

In these, studied residual H_2O_2 concentration is not measured. But, considering is possible that hydrogen peroxide concentration is decreasing significantly before 60 min, Fe^{3+} regeneration in Fe^{2+} will be interrupted because the photo-Fenton reaction is ceased. Thus, the oxidants present degrade the organic matrix through the UV and Fe^{2+} radiation remaining [18].

Standard reduction potential of Fe^{2+} is −0.44 V and H_2O_2 has +2.77 V. Thus, photo-Fenton reaction (Eq. 2) is catalyzed with the light effect and the presence of Fe^{2+} acting as another reducer of hydrogen peroxide. In this way, the formation of hydroxyl radical is more possible in the presence of these two reducers (UV radiation and Fe^{2+}). But, photo-Fenton (Eq. 2) has a rate constant ($k_1 = 76\ M^{-1}\ s^{-1}$) and quantum yield (0.14 in 313 nm) much lower than reaction represented in Eq. 9 between hydroxyl radical and hydrogen peroxide [7, 17, 19].

$$H_2O_2 + OH^{\cdot} \rightarrow HO_2^{\cdot} + H_2 \qquad k_3 = 2.7 \times 10^7 M^{-1}\ s^{-1}. \quad (9)$$

Considering Eq. 9, the excess of hydrogen peroxide for photo-Fenton system is not good to degradation global efficiency, because the reagent acts how hydroxyl radical scavenger. In these conditions, Fe^{2+} concentration will be much less than Fe^{3+} generating reaction represented in Eq. 9 and growing hydroperoxyl radical concentration (with less oxidant power than hydroxyl radical) [17, 20].

Iron speciation is an important factor for process. In this study, it uses ferrous sulfate (II) with acid oxalic. Acid oxalic should be forming oxalate iron complex that reacts more rapidly with hydrogen peroxide accelerating degradation process (Eqs. 10 and 11). This complex assists LMCT phenomenon that reduces Fe^{3+} to Fe^{2+} [7]. The presence of oxalate increases radiation absorption to the visible domain that is 500 nm, with presenting a high quantum yield, $\Phi_{Fe}^{2+} = 0.86$. The influence of better absorption is not unique. Oxalate helps in continuous generation of Fe^{2+}, enabling a cyclical system in conjunct with the UV radiation. But this is only possible when the substrate is easily oxidizable. Otherwise, it will be competing with hydrogen peroxide for Fe^{2+} or it will be dimerized by converting again to the original compound [5, 6].

Table 3 Results photo-Fenton system in different pHs

pH	Time (min)	Degradação (%)
1.5–2	110	46
2.5–3	60	70
3–3.5	120	3
4–4.5	60	35
5–5.5	75	26
7.5–8	110	30

Kim and Vogelpohl [21] achieved same results utilizing two types of iron speciation. In one experiment, they used oxalate and ferric sulfate and in other just ferrous sulfate (besides UVA and H_2O_2). Differences in results were less iron concentration and minor velocity utilized when ferric sulfate (III) and oxalate are used [18].

$$Fe^{III}(C_2O_4)_3^{3-} + h\nu \rightarrow Fe^{2+} + CO_2 + 2.5C_2O_4^{2-}, \tag{10}$$

$$Fe^{2+} + H_2O_2 + 3C_2O_4^{2-} \rightarrow Fe^{III}(C_2O_4)_3^{3-} + HO^- + HO^{\cdot}. \tag{11}$$

In another way, product of radiation $Fe^{III}(C_2O_4)_3^{3-}$ is a $C_2O_4{}^{2-}$ ion with dissociates in aqueous medium forming carbonic dioxide in solution, increasing CO_2 availability which is demonstrated in Eq. 10 [5]. When carbon dioxide disponibility is in larger quantity, acid carbonic concentration is growing too which has been described in Eqs. 7 and 8.

From the better degradation result in Table 2 (70% degradation and 1.4 molar rate), pH was varied to evaluate TOC comportment. Results are shown in Table 3.

In different ranges for pH, the better result is obtained in 2.5–3. For the greatest acid condition (pH between 1.5–2). To neutral conditions (pH = 5–5.5 and 7.5–8) a 26–30% of TOC removal is achieved. Worst value for removal appears in the range maintained between 3–3.5 with just 3%.

Figure 5 shows degradation variation in 2 h reactions in function of pH. In 2.5–3 pH range value, degradation is much higher than in another range. Just one peak with 27% degradation appears in 4–4.5 pH, after this discrete growing is presented. The minor value is obtained in 3–3.5 range (2% degradation), but was studied with much significant value in this pH. The study with landfill leachate degradation obtained 75% in the same conditions [18].

Photo-Fenton reaction works with ferrous ions and a small pH range. Studies say the better range to photo-Fenton comprehend is 2.5–3 because of the iron comportment. For solution with pH above 3, ferrous ions precipitate decreasing UV radiance interaction with substrate and hydrogen peroxide molecule. When pH under 2.5 exists a lot of H^+ ions are in solution with aging hydroxyl scavenger [1, 6].

Fatty amines, like as ether monoamines, has a propriety to be dissociated from 8 pH [1]. But, for this pH hydrogen peroxide behaves like a reducer and not as oxidant decreasing hydroxyl radical production and global efficiency processes [13].

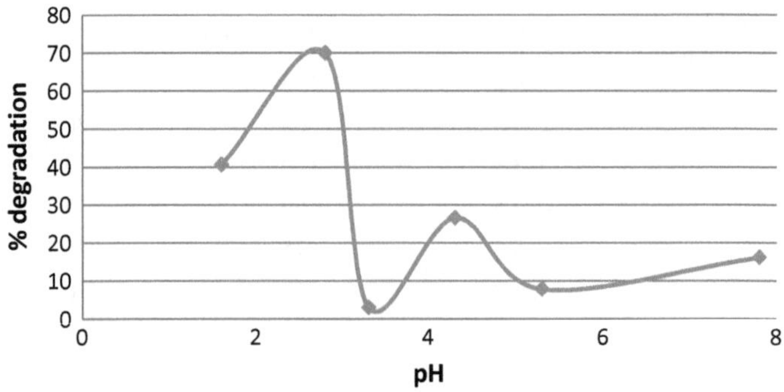

Fig. 5 Degradation as a pH function fixed in final of 2 h reaction

Conclusion

Ether monoamine mixture is a persistent substance, requiring a stoichiometry rate between Fe^{2+} and H_2O_2 adjusted in pH 1.4 and pH 2.5 to obtain 70% degradation in 60 min. In point of this, ferrous presence is a significant importance for degradation since in presence of just hydrogen peroxide and UV radiation better result for degradation achieved is 10% in 2 h using 15 mM. A growing concentration for hydrogen peroxide does not significant a growing in degradation, and the same a growing concentration for photo-Fenton reagents too. Variation in pH demonstrates better result in 2.5–3 range and this variation is not favorable to improve efficiency processes. A recommendation for future studies is the evaluation of hydrogen peroxide reminiscent, which verifies kinetic production of hydrogen peroxide efficiency, and also verify temperature and aeration influences. For reactor configuration, it is recommended to evaluate the efficiency when mercury lamp adjusted in 200–204 nm is used.

References

1. Pignatello JJ, Oliveros E, MacKay A (2006) Advanced oxidation processes for organic contaminant destruction based on the fenton reaction and related chemistry. Crit Rev Environ Sci Technol 36(1):1–84. https://doi.org/10.1080/10643380500326564
2. Bach L, Nørregaard RD, Hansen V, Gustavson K (2016) Review on environmental risk assessment of mining chemicals used for mineral separation in the mineral resources industry and recommendations for Greenland. http://dce2.au.dk/pub/SR203.pdf. Accessed 21 Aug 2018
3. Parsons S (2005) Advanced oxidation processes for water and wastewater treatment. IWA Publishing, Londres
4. Scott MJ, Jones MN (2000) The biodegradation of surfactants in the environment. Biochim Biophys Acta Biomembr 1508(1–2):235–251. https://doi.org/10.1016/S0304-4157(00)00013-7
5. Leja J (1982) Surface chemistry of froth flotation. Plenum Press, New York, pp 205–338

6. Pupo Nogueira RF, Trovó AG, Da Silva MRA, Villa RD, De Oliveira MC (2007) Fundamentos e aplicaçõ es ambientais dos processos Fenton e foto-Fenton. Quim Nova 30(2):400–408. https://doi.org/10.1590/S0100-40422007000200030
7. Pignatello JJ, Oliveros E, MacKay A (2006) Advanced oxidation processes for organic contaminant destruction based on the fenton reaction and related chemistry. Crit Rev Environ Sci Technol 36(1):1–84. https://doi.org/10.1080/10643380500326564
8. Yang M, Hu J, Ito K (1998) Characteristics of Fe^{2+}/h2o2/uv oxidization process. Environ Technol (United Kingdom) 19(2):183–191. https://doi.org/10.1080/09593330.1998.9618629
9. Garrido-Ramírez EG, Theng BKG, Mora ML (2010) Clays and oxide minerals as catalysts and nanocatalysts in Fenton-like reactions—a review. Appl Clay Sci 47(3–4):182–192. https://doi.org/10.1016/j.clay.2009.11.044
10. Clariant S/A (2009) FISPQ—Ficha de Informação de Segurança de Produtos Químicos em acordo com a NBR-14725. Syngenta, Paulinia
11. Barbeni M, Minero C, Pellzzetti E, Borgarello E, Serpone N (1987) Chemical degradation of chlorophenols with Fenton's reagent (Fe^{2++} H_2O_2), 16. https://ac.els-cdn.com/0045653587902815/1-s2.0-0045653587902815-main.pdf?_tid=8d1ba1ff-784d-470d-8802-db5e167c9e84&acdnat=1533510829_262fe690b57eca28583b816a1bc10ed0. Accessed 5 Aug 2018
12. Monteagudo JM, Durán A, Culebradas R, San Martín I, Carnicer A (2013) Optimization of pharmaceutical wastewater treatment bysolar/ferrioxalate photo-catalysis. J Environ Manage 128:210–219. https://doi.org/10.1016/j.jenvman.2013.05.013
13. Mierzwa JC, Rodrigues R, Teixeira ACSC (2018) UV-hydrogen peroxide processes. In: Advanced oxidation processes for waste water treatment. Elsevier 13–48. https://doi.org/10.1016/b978-0-12-810499-6.00002-4
14. Acosta AML (2016) Processos de tratamento não convencionais para degradação do antibiótico sulfadiazina em meio aquoso
15. Pignatello JJ (1992) Dark and photo assisted Fe^{3+} catalyzed degradation of chlorophenoxy herbicides by hydrogen peroxide. Environ Sci Technol 26(5):944–951. https://pubs.acs.org/sharingguidelines. Accessed 23 Aug 2018
16. Skoog DA, Holler FJ, Nieman TA (2013) Fundamentos de quimica analítica, vol 53. 8th edn. Thomson. https://doi.org/10.1017/cbo9781107415324.004
17. Oppenländer T (2003) Photochemical purification of water and air. Wiley-VCH, Villingen-Schwenningen, pp 101–183
18. Kim S-M, Vogelpohl A (1998) Degradation of organic pollutants by the photo-Fenton-process. Chem Eng Technol 21(2):187–191. https://doi.org/10.1002/(SICI)1521-4125(199802)21:2%3c187:AID-CEAT187%3e3.0.CO;2-H
19. Giri RR, Ozaki H, Takayanagi Y, Taniguchi S, Takanami R (2011) Efficacy of ultraviolet radiation and hydrogen peroxide oxidation to eliminate large number of pharmaceutical compounds in mixed solution. Int J Environ Sci Tech 8(1):19–30. http://www.bioline.org.br/pdf?st11002. Accessed 22 Aug 2018
20. Rosenqvist, T (2004) Principles of extractive metallurgy. Tapir Academic Press, pp 238–255
21. Kim S-M, Vogelpohl A (1998) Degradation of organic pollutants by the photo-Fenton-process. Chem Eng Technol 21(2):187–191. https://doi.org/10.1002/(SICI)1521-4125(199802)21:2%3c187:AID-CEAT187%3e3.0.CO;2-H

Influence of Metallic Impurities on Solvent Extraction of Cobalt and Nickel from a Laterite Waste Liquor

Paula Aliprandini, Mónica M. Jiménez Correa, Jorge A. Soares Tenório and Denise Crocce Romano Espinosa

Abstract Limonite is the surface zone of lateritic ore and it was considered an ore waste. However, because of the demand, limonite is now considered an important resource of nickel and cobalt. The presence of impurities in the ore such as magnesium, manganese, and chromium makes it difficult for the recovery of nickel and cobalt. The separation of nickel and cobalt can be done by solvent extraction technique using Cyanex 272 as the extractant. In order to study the extraction of cobalt and nickel from a solution containing chromium, magnesium, and manganese, the pH range between 3.5 and 5.5 was investigated. However, precipitation of chromium, magnesium, manganese, and nickel was observed at pH 5.5. By increasing the pH value, the extraction of cobalt, magnesium, manganese, and chromium was increased. Nickel was not extracted during the study. It was not possible to obtain a solution containing only nickel or cobalt without the presence of the other metals in solution.

Keywords Limonite ore · Cyanex 272 · Purification · Impurities

Introduction

Nickel laterite ore is considered an important resource of nickel. The ore is divided into three different zones [1]. Each zone has a specific composition. Limonite is the surface zone [2]. Until recently, limonite was considered as a waste zone from the ore. However, because of the demand for specific metals, previously waste ores are now considered.

In this zone, nickel is the main value metal and its concentration is between 1.0 and 1.8% [3]. The major use of nickel is the stainless steels, electronics equipment, and batteries [2, 4, 5]. In addition to nickel, the metal of commercial interest is cobalt. The concentration of cobalt is between 0.05 and 0.3% in the limonite ore

---

P. Aliprandini (✉) · M. M. Jiménez Correa · J. A. Soares Tenório · D. C. R. Espinosa
Department of Chemical Engineering, Polytechnic School of the University of São Paulo, R. Do Lago, 250, São Paulo 05424-970, Brazil
e-mail: paulaaliprandini@usp.br

G. Gaustad et al. (eds.), *REWAS 2019*, The Minerals, Metals & Materials Series,
https://doi.org/10.1007/978-3-030-10386-6_16

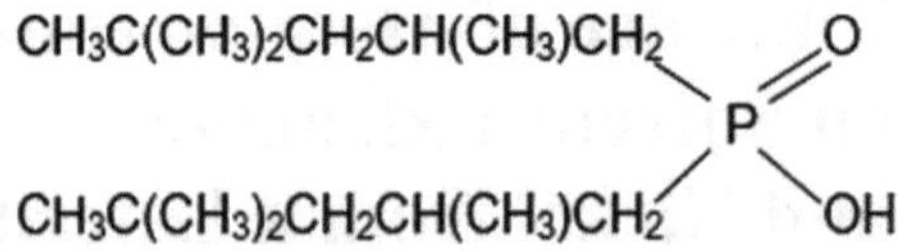

Fig. 1 Commercial extractant Cyanex 272 (bis (2, 4, 4-trimethylpentyl) phosphinic acid)

[2, 3, 6, 7]. Cobalt is used in superalloys. Therefore, recovery of both metals is of great commercial interest [7].

However, because of their chemical characteristics, it is difficult to separate cobalt from nickel. In addition, the presence of impurities affects the recovery of metals of interest as nickel and cobalt [6, 8].

Solvent extraction is an industrial technique used to separate nickel and cobalt. The solvent extraction technique uses two immiscible solutions: an aqueous phase and an organic phase [2, 9, 10]. When the solutions are in contact, and under specific conditions, the metal ions can be transferred between the phases [2].

Cyanex 272 is the extractant used to separate nickel and cobalt. The active component is bis-(2, 4, 4-trimethylpentyl) phosphinic acid (Fig. 1) [11].

During the extraction process, the extractant interacts with metals to form a hydrophobic complex. Specific parameter, such as pH value, can make the extractant selective for one metal relative to another.

The presence of other metal in solution can affect the extraction process of cobalt and nickel.

A solution containing nickel and cobalt, in addition to other metals considered impurities, was used in this study. Chromium, magnesium, and manganese are typically metals present in leaching solution from limonite ore. During the process, it is expected to obtain pure solutions of each interest metal.

Methodology

A synthetic solution was in contact with an organic solution. The metal concentrations used are shown in Table 1. All the metals were used as sulfate salts of analytical grade.

The solution composition was based on a previous study. The composition of the solution is the remainder after the removal of most impurities such as iron, aluminum, and zinc [12].

Cyanex 272 was used as extractant and kerosene as diluent. The organic solution was prepared by dissolving 20% v/v of Cyanex 272 in kerosene.

The effect of equilibrium pH was carried out by varying the pH of solution in a range from 3.5 to 5.5, with an increase of 0.5 by adding sodium hydroxide (NaOH).

Table 1 Metal concentrations in synthetic leach solution

Metals	Co	Cr	Ni	Mg	Mn
Concentration (g/L)	0.05	0.16	2.52	6.35	0.36

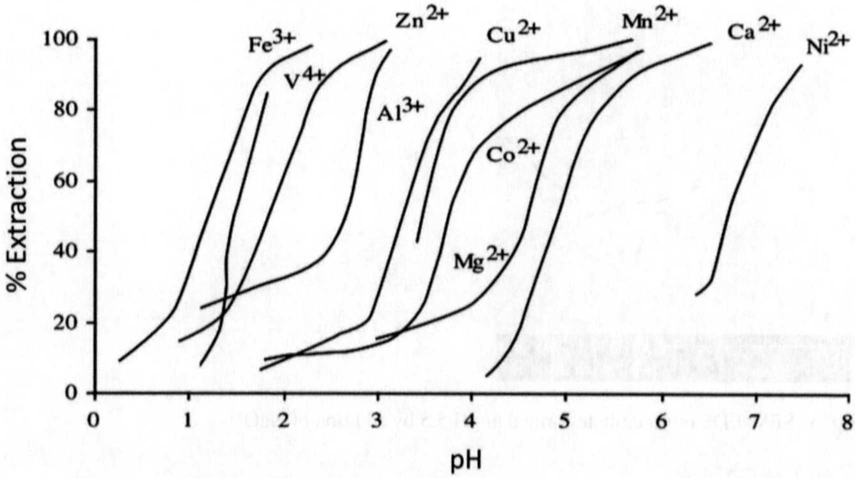

Fig. 2 Extraction of different metals on range pH values in solvent extraction using Cyanex 272 [11]

The pH range was based on the extraction behavior of metals during extraction with Cyanex 272 (Fig. 2). From the diagram, the extraction of metal ions from the synthetic solution begins at pH values greater than 3.

First, 40 mL of the aqueous solution was adjusted to target pH and a similar volume of organic solution was added to the aqueous solution. The system was stirred for 20 min while the pH was maintained constant by adding NaOH. The experiments were performed at room temperature.

Energy-dispersive X-ray spectroscopy (EDX) was used to determine the metals concentration remaining in the aqueous solution after the solvent extraction process.

Results

The study of solvent extraction of cobalt, chromium, magnesium, and manganese was carried out between pH 3.5 and 5.5. However, the precipitation of chromium began at pH 5.5. Figure 3 shows the microscopy of precipitate obtained after drying in oven. The precipitate was analyzed using a scanning electron microscopy with energy-dispersive X-ray spectroscopy (SEM-EDS).

The presence of chromium, magnesium, manganese, and nickel was identified. Therefore, the extraction tests were performed up to pH 5.0. Figure 4 shows the extraction percentage for each metal when pH values are changed between 3.5 and 5.0 using Cyanex 272.

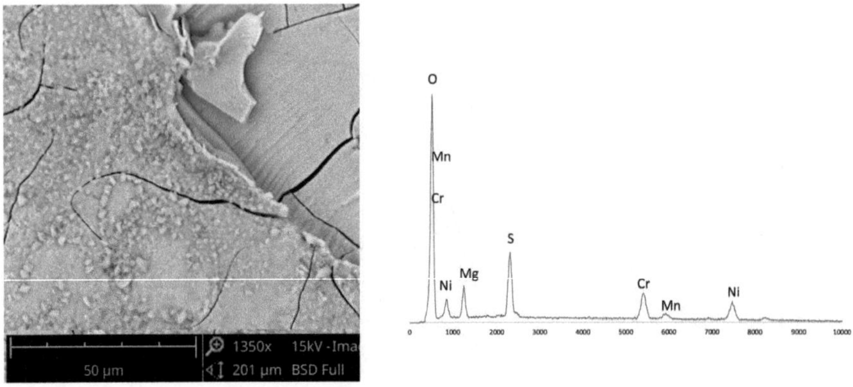

Fig. 3 SEM-EDS of precipitate formed at pH 5.5 by addition of NaOH

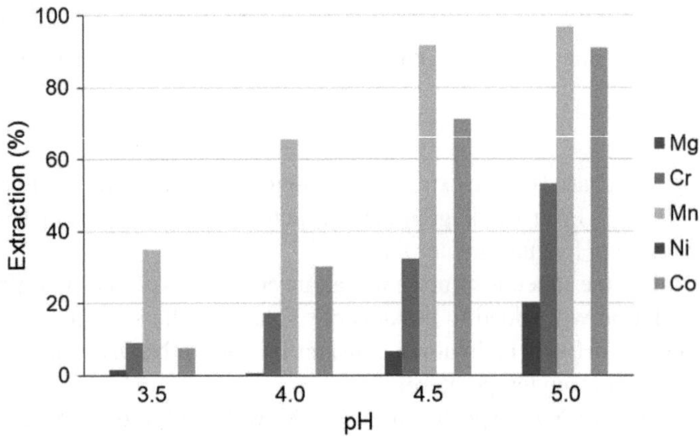

Fig. 4 Extraction percentage of metals during the solvent extraction using Cyanex 272

The nickel was not extracted at any pH value. On the other hand, cobalt extraction was 8% at pH 3.5 and 91% at pH 5. This shows the efficiency of the extractant to separate the nickel from cobalt.

The percentage of chromium extracted was from 9 to 53% between pH 3.5 and 5. For the same pH range, manganese and magnesium were extracted in the percentages 35–97% and 1–20%, respectively.

As can be seen in Fig. 2, the selectivity pH of the extractant for cobalt, manganese, and magnesium is very close. This makes the separation difficult. According to Agatzini Leonardou et al. [13], the extraction of all these metals occurs together and allows only nickel in solution. However, the same author indicates pH 5.5 as being ideal for extracting cobalt and magnesium from nickel. Because of presence of chromium, this pH value could not be achieved. For the proposed working conditions, it was not possible to separate nickel from other metals. Even, it was not possible to separate cobalt.

Therefore, the differences in extraction between the metals made it difficult to obtain pure solutions of nickel or cobalt. Therefore, other work steps will be required to obtain solutions containing only the metals of interest.

Conclusion

- The presence of chromium in solution affected the pH value. The extraction pH value was limited to 5.0. At pH values greater than 5, the precipitation started with chromium, magnesium, manganese, and nickel.
- The extraction of cobalt, magnesium, manganese, and chromium increased with increasing pH value.
- The higher extraction percentage for cobalt was 91%; for chromium was 53%; for magnesium was 20%; and for manganese was 97% at pH 5.0.
- Nickel was not extracted in the studied pH range.
- It was not possible to separate nickel from other metals on the chosen work conditions.
- It was not possible to separate cobalt from other metals.
- The near pH value for extraction of metals makes the selective extraction process difficult.
- Additional work will be needed to evaluate the separation between cobalt and nickel.

Acknowledgements Financial support for this study was supplied from the Coordination for the Improvement of Higher Education Personnel (CAPES), the Counsel of Technological and Scientific Development (CNPq), and the São Paulo Research Foundation (FAPESP—grant 2012/51871-9). The authors thank Ana Carolina Fadel Dalsin for her help on the project.

References

1. Kesler SE, Simon AC (2015) Mineral resources, economics and the environment. 2nd edn. Cambridge University Press
2. Gupta CK (2003) Chemical metallurgy principles and practice. Wiley-VCH Verlag GmbH & Co., KgaA
3. Neudorf D (2006) Method for nickel and cobalt recovery from laterite ores by reaction with concentrated acid and water leaching. US2006/0002835
4. Guo X, Li D, Park KH, Tian Q, Wu Z (2009) Leaching behavior of metals from a limonitic nickel laterite using a sulfation-roasting-leaching process. Hydrometallurgy 99:144–150
5. Mudd GM (2010) Global trends and environmental issues in nickel mining: sulfides versus laterites. Ore Geol Rev 38(1–2):9–26
6. Chang Y, Zhai X, Li B, Fu Y (2010) Removal of iron from acidic leach liquor of lateritic nickel ore by goethite precipitate. Hydrometallurgy 101(1–2):84–87

7. Kyle J (2010) Nickel laterite processing technologies—where to next? In: ALTA 2010 nickel/cobalt/copper conference, pp 1–36
8. McDonald RG, Whittington BI (2008) Atmospheric acid leaching of nickel laterites review. Part I Surphuric Acid Technol Hydrometal 91(1–4):35–55
9. Free M (2013) Hydrometallurgy: fundamentals and applications. Wiley
10. Kislik VS (2012) Solvent extraction: classical and novel approaches. Elsevier
11. CYTEC (2008) CYANEX 272 extractant. Construction materials: 16. https://www.cytec.com/sites/default/files/datasheets/CYANEX%20272%20Brochure.pdf. Accessed 21 Feb 2016
12. Aliprandini P (2017) O uso da extração por solventes para tratamento de licor de lixiviação de minério limonítico de níquel (master's thesis). University of São Paulo, São Paulo, Brazil
13. Agatzini-Leonardou S, Tsakiridis PE, Oustadakis P, Karidakis T, Katsiapi A (2009) Hydrometallurgical process for the separation and recovery of nickel from sulphate heap leach liquor of nickeliferrous laterite ores. Miner Eng 22(14):1181–1192

Iron Recovery from Nickel Slag by Aluminum Dross: Viscosity Evolution in Different Periods

Guangzong Zhang, Nan Wang, Min Chen, Ying Wang and Hui Li

Abstract Nickel slag can be recycled as one of the excellent secondary sources due to valuable iron resource. Slag viscosity during iron recovery from nickel slag by aluminum dross was studied from 1773 to 1873 K, and the viscosity evolution in different periods was systemically discussed. Experimental results showed that slag viscosity increased suddenly after "FeO"/("FeO" + Al_2O_3) ratio lower than 0.53, signifying the transition of dominant role from "FeO" to Al_2O_3. Structure investigation by Raman spectra showed the center of curves transferred to higher shift, which indicated the formation of many more complicated structural units during the reduction process. Furthermore, the increasing units ratio $(Q^2 + Q^3)/(Q^0 + Q^1)$ in $[SiO_4]$-tetrahedra showed a good agreement with the variation of apparent activation energy.

Keywords Nickel slag · Aluminum dross · Viscosity · Slag structure

Introduction

Nickel slag is the by-product from nickel industries containing valuable iron resource. The content of iron in nickel slag can typically reach up to 50 wt%. Since 2008, China has been the largest producer and consumer of nickel in the world on account of the large reserves of nickel sulfide [1]. It is estimated that about 160 tons of granulated nickel slag will be discharged per year and more than 33 million tons of the nickel slag have been accumulated up to now. Generally, so many nickel slags are stockpiled in heaps and only part of them are utilized in road construction or cement manufacture. The negligence of utilization undoubtedly leads to the wastes of resources, and therefore a scientific disposal methodology is really essential to recycle the nickel slag as an excellent secondary source.

Aluminothermic reduction attracts the authors' attention due to its exothermic characteristic and in the present work aluminum dross is selected as the reductant. As

G. Zhang · N. Wang (✉) · M. Chen · Y. Wang · H. Li
School of Metallurgy, Northeastern University, Shenyang 110819, Liaoning, China
e-mail: wangn@smm.neu.edu.cn

G. Gaustad et al. (eds.), *REWAS 2019*, The Minerals, Metals & Materials Series,
https://doi.org/10.1007/978-3-030-10386-6_17

it is known, aluminum dross is also the metallurgical solid waste from the aluminum industries, and the content of Al in the dross is typically as high as 35 wt%. Recycling nickel slag by aluminum dross can realize the full use of the metallurgical solid wastes as well as guarantee the fluidity of slag during reduction process. However, the thermal physical properties of slag during reduction process such as the viscosity, surface tension, and so on are not clear.

Viscosity as one of the most important properties of slag has a significant effect on the reduction proceeding. As stated by Guo et al., slag viscosity directly results in the change of kinetic condition of reduction. In past years, many researches have been conducted to investigate the viscosity as a function of slag components and slag temperature [2–5].

The current slag is $CaO–SiO_2$–"FeO"–Al_2O_3–MgO system. Based on the studies by Park et al. and other researchers, slag basicity, slag components such as "FeO", Al_2O_3, and SiO_2 are all the factors that are affecting viscosity. Nevertheless, the previous studies mainly focus on the viscosity variation under a fixed "FeO" content or Al_2O_3 content, and few works have been carried out during reduction process. Especially for the current slag, "FeO" content decreases and Al_2O_3 content increases with the addition of aluminum dross [6, 7].

The objective of this work is to systematically study the viscosity evolution during iron-exacting process from nickel slag by aluminum dross. It is speculated that the dominant role has changed from "FeO" to Al_2O_3, and therefore the viscous behavior of slag under different reduction periods, i.e., the early and the later periods, are further investigated. In addition, slag structure by Raman spectroscopy is analyzed to reveal the mechanism of viscosity evolution in the microscopic view.

Experimental

Slag Preparation

Based on the material balance, chemical compositions of the investigated slags were calculated and listed in Table 1. Slags A to G were represented as the compositions during reduction process, and slags S1–S4 were carried out to make the comparison with slag A and D, respectively. To obtain a higher activity of "FeO", the basicity of the initial slag was adjusted to be 1.0 upon the thermodynamic calculation.

Reagent-grade chemicals $CaCO_3$, FeC_2O_4, SiO_2, MgO, and Al_2O_3 with the purity more than 99.9% were adopted to prepare the investigated slags, where $CaCO_3$ and FeC_2O_4 were the suppliers of CaO and "FeO", respectively. $CaCO_3$ powder was calcined at 900 °C for 24 h before mixing the chemicals together. During the heating process, 180 g of the mixture of FeC_2O_4 and other components (CaO, SiO_2, MgO, and Al_2O_3) were charged into a molybdenum crucible and placed in the high-temperature zone in a tubular electric furnace. It should be noted that the crucible was held at 873 K for 2 h to decarburize FeC_2O_4 absolutely. When slag temperature

Table 1 Experimental conditions of the investigated slags

Slag no.	Temperature (K)	*w* (SiO_2) (wt%)	*w* (Al_2O_3) (wt%)	*w* (CaO) (wt%)	*w* (MgO) (wt%)	*w* (FeO) (wt%)	"FeO"/("FeO" + Al_2O_3)
A	1773 1823 1873	26.39	7.34	25.72	7.39	33.17	0.82
B		27.04	14.61	25.71	7.92	24.72	0.63
C		27.37	18.20	25.71	8.18	20.54	0.53
D		27.70	21.77	25.71	8.44	16.36	0.43
E		28.02	25.32	25.71	8.70	12.23	0.33
F		28.35	28.85	25.71	8.96	8.13	0.22
G		28.98	35.81	25.72	9.49	0.00	0.00
S1		32.78	7.34	31.95	7.39	20.54	0.74
S2		20.88	18.20	20.35	7.39	33.17	0.65
S3		31.98	21.77	29.68	8.44	8.13	0.27
S4		24.04	28.85	22.31	8.44	16.36	0.36

reached up to 1873 K, the crucible should be maintained for about 1.5 h to make the slag homogeneous. In the whole slag preparation procedure, ultrahigh-purity argon (99.9999%) with the flow rate of 0.5 L min^{-1} was guaranteed to avoid the oxidation of slag components and the crucible. After then, the crucible was taken out rapidly and quenched in the liquid nitrogen (with the cooling rate more than 500 K s^{-1}). The quenched slags were then examined by X-ray diffraction (XRD, D8 Advance, Bruker AXS Gmbh, Germany) and X-ray fluorescence (XRF, ZSX100e, Rigaku, Japan).

No apparent characteristic peaks were observed according to the XRD results, which indicated that the slags were all amorphous. In addition, the composition variation between the initial and final slags was negligible, signifying the samples obtained this way could be used for viscosity measurement.

Viscosity Measurement

Rotating cylindrical method was adopted in viscosity measurement. The details of experimental setup with a rotating viscometer (RTW-10, China) were shown in Fig. 1. The temperature of the sample was continuously detected using a $PtRh_{30}$–$PtRh_6$ thermocouple. To keep the valence of Fe^{2+}, a piece of electrolytic iron plate (about 2.5 g) was placed on the bottom of the crucible. For each measurement experiment, 150 g of the pre-melted slag was charged into the Mo crucible. To ensure the melt homogenous, the crucible was maintained for more than 30 min when the temperature reached 1873 K. After then, the spindle with rotating speed of 200 rpm was immersed into the molten slag and the viscosity measurement was carried out under slag temperature of 1873 K, 1823 K, and 1773 K, respectively. The equilibration time

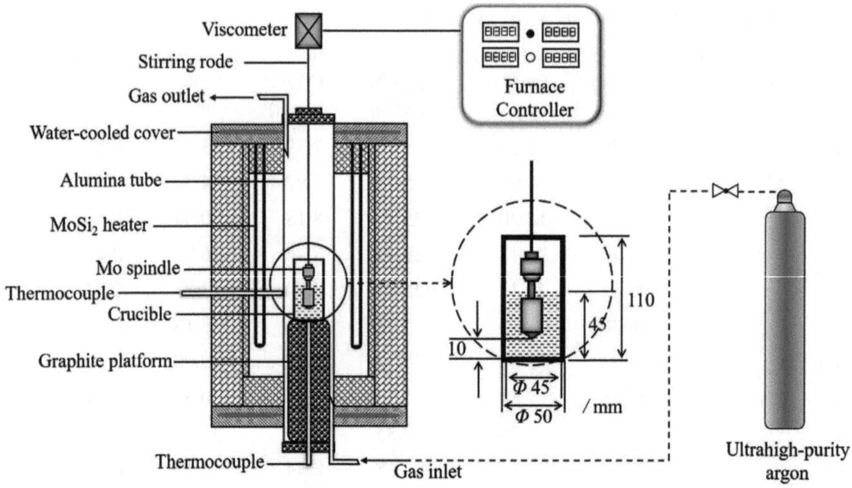

Fig. 1 Schematic diagram of experimental setup

was approximately 30 min for each temperature. Similarly, ultrahigh-purity argon was also used as the protective gas to prevent the oxidation of the crucible, spindle, and the flux. The average viscosity values were determined through three repeated viscosity experiments for each slag component.

After finishing the measurement, slags were reheated to 1873 K and then rapidly taken out and quenched in liquid nitrogen. Due to the negligible proportion of Fe^{3+} in slags, it was assumed that iron ions were fully divalent in all the samples. XRF results also showed the content of MoO_3 should be ignored, indicating that the slag samples obtained this way can be used for structure measurement. More details for the viscosity measurement could be found in our previous works [8, 9].

Raman Spectroscopy

Structural units of the quenched slags were quantitatively analyzed by Raman spectroscopy (Renishaw inVia, Renishaw, England). The diode laser at 532 nm was used as the laser source and the measured shift range was from 2000 to 100 cm^{-1}. The slag structure measured at room temperature was assumed the same as that at 1873 K based on the study by Mysen and Richet [10].

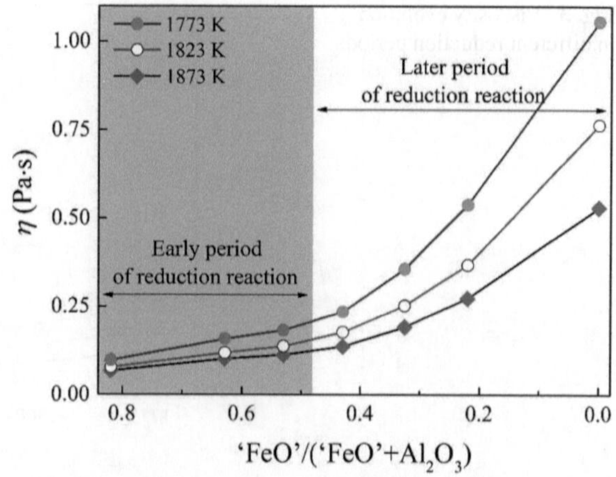

Fig. 2 Viscosity variation under different "FeO"/("FeO" + Al_2O_3) ratios at 1773–1873 K

Results and Discussion

Viscosity Variation

As shown in Table 1, "FeO" content decreases from 33.17 wt% to zero combined with the increasing Al_2O_3 content from 7.34 to 35.81 wt%, and the "FeO"/("FeO" + Al_2O_3) ratio changes from 0.82 to 0.00 with the reduction process. Figure 2 shows the viscosity variation under different "FeO"/("FeO" + Al_2O_3) ratios at 1773–1873 K. It could be found that viscosity decreases with the increasing temperature as well as the increasing "FeO"/("FeO" + Al_2O_3) ratio. Under the fixed "FeO"/("FeO" + Al_2O_3) ratio of 0.22, for instance, viscosities are 0.538 Pa s and 0.273 Pa s at 1773 K and 1873 K, respectively. Whereas, the dependence of slag viscosity on "FeO"/("FeO" + Al_2O_3) ratio reflects an increasing viscosity with the reduction process. In the early period of reduction (the "FeO"/("FeO" + Al_2O_3) ratio is higher than 0.53), slag viscosity changes slowly, followed by a sudden increase then. In view of the fact that the ternary slag basicity ($R = (CaO + MgO)/SiO_2$) could be ignored, the significant evolution of viscosity is speculated to be caused by the dominant role of "FeO" and Al_2O_3. To take a knowledge of the viscosity evolution in different periods, slags A–D and S1–S4 are further studied.

Viscosities of slags A–D and S1–S4 are shown in Fig. 3. Comparing slags A, S1, and S2, it could be found that the effects of "FeO" and Al_2O_3 on viscosity are different. In the early reduction process, the viscosity under a constant slag temperature changes more significantly with the decreasing "FeO" content. For example, the viscosity of slag with 33.17 wt pct "FeO" and 7.34 wt pct Al_2O_3 is 0.097 Pa s at 1773 K, and it increases to 0.14 Pa s as "FeO" decreases, which is much higher than the value of 0.116 Pa s led by the increment of Al_2O_3. Therefore, "FeO" plays the dominant role.

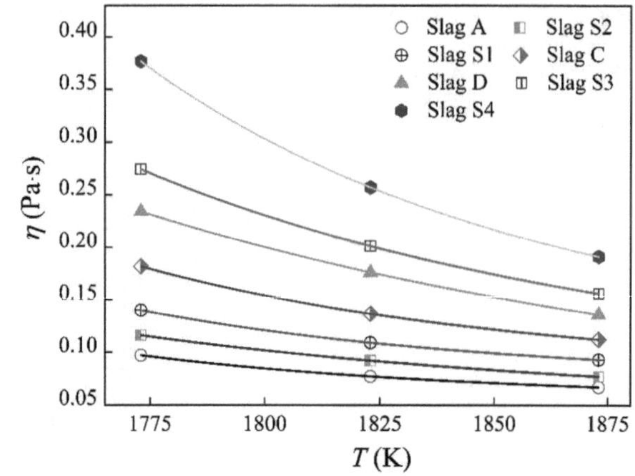

Fig. 3 Viscosity evolution in different reduction periods

Nevertheless, when the "FeO"/("FeO" + Al_2O_3) ratio is lower than 0.53, the influences of these two components on viscosity tend to be contrary. As slags D, S3, and S4 are shown, viscosity changes more evidently in the later period, which should be attributed to the coupling effect of "FeO" and Al_2O_3. What is more, the increment of Al_2O_3 from 21.77 to 28.85 wt% results in a higher value when compared with the decrement of "FeO". The evident increase in viscosity precisely demonstrates the dominant role of Al_2O_3. In terms of the fact that slag viscosity is correlated with its structure, hence the structures for these slags are further studied.

Structure Investigation

Figure 4 represents the Raman spectra of the quenched slags. Based on the related works [10–13], different Raman bands are related to the different structural units, which are listed in Table 2. The main bands between 850 and 1160 cm^{-1} can be separated into four regions, i.e., 850–880 cm^{-1}, 900–930 cm^{-1}, 950–980 cm^{-1}, and 1040–1060 cm^{-1}, reflecting the stretching vibrations of silicate structural units Q^0, Q^1, Q^2, and Q^3, respectively. The superscripts 0, 1, 2, and 3 are the numbers of the bridging oxygen per [SiO_4]-tetrahedra. In Fig. 4, the structural units have been marked at the relevant positions. From the tendency of curves, the center of Raman spectra has been transferred to higher shift. In addition, with the increasing viscosity, the bands change to be more announced, indicating that many more complicated structures have been generated.

In order to analyze the structure evolution more quantitatively, Raman spectra of slags A–D and S1–S4 are deconvolved using Gaussian-deconvolution function in software Origin 8.0, and Fig. 5 presents the results. In the present work, the relative area fraction of individual unit is used for quantitative estimation, which is obtained by dividing the integrated area of all the structural units.

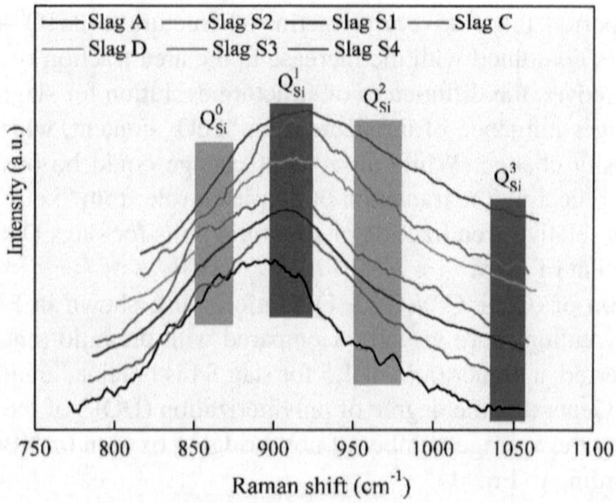

Fig. 4 Raman spectra of the quenched slags

Table 2 Assignments of characteristic bands in Raman spectra

Spectra-centered shift (cm^{-1})	Structural units (Q notation)	Unit types
850–880	SiO_4^{4-}(Q^0)	Monomers
900–930	$Si_2O_7^{6-}$(Q^1)	Dimers
950–980	SiO_3^{2-}(Q^2)	Chains or rings
1040–1060	$Si_2O_5^{2-}$(Q^3)	Sheets

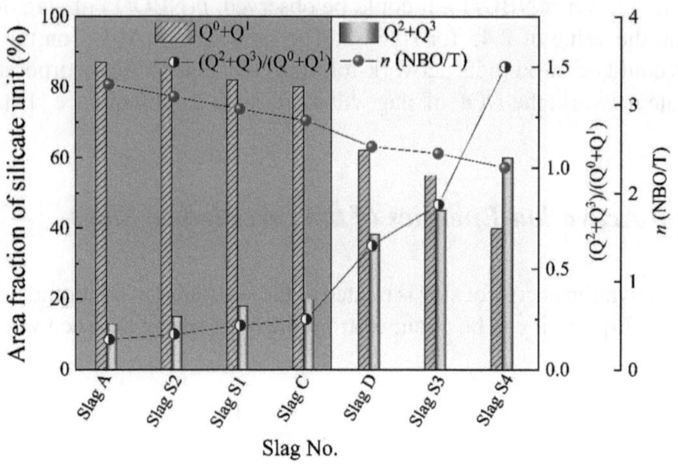

Fig. 5 Deconvolved results of Raman spectra

In the early period, the relative area fraction of structural units ($Q^0 + Q^1$) decreases from 87 to 80%, combined with the increase in the area fraction of structural units ($Q^2 + Q^3$). Moreover, the differences of structure evolution for slags A, S1, and S2 suggest a stronger influence of the decreasing "FeO" content, which is consistent with the viscosity change. While an evident change could be observed between slags C and D, reflecting the transition of dominant role from "FeO" to Al_2O_3. The development of relative area fraction of structural units for slags D, S3, and S4 also confirms this point of view.

The evolution of ($Q^2 + Q^3$)/($Q^0 + Q^1$) ratio is also shown in Fig. 5 to exhibit the structure evolution more visually. Compared with the ratio range from 0.14 to 0.25 in early period, a higher ratio of 1.5 for slag S4 is obtained under the dominant role of Al_2O_3. Generally, the degree of polymerization (DOP) of the silicate melt is characterized by the average number of non-bridging oxygen (*n*(NBO/T)), and it is evaluated according to Eq. (1).

$$n(\mathrm{NBO/T}) = \sum (4 - i) \cdot X_i, \tag{1}$$

where i is the number of bridging oxygen of each structural unit and X_i is the relative area fraction of individual unit.

As slags A and S1 presented in Fig. 5, *n*(NBO/T) decreases from 3.23 to 2.95 with the "FeO" content ranging from 33.17 to 20.54 wt%. The influence of "FeO" on structure evolution should be attributed to the decreasing number of free Fe^{2+} and O^{2-} irons, which induces the generation of complex structural units [14]. However, after "FeO"/("FeO" + Al_2O_3) ratio lower than 0.53, further increasing Al_2O_3 content will lead to a lower *n*(NBO/T). It could be observed, *n*(NBO/T) of slag S4 is 2.29, lower than the value of 2.45 for slag S3. The influence of Al_2O_3 on the structure evolution could be owed to its network-forming role. Due to the incorporation of Al into silicate network, the DOP of slag will increase as a consequence [15].

Apparent Activation Energies of the Investigated Slags

Apparent activation energy of slag is treated as the resistance of the liquid to shearing. As shown in Eq. (2), it can be obtained from the relationship between viscosity and slag temperature.

$$\ln \eta = \ln A + \frac{E_v}{R} \cdot \frac{1}{T}, \tag{2}$$

where η is the slag viscosity (Pa s), A is the pre-exponential factor, E_v is the apparent activation energy (kJ mol^{-1}), R is the gas constant (8.314 J mol^{-1} K^{-1}), and T is the slag temperature (K).

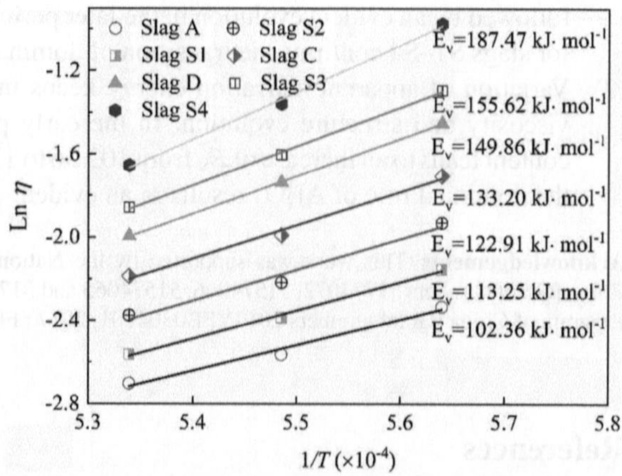

Fig. 6 Relationship between viscosity and reciprocal temperature

Figure 6 shows the dependence of viscosity on reciprocal temperature, also, the results of apparent activation energy are exhibited in the figure. Either in the early or the later period of reduction, the change of apparent activation energy shows the same tendency with those of viscosity and structure evolution. In the early period, the decreasing "FeO" content leads to an increase of E_v from 102.36 to 122.91 k J mol^{-1}. While, after the transition from "FeO" dominant to Al_2O_3 dominant, an evident change in apparent activation energy happens, which is confirmed by the values of 155.61 and 187.46 k J mol^{-1} for slags S3 and S4, respectively. Therefore, it can be concluded that the variation of apparent activation energy shows a good agreement with the evolution of viscosity and structure.

Conclusions

Viscous behavior of slags during iron recovery from nickel slag by aluminum dross has been studied from 1773 to 1873 K. Additionally, the evolution of viscosity and structure in early and later periods of reduction process has been discussed, respectively. The following conclusions can be drawn based on the experimental results.

(1) The influences of "FeO" and Al_2O_3 on viscosity evolution are different. In the early period, "FeO" plays the major impact. While after the "FeO"/("FeO" + Al_2O_3) ratio lower than 0.53, slag viscosity depends on the dominant role of Al_2O_3.

(2) Structure investigation indicates the bands of Raman spectra not only transfer to higher shift, but also change to be more announced, indicating many more complicated structures have been generated. What is more, the relative area fractions of structural units (Q^0 + Q^1) and (Q^2 + Q^3) develop slowly first,

followed by an evident evolution in the later period. Development of n(NBO/T) for slags S1–S4 confirms the transition of dominant role from "FeO" to Al_2O_3.

(3) Variation of apparent activation energy keeps in a good accordance with the viscosity and structure evolution. In the early period, the decreasing "FeO" content leads to an increase of E_v from 102.36 to 122.91 $kJ\,mol^{-1}$. Nevertheless, the dominant role of Al_2O_3 results in an evident change.

Acknowledgements This work was supported by the National Natural Science Foundation of China [Grant numbers 51774072, 51574066, 51574065 and 51774073] and the National Key R&D Program of China [Grant numbers 2017YFB0304201, 2017YFB0304203 and 2016YFB0300602].

References

1. Xu K (2008) Iron and steel industry development and technological innovation in China. Iron Steel 43:1–13
2. Guo H, Yin SH, Yu QJ, Yang X, Huang HL, Yang Y, Gao F (2018) Iron recovery and active residue production from basic oxygen furnace (BOF) slag for supplementary cementitious materials. Resour Conserv Recycl 129:209–218
3. Mills KC, Sridhar S (1999) Viscosities of ironmaking and steelmaking slags. Ironmak Steelmak 26:262–268
4. Starodub K, Kuminova Y, Dinsdale A, Cheverikin V, Filichkina V, Saynazarov A, Khvan A, Kondratiev A (2016) Experimental investigation and modeling of copper smelting slags. Metall Mater Trans B 47B:2904–2918
5. Chen M, Zhao BJ (2016) Viscosity measurements of the SiO_2–K_2O–CaO system relevant to biomass slags. Fuel 180:638–644
6. Park JH, Min DJ, Song HS (2004) Amphoteric behavior of alumina in viscous flow and structure of CaO–SiO_2 (–MgO)–Al_2O_3 slags. Metall Mater Trans B 35B:269–275
7. Lee YS, Min DJ, Jung SM, Yi SH (2004) Influence of basicity and FeO content on viscosity of blast furnace type slags containing FeO. ISIJ Int 44:1283–1290
8. Zhang GZ, Wang N, Chen M, Li H (2018) Viscosity and structure of CaO–SiO_2–'FeO'–Al_2O_3–MgO system during iron-extracting process from nickel slag by aluminum dross. Part 1: coupling effect of 'FeO' and Al_2O_3. Steel Res Int 89:1800272
9. Zhang GZ, Wang N, Chen M, Wang Y (2018) Viscosity and structure of CaO–SiO_2–'FeO'–Al_2O_3–MgO system during iron-extracting process from nickel slag by aluminum dross. Part 2: influence of Al_2O_3/SiO_2 ratio. Steel Res Int 89:1800273
10. Mysen BO, Richet P (2005) Silicate glasses and melts: properties and structure, Henderson P (ed). Amsterdam, The Netherlands, pp 267–284
11. Huang WJ, Zhao YH, Yu S, Zhang LX, Ye ZC, Wang N, Chen M (2016) Viscosity property and structure analysis of FeO–SiO_2–V_2O_3–TiO_2–Cr_2O_3 slags. ISIJ Int 56:594–601
12. Gao JX, Wen GH, Huang T, Tang P, Liu Q (2018) Effects of the composition on the structure and viscosity of the CaO–SiO_2–based mold flux. J Non-Cryst Solids 490:33–39
13. Maeharay T, Yano T, Shibata S, Yamane M (2004) Structure and phase transformation of alkali silicate melts analysed by Raman spectroscopy. Philos Mag 84:3085–3099
14. Toop GW, Samis CS (1962) Activities of ions in silicate melts. Trans Met Soc AIME 224:878–887
15. Toplis MJ, Dingwell DB (2004) Shear viscosities of CaO–Al_2O_3–SiO_2 and MgO–Al_2O_3–SiO_2 liquids: implications for the structural role of aluminium and the degree of polymerisation of synthetic and natural aluminosilicate melts. Geochim Cosmochim Acta 68:5169–5188

Isolation of Cyanide-Degrading Bacteria from Cassava-Processing Effluent

Amzy Tania Vallenas-Arévalo, Carlos Gonzalo Alvarez Rosario, Marcela dos Passos Galluzi Baltazar, Denise Crocce Romano Espinosa and Jorge Alberto Soares Tenório

Abstract Cyanidation is a widely used process for gold leaching where cyanide-containing solutions are used to extract metals from ores. However, cyanidation produces toxic wastes that must be treated prior to discharge to the environment. In this context, cyanide biodegradation has appeared as an environmental friendly and alternative technology. In this study, cyanide-degrading bacteria was isolated from cassava-processing effluents containing 300 ppm of free cyanide using a selective media with cyanide and glycerol. From the effluent, four isolated strains and one consortium were obtained and were tested to degrade cyanide in alkaline medium containing 150 mg L^{-1} free cyanide and 0.5 g L^{-1} scratch for 72 h in orbital agitation. Two strains and the consortium showed the biggest difference from negative control degrading 27–30% of free cyanide in solution proving their potential use in cyanide treatment. Finally, microscopy analysis showed rod-shaped cells in selected samples and classified isolated strains as gram negative and the consortium as gram positive.

Keywords Cyanide · Biodegradation · Bacteria · Cassava processing · Isolation

Introduction

Cyanide salts are used in the leaching of gold and silver from ores. During this process, cyanide complexes $[Ag(CN)_2]^-$ and $[Au(CN)_2]^-$ are formed extracting these precious metals from ores and making them soluble in water in alkaline conditions [1]. Cyanidation of gold works according to the Elsner's equation which is given below:

$$4Au + 8NaCN + O_2 + 2H_2O \rightarrow 4Na\left[Au(CN)_2\right] + 4NaOH.$$

A. T. Vallenas-Arévalo (✉) · C. G. A. Rosario · M. dos Passos Galluzi Baltazar · D. C. R. Espinosa · J. A. S. Tenório
Chemical Engineering Department, Escola Politécnica, Universidade de São Paulo (USP), Rua do Lago, 250, São Paulo CEP 05508-080, Brazil
e-mail: avallenas@usp.br

G. Gaustad et al. (eds.), *REWAS 2019*, The Minerals, Metals & Materials Series,
https://doi.org/10.1007/978-3-030-10386-6_18

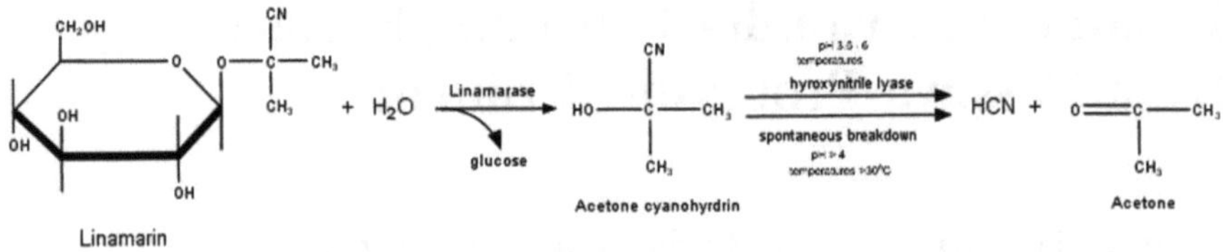

Fig. 1 Free cyanide formation from linamarin, a cyanogen glycoside present in cassava

The solution containing the soluble gold–cyanide complexes is called "pregnant solution" and the solids are discarded to a tailing pond. Gold cyanidation wastes usually contain free cyanide (CN^- and HCN), which is the most toxic form of cyanide and cyanide complexes. Cyanide-containing wastes can also be produced in other industrial processes such as plastic manufacturing, pesticides production, electroplating, and cassava-processing industries [2].

In cassava-processing industries, cyanide is present in effluents due to its liberation from cassava pulp. Cyanogenic glycoside present in cassava, linamarin, is hydrolyzed by the linamarase enzyme (Fig. 1). These components are both present in cassava cells, but are located in different parts of the cell, so they are together only after the pulp-grating process. The reaction produces α-hydroxynitriles and liberates glucose; then the hydroxynitrile lyase enzyme catalyzes the transformation of α-hydroxynitriles in HCN and acetone in a process called "cyanogenesis" [3]. Because of this, cassava-processing industries produce effluents that represent a risk to the environment if not treated properly.

There is a variety of methodologies for cyanide degradation including chemical, physical, and natural processes [4]. In industries, cyanide can be treated using oxidation technologies that are often expensive and complex to operate or can produce toxic by-products [5]. In this context, biotreating cyanide-containing effluents using native bacteria could represent several economical and technical advantages.

Cyanide can be treated by bacteria in a viable process of cyanide-containing effluents [5]. Commonly, *Pseudomonas sp.* and *Bacillus sp.* are used to treat cyanide-containing effluents [6, 7] but other bacteria, fungus, and algae have also been reported [8]. Therefore, the purpose of this study is to isolate from a cassava-processing industry effluent, cyanide-degrading bacteria with potential use in the biotreatment of cyanide-containing effluents.

Methodology

A cassava-processing effluent sample was collected from a factory located in Santa Maria da Serra, São Paulo, Brazil. Effluent was characterized for pH and free-cyanide concentration. Samples were taken, transported, and manipulated according to the Standard Methods for the Examination of Water and Wastewater [9].

Selective Enrichment and Strain Isolation

A selective media for cyanide-degrading bacteria selection was prepared. This media was modified from [10] and contained NH_4Cl 1 g L^{-1}, KH_2PO_4 0.5 g L^{-1}, K_2HPO_4 1.5 g L^{-1}, $MgSO_4$ 0.2 g L^{-1}, NaCl 0.5 g L^{-1}, scratch 0.5 g L^{-1} and was supplemented in 1% (v/v) with a cyanide concentrated solution (37.7 mg NaCN mL^{-1} 75% purity in 0.25% (w/v) NaOH) to obtain a final pH of 8 and a final concentration of 150 mg L^{-1} of free cyanide. Selective media was sterilized by filtration using a 0.22 μm pore filter. Solid selective media was prepared by adding 18 g L^{-1} of agar sterilized separately by autoclaving at 121 °C for 30 min.

An aliquot of 100 μL of effluent was inoculated in selective media plates and incubated at 30 °C and checked daily for colony formation during 5 days. Then, each colony with different morphologies was isolated and inoculated in a new plate by streaking technique twice to assure a single species was isolated.

Cyanide-Degrading Screening

After isolated strains were obtained from cassava-processing effluent, they were tested individually for their ability to degrade free cyanide. In order to achieve this, each strain was cultivated in LB solid medium plates (15 g L^{-1} peptone, 10 g L^{-1} NaCl, 5 g L^{-1} yeast, and 15 g L^{-1} agar) and incubated in stove at 30 °C overnight. Then, a colony of each strain was transferred to a 50 mL falcon tube containing 15 mL of LB liquid medium (15 g L^{-1} peptone, 10 g L^{-1} NaCl, and 5 g L^{-1} yeast) and incubated at 30 °C 185 RPM until optical density at $\lambda = 600$ nm (OD600) was 0.8. Each tube was then centrifuged at 8000 RPM for 5 min and washed twice using a saline solution (NaCl 0.9%). Then, tubes were centrifuged once more and pellet was resuspended in solution containing 0.9% NaCl and 0.1% scratch and supplemented with NaCN to a final concentration of 150 mg L^{-1} CN^-. Tubes were incubated in orbital agitation at 30 °C 185RPM for 72 h. A negative control containing no bacteria was carried out under the same incubation conditions. All assays were executed in triplicates.

After incubation period, free cyanide in solution was measured using a polarographic technique with a voltammetric analyzer (VA) Computrace Control 797, Metrohm. In this technique, an electrolyte solution with 0.2 M boric acid (H_3BO_3) and a 0.17 M potassium hydroxide (KOH) solution were used. Cyanide degradation was calculated in % according to Eq. 1. Isolates with more degradation potential were selected for further characterization.

$$\%[\mathrm{CN}]d = \frac{[\mathrm{CN}]_i - [\mathrm{CN}]_f}{[\mathrm{CN}]_i} \times 100\%, \qquad (1)$$

where

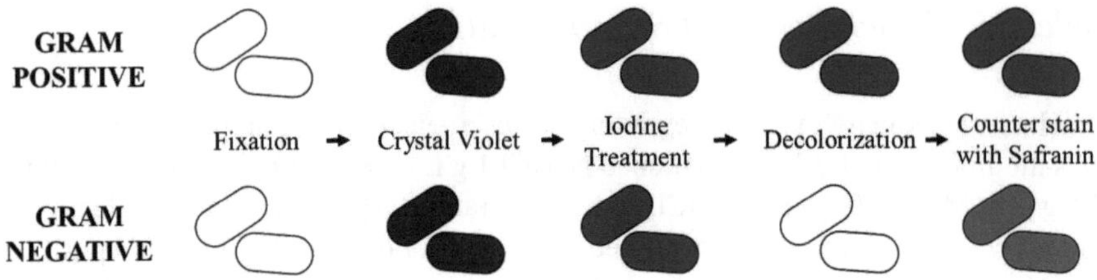

Fig. 2 Gram stain technique

%[CN] Percentage of free-cyanide degraded.
$[CN]_i$ Initial concentration of free cyanide in solution (mg L^{-1}).
$[CN]_f$ Final concentration of free cyanide in solution (mg L^{-1}).

Gram Stain

Selected strains were characterized by the gram stain technique, which is based on the ability of the cells in retaining the crystal violet dye due to the presence of certain structures in the cell wall of the microorganism.

First, reagents were prepared. For gram crystal violet solution, 1 mL of crystal violet stock solution (20 g of crystal violet dissolved in 100 mL of ethanol) was mixed with 40 mL of oxalate stock solution (1 g of ammonium oxalate in 100 mL of water) and 10 mL of water. For gram iodine solution, 1 g of iodine, 2 g of potassium iodide, and 3 g of sodium bicarbonate are dissolved in 300 mL of water. For decolorizer solution, equal volumes of 95% ethanol and acetone are mixed. Finally, for gram safranin solution, 1 mL of safranin stock solution (2.5 g of safranin in 100 mL of 95% ethanol) is mixed with 5 mL of water.

Selected strains were cultivated in LB plates (15 g L^{-1} peptone, 5 g L^{-1} yeast, 10 g L^{-1} NaCl, and 15 g L^{-1} agar) and incubated in stove at 30 °C overnight. A 20 μL of sterile ultra-pure water was added on a clean microscope slide and a colony was transferred from the plate using an inoculation loop, spread and left to air dry in a sterile environment.

The gram stain technique consists of five steps illustrated in Fig. 2. First, two drops of gram crystal violet solution were added to the fixed culture. After 1 min, the slide was gently rinsed using ultra-pure water. Then, two drops of gram iodine solution, or enough to cover the culture, was poured. After 1 min, the slide was rinsed with ultra-pure water carefully to not wash off the culture from the slide. Then, a few drops of the decolorizer solution were added and the slide was rinsed with ultra-pure water after 5 s. Finally, culture was counterstained with three drops of safranin solution and washed off using ultra-pure water after 30 s. The microscope slide was let to air dry and then analyzed by optical microscopy (Leica DM2700M).

Results and Discussion

A sample of a pressing-process effluent from a cassava-processing industry was collected in Santa Maria da Serra, São Paulo, Brazil. Effluent contained 304.8 mg L^{-1} of free cyanide and a pH value of 5 which confirms the spontaneous formation of free cyanide during cassava processing previously reported [3].

Strain Isolation

The plates containing the selective medium were incubated at 30 °C for 5 days. Cyanide-resistant colonies with different morphologies were selected and isolated to pure culture by the streaking technique. This procedure was repeated twice in fresh LB plates to obtain single-microorganism cultures. From the cassava-processing effluent, eight individual strains and one consortium (containing a mixture of all the individual strains) were isolated and categorized as cyanide resistant. Strains were designated as M1, M2, M3, M4, M5, M6, M7, M8, and C1, respectively, for the eight individual strains and the consortium. All the isolated strains and the consortium were cultivated in LB liquid medium and mixed with glycerol to store at −80 °C for future assays.

Cyanide-Degrading Screening

Cyanide is toxic for most living organisms since it inhibits the cytochrome c oxidase, a key enzyme in the electron transport chain (ETC), during cellular respiration [11]. However, some microorganisms have developed alternative pathways to survive cyanide toxicity and others have developed metabolic pathways that use cyanide as a nitrogen source, degrading cyanide into less toxic compounds [8]. Cyanide-resistant microorganisms are thus not necessarily cyanide-degrading organisms too. Therefore, a screening assay for general degradation was performed with all isolates to assess their potential ability to degrade cyanide.

In order to achieve this, cyanide and scratch were added as the sole nitrogen and carbon source, respectively. Scratch was selected as a suitable carbon source since it is a main component in cassava pulp and cassava-processing effluents will surely have it in its composition. Because of final pH in assay solution was eight, cyanide is highly probable to volatilize or oxide when in contact with the oxygen of the air. This is why it is important to evaluate the microorganisms performance comparing it to a negative control.

Free-cyanide degradation in each sample is shown in Fig. 3. It also shows the cyanide reduction by natural degradation in the negative control. Every sample

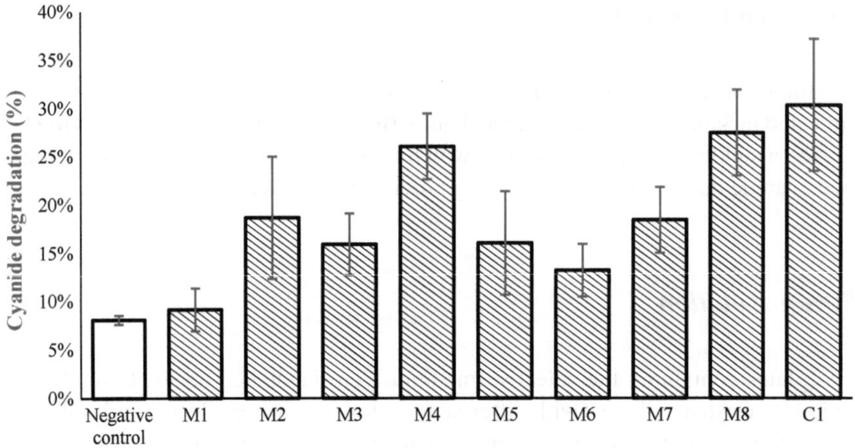

Fig. 3 Free-cyanide degradation of native strains isolated from cassava-processing industry effluent. Percentage degraded after 72 h of incubation at 30 °C 185RPM. Error bars show difference between triplicates

showed a bigger degradation rate than the negative control, which indicates that all the studied strains consume cyanide in their metabolism in some level.

The consortium C1 was the sample that had the biggest degradation rate reducing cyanide by 30%. Within the individual isolates, strains M4 and M8 showed the biggest degradation rates, reducing cyanide by 26% and 27.5%, respectively. Negative control showed a free-cyanide reduction of less than 10%, which is acceptable considering the final pH of the assay solution. After cyanide-degrading screening, isolates M4 and M8, and consortium C1 were chosen for further analyses and characterization.

In consortia, several species work in synergy, co-metabolism, and mutualism to adapt or survive together. Usually, the by-products of one strain's metabolism are used in another strain's metabolism. Because of this, it is expected for the consortium to show a greater degradation rate than individual strains. However, using consortia in effluent treatment can bring other challenges like maintaining the same microbiological quality of the inoculum during assays. As these populations are constantly changing, they tend to produce results that are more variable.

Gram Stain and Optical Microscopy Characterization

Gram stain and optical microscopy characterization is the first step to identify a microbial strain. In this study, the two individual strains with more degradation potential were analyzed by optical microscopy after gram stain. The difference between gram-positive and gram-negative bacteria is the composition of the cellular wall.

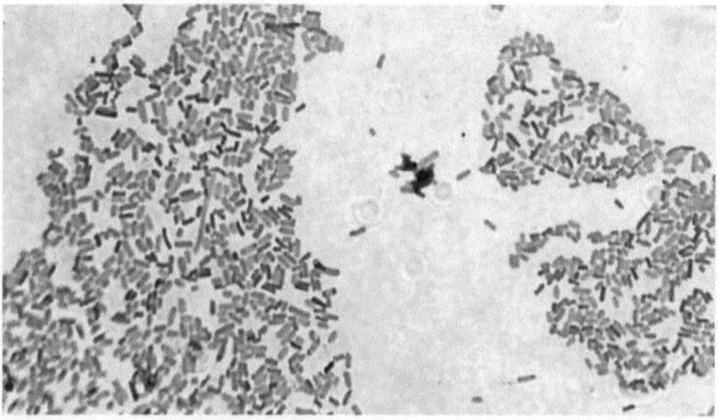

Fig. 4 M4 strain grown in LB medium plate, incubated at 30 °C for 24 h. Gram stain

Gram positive possess a thick cell wall compared to gram-negative bacteria that have a thinner cellular wall and at the same time possess an additional outer membrane, which gives cells different characteristics when responding to stress, heat, or antibiotics [12].

Table 1 shows the results for strain characterization through gram stain and optical microscopy. M4 and M8 strains were classified as gram-negative organisms, while C1 consortium presented most cells with a gram-positive morphology. All strains have a rod-shaped morphology. Figures 4, 5, and 6 show the optical microscopy images of the studied strains. M4 and M8 strains presented cells with a length between 1 and 3 μm and presented cylindrical form. Furthermore, consortium C1 presented rod-shaped cells with lengths between 1.5 and 5 μm (Fig. 6).

Rod-shaped bacteria has been previously reported as cyanide-degrading organisms, including gram-positive species belonging to the *bacillus* and *corynobacterium* genres [10, 13–15], and gram-negative species belonging to the *pseudomonas* [7, 16–18], *klebsiella* [19, 20], and *enterobacter* [21] genres. A molecular identification assay will be needed to identify specifically the studied strains in order to work toward a better understanding of cyanide-degrading mechanisms in bacteria and its application to effluent treatment.

Table 1 Strain characterization

Strain	Gram stain	Cell morphology
M4	–	Rod shaped
M8	–	Rod shaped
C1	+	Rod shaped

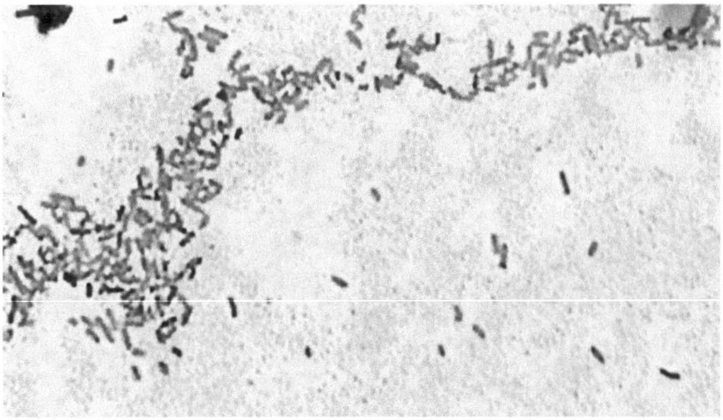

Fig. 5 M8 strain grown in LB medium plate, incubated at 30 °C for 24 h. Gram stain

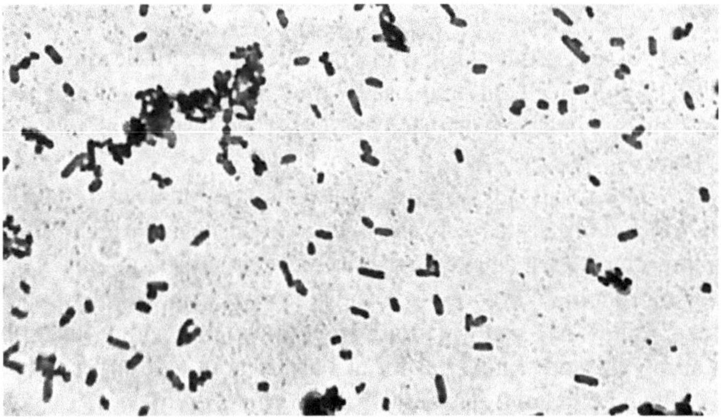

Fig. 6 C1 consortium grown in LB medium plate, incubated at 30 °C for 24 h. Gram stain

Conclusions

Cyanide is a highly toxic compound and effluents containing cyanide must be treated prior to discharge to the environment. Cassava-processing industries produce cyanide-containing effluents due to the cyanogen glycosides present in cassava pulp. From the eight individual strains isolated from a cassava-processing effluent containing 305 mg L^{-1} of free cyanide, strains M4 and M8 showed the greater potential for cyanide biotreating and were selected for further characterization. Furthermore, the isolated consortium achieved the greatest degradation rate, most probably due to synergetic metabolism between species. This study showed the importance and the potential for cyanide biotreating using native bacteria isolated from cyanide-containing environments. Native isolated strains were able to reduce cyanide concentration between 10 and 27.5% after 72 h incubation indicating their potential use in cyanide treatment.

Acknowledgements The authors would like to thank the Brazilian Coordination for the Improvement of Higher Education Personnel (CAPES) and the Brazilian National Council for Scientific and Technological Development (CNPq) for the financial support provided for this work.

References

1. Kuyucak N, Akcil A (2013) Cyanide and removal options from effluents in gold mining and metallurgical processes. Miner Eng 50–51:13–29
2. Mudder T, Botz M (2004) Cyanide and society: a critical review. Tha Eur J Miner Process Environ Prot 4(1):62–74
3. Chisté RC, Cohen KDO, Mathias EDA, Oliveira SS (2010) Quantificação de cianeto total nas etapas de processamento das farinhas de mandioca dos grupos seca e d'água. Acta Amaz 40(1):221–226
4. Akcil A (2003) Destruction of cyanide in gold mill effluents: biological versus chemical treatments. Biotechnol Adv 21(6):501–511
5. Botz M (1999) Overview of cyanide treatment methods. Gold Inst 11
6. Huertas MJ et al (2010) Alkaline cyanide degradation by Pseudomonas pseudoalcaligenes CECT5344 in a batch reactor. Influence of pH. J Hazard Mater 179(1–3):72–78
7. Luque Almagro VM et al (2005) Bacterial degradation of cyanide and its metal complexes under alkaline conditions. Appl Environ Microbiol 71(2):940–947
8. Dash RR, Gaur A, Balomajumder C (2009) Cyanide in industrial wastewaters and its removal: a review on biotreatment. J Hazard Mater 163(1):1–11
9. APHA/AWWA/WEF (2012) Standard methods for the examination of water and wastewater. Stand Methods 541
10. Wu CF et al (2014) An effective method for the detoxification of cyanide-rich wastewater by Bacillus sp CN-22. Appl Microbiol Biotechnol 98(8):3801–3807
11. ATSDR (2006) Toxicological Profile for Cyanide
12. Mai-prochnow A, Clauson M, Hong J, Murphy AB (2016) Gram positive and Gram negative bacteria differ in their sensitivity to cold plasma. Nat Publ Gr 1–11
13. Mekuto L, Jackson VA, Ntwampe KS, Obed (2013) Biodegradation of free cyanide using bacillus sp consortium dominated by Bacillus safensis, Lichenformis and Tequilensis Strains: a bioprocess supported solely with whey. Bioremediat Biodegrad 18:1–7
14. Lee C, Kim J, Chang J, Hwang S (2003) Isolation and identification of thiocyanate utilizing chemolithotrophs from gold mine soils. Biodegradation 14(3):183–188
15. Gupta N, Balomajumder C, Agarwal VK (2010) Enzymatic mechanism and biochemistry for cyanide degradation: a review. J Hazard Mater 176(1–3):1–13
16. Singh N, Balomajumder C (2016) Simultaneous biosorption and bioaccumulation of phenol and cyanide using coconut shell activated carbon immobilized pseudomonas putida (MTCC 1194). J Environ Chem Eng 4(2):1604–1614
17. Castillo F, Rold MD, Moreno-vivi C (2005) Alkaline cyanide biodegradation by Pseudomonas pseudoalcaligenes CECT5344 33:168–169
18. Watanabe A, Yano K, Ikebukuro K, Karube I (1998) Cyanide hydrolysis in a cyanide-degrading bacterium, pseudomonas stutzeri AK61, by cyanidase. Microbiology 144(6):1677–1682
19. Kao CM, Liu JK, Lou HR, Lin CS, Chen SC (2003) Biotransformation of cyanide to methane and ammonia by Klebsiella oxytoca. Chemosphere 50(8):1055–1061
20. Chen CY, Kao CM, Chen SC (2008) Application of Klebsiella oxytoca immobilized cells on the treatment of cyanide wastewater. Chemosphere 71(1):133–139
21. Mekuto L, Ntwampe SKO, Akcil A (2016) An integrated biological approach for treatment of cyanidation wastewater. Sci Total Environ 571:711–720

Part V
REWAS 2019: Secondary and Byproduct Sources of Materials, Minerals, and Metals: Secondary and Byproduct Beneficial Use

Introducing the Extraordinary Leuven Cement: Raw Materials, Process, Performance, and First Real-Life Applications

Yiannis Pontikes

Abstract The development of alternative cementitious binders has been primarily fueled by the need to reduce the environmental footprint of Ordinary Portland Cement (OPC). It would be reasonable to expect though that these new binders to come should be on one hand better in terms of environmental footprint, and on the other hand, at least comparable in terms of performance and availability to society. The latter, in engineering terms, translates into a production process that is relatively easy to erect, technically and financially, and robust during operation. Moreover, the binder itself should be composed of abundant elements so as to empower construction and growth for all. The most abundant elements in earth's crust are oxygen, silicon, aluminum, iron, calcium, and sodium. Assuming a high-temperature process is employed, then the parent minerals where these elements are present are of little interest: they will all melt, and by adjusting the chemistry and the cooling conditions, the solidified product will be the glass precursor to be used as the main component in the cementitious binder thereafter.

In the work herein, we present our approach to meet all the above. We try to translate intentions (i.e., develop a binder of low environmental footprint and abundant to all) into actions (i.e., develop and test the process and the materials), respecting the obvious boundaries of the system (i.e., thermodynamics).

The work is structured around the production process of this new binder. It starts with an overview of the different steps (i.e., unit operations) and continues with in-depth presentation of raw materials, firing conditions, milling, additives, (…), concluding to the two lines of binders that have been developed over the past years, one with OPC (blended) and the other one without. In addition to OPC, these formulations can also integrate other materials, namely, ground granulated blast furnace slag, fly ash, calcined clays, and more, thus, resemble to a great extent the family of blended cements in the market today. The data communicated refer to work conducted at both laboratory and pilot-plant scale, and extend from the atomic structure of these new binders to mortar formulations and ultrahigh strength concrete.

Y. Pontikes (✉)
KU Leuven Department of Materials Engineering, Kasteelpark Arenberg 44, 3001 Leuven, Belgium
e-mail: yiannis.pontikes@kuleuven.be

G. Gaustad et al. (eds.), *REWAS 2019*, The Minerals, Metals & Materials Series,
https://doi.org/10.1007/978-3-030-10386-6_19

The name of this new binder is Extraordinary Leuven Cement. It is abbreviated as ELCE, suggesting that "elce", an obsolete form of "else", i.e., something different is a possibility even for one of the most massively used materials today.

Keywords Cement · Binder · Sustainable

Ferroalloy Production from Spent Petroleum Catalysts by Reductive Smelting and Selective Oxidation Processes

Jong-Jin Pak, Do-Hyeong Kim, Min-Kyu Paek and Yong-Dae Kim

Abstract Solid spent catalysts from the hydroprocessing units of petroleum refining industries contain valuable metals of V, Ni, and Mo in appreciable concentration together with a significant amount of sulfur in an alumina supporting material. A recycling process for this resource has been developed and commercialized by reductive smelting and selective oxidation processes to recover these metals in the form of ferroalloys. The carbon saturated Fe–V–Ni–Mo melt recovered from the reductive smelting was oxidized to separate vanadium into a slag which can be processed for Fe–V alloy production. Then Fe–V and Fe–Ni–Mo alloys were successfully manufactured with high recovery ratios of over 90% and 95%, respectively.

Keywords Spent catalyst · Reductive smelting · Ferroalloy · Vanadium · Nickel · Molybdenum

Introduction

The spent catalysts discarded from the petroleum refinery operations contain valuable metals such as vanadium, nickel, and molybdenum in the form of oxides and sulfides up to 30 mass% in total. These metals are highly valuable and are used as alloying elements extensively in the steel industry. These metals are usually manufactured from the ores and minerals containing them. Spent catalysts could be used as a cheap source for these valuable metals [1]. However, the spent catalysts also contain a very high sulfur content of more than 10 mass% derived from the hydrodesulfurization processes of heavy oils.

J.-J. Pak (✉) · D.-H. Kim
Department of Materials Engineering, Hanyang University, ERICA, Ansan 426-791, South Korea
e-mail: jjpak@hanyang.ac.kr

M.-K. Paek
Department of Chemical and Metallurgical Engineering, Aalto University, Kemistintie 1, 02150 Espoo, Finland

Y.-D. Kim
Golden River Co, Gunsan 54002, South Korea

G. Gaustad et al. (eds.), *REWAS 2019*, The Minerals, Metals & Materials Series,
https://doi.org/10.1007/978-3-030-10386-6_20

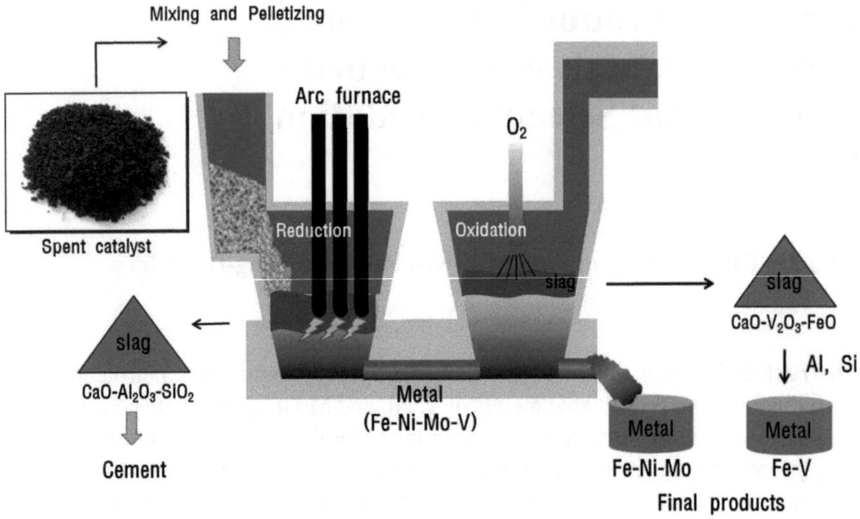

Fig. 1 Process route for recycling spent catalysts

Several industrial-scale wet processes such as chlorination, acid leaching, alkali leaching, roasting with soda salts, etc., have been developed for the recovery of Mo, Ni, and V from the spent catalysts [1]. However, the roasting process for sulfur removal from the catalysts and the leaching processes produce unattractive by-products such as toxic gases, liquid streams containing strong acids, bases, chlorine, etc., which require a high level of environmental precautions and costs.

Recently, the reductive smelting process of this resource using the carbothermic reaction in an electric arc furnace and the selective vanadium oxidation process was developed to recover these metals in the form of Fe–Ni–Mo and Fe–V alloys as shown in Fig. 1 [2]. The process route starts with dry catalysts that are melted with cokes in an EAF at temperatures around 1500–1600 °C. Heavy metals sink to the bottom as carbon saturated molten ferroalloys and are separated from liquid slag containing high alumina. Alumina was originally present in the catalyst as a supporting material. The Fe–V–Ni–Mo melt is then oxidized to separate vanadium into a slag which can be processed for Fe–V alloy production. The remaining Fe–Ni–Mo alloy melt after the vanadium oxidation can be used as a raw material for stainless steelmaking. It has been shown that this pyrometallurgical route has the advantage of a complete recycling scheme: metals for manufacturing ferroalloys and slags for manufacturing special cements. The process has advantages over the conventional hydrometallurgical route in terms of cost-effectiveness and environmental issues by the elimination of roasting and leaching processes of spent catalysts.

Table 1 Composition of spent catalysts (mass%)

Spent catalysts	V	Ni	Mo	Al_2O_3	C	S
SC1	19.9	7.9	2.78	47.6	8.49	11.2
SC2	–	5.38	5.39	70.7	4.39	0.42
SC3	10.0	–	–	74.3	0.39	0.16

For successful commercialization of this process, high metal recovery ratio and sulfur control in liquid Fe–V–Ni–Mo alloys in the reductive smelting process and the effective separation of vanadium from this alloy are very important. In the present work, some experimental results of our process for recycling spent catalysts will be presented together with thermodynamics of sulfur in these special alloys.

Experimental

A pilot-scale 350 kVA EAF and a commercial-scale 2,000 kVA EAF were utilized for a reductive smelting of spent catalysts to produce carbon saturated Fe–V, Fe–Ni–Mo and Fe–V–Ni–Mo alloy melts. In case of Fe–V–Ni–Mo melt, the selective oxidation of V from melt to slag was carried out using a 500 kW induction furnace. Table 1 shows the analysis of spent catalysts used in the present study after the de-oil process.

Figure 2 shows photos of the 2000 kVA EAF operation. The reductive smelting process comprises melting a charge containing de-oiled spent catalysts, coke, iron source, and sub-materials like lime and silica. The inner lining of the furnace is a carbonaceous material to maintain a strong reducing condition for high metal recovery ratio and desulfurization efficiency. Carbon saturated liquid iron containing V, Ni, and Mo can be produced by a carbothermic reduction of their oxides and sulfides in the catalysts at a high temperature above 1500 °C. The Fe–V–Ni–Mo alloy recovered from spent catalysts was treated for selective vanadium oxidation by oxygen blowing and iron ore addition to produce vanadium-rich slag. Fe–V alloy was produced by an aluminothermic process of vanadium-rich slag as shown in Fig. 3.

Results and Discussion

Reductive Smelting of Spent Catalysts

Figure 4 shows a photo of metals recovered from spent catalysts and the XRD data of slag produced in EAF. Table 2 shows the compositions of metals recovered from different catalysts. The remaining V, Ni, and Mo in slag was negligible showing the high recovery ratios of metallic elements (>95% for V and ~100% for Ni and

Fig. 2 Reductive smelting process by EAF

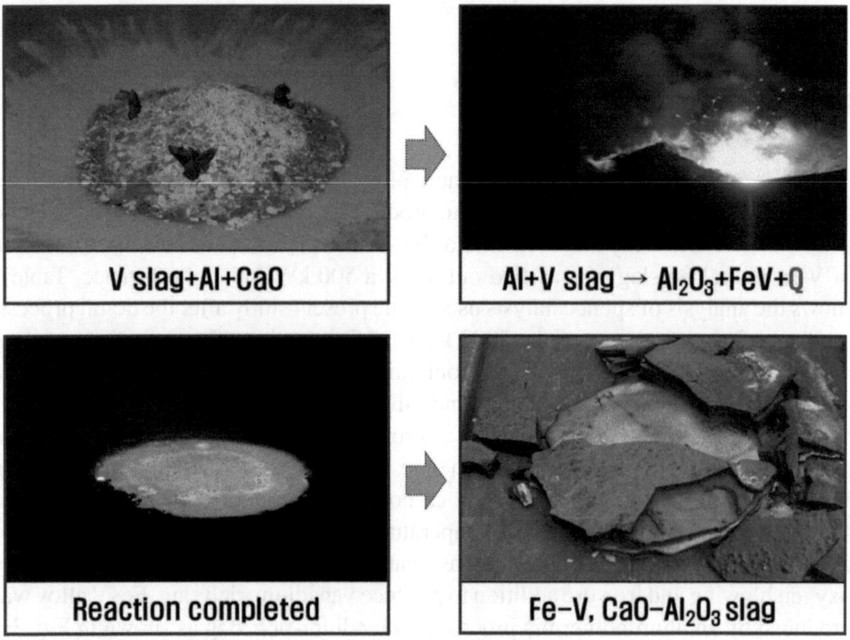

Fig. 3 Aluminothermit process for Fe–V alloy production

Mo) from spent catalysts during reductive smelting. The composition of metal can be controlled by the selection of catalyst/metal ratio in EAF operation. The slag generated from high sulfur catalyst (SC1 in Table 1) contained a very high sulfur up to 7 mass%, and the slag was saturated with CaS as shown in Fig. 4b.

Impurity elements of P and S in metals were less than 0.03 mass%. A very high sulfur distribution ratio between slag and metal, (%S)/[%S], was obtained under carbon saturated condition in EAF, and it allowed the elimination or the reduction of a cost-push roasting process of spent catalysts for sulfur removal. The slag generated

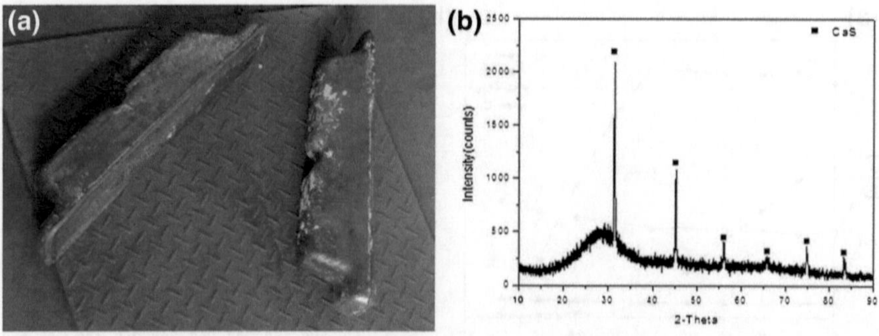

Fig. 4 **a** Fe–V–Ni–Mo alloy recovered from EAF and **b** XRD data of EAF slag

Table 2 Composition of metals recovered from different catalysts (mass%)

Catalyst	Metal					
	V	Ni	Mo	C	P	S
SC1	16.3	5.67	3.5	7.25	0.02	0.023
SC2	–	15.8	15.8	4.90	0.01	0.011
SC3	27.4	–	–	8.87	0.03	0.012

from the smelting process is free from leachable heavy metals, and it can be utilized for raw materials for other valuable products such as special cements.

Selective Vanadium Oxidation

The Fe–V–Ni–Mo alloy recovered from spent catalysts was treated for a selective vanadium oxidation using an induction furnace. Figure 5 shows the variation of melt composition during oxygen blowing and iron ore addition. In case of oxygen blowing, the decarburization reaction occurred first and the vanadium oxidation started when [C] content decreased to about 1 mass% in the melt. In case of iron ore addition, the vanadium oxidation started at higher [C] content. Nickel and molybdenum were not oxidized until the end of oxidation process. Table 3 shows the composition of vanadium-rich slag obtained after the oxidation processes. The mass ratio of V/Fe in slag was higher in case of iron ore addition. Fe–50–60%V alloy was manufactured directly from the vanadium-rich slag by an aluminothermic process shown in Fig. 3. Figure 6 shows a vanadium-rich slag (a) and a Fe–60%V ferroalloy (b) produced therefrom. The overall recovery ratio of V from spent catalyst via the reductive smelting and the selective oxidation was over 90%. The recovery ratios for Ni and Mo were over 95%.

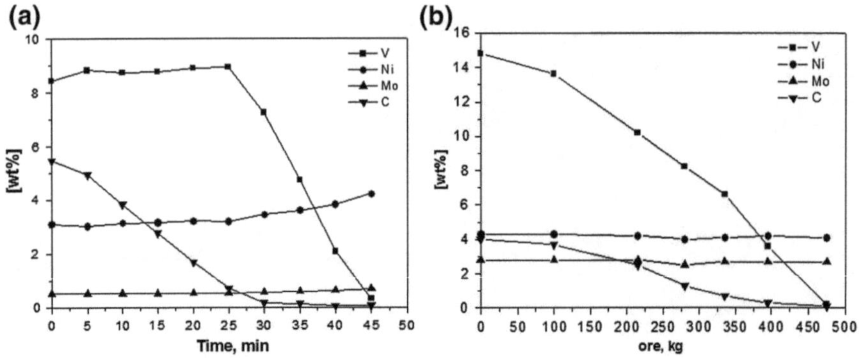

Fig. 5 Change of melt composition during V oxidation **a** by oxygen blowing and **b** by addition of Fe ore

Table 3 Composition of V-rich slag after V oxidation (wt%)

Oxidation	V	Fe	Ni	Mo	V/Fe
Oxygen	17.1	17.2	0.40	0.29	1.0
Iron ore	28.6	17.5	0.13	0.15	1.64

Fig. 6 **a** V-rich slag and **b** Fe–60%V alloy

Thermodynamics of Sulfur in Carbon Saturated Fe–V–Ni–Mo Alloys

Thermodynamics of sulfur in carbon saturated Fe–V–Ni–Mo alloy is very important for producing low sulfur ferroalloys from spent catalysts. In the present study, thermodynamic interactions between those alloying elements and sulfur in liquid iron were studied using the slag/metal equilibration technique at 1600 °C. The equilibrium sulfur distributions between a slag with a known sulfide capacity and carbon saturated liquid Fe–V–Ni–Mo alloys, (%S)/[%S], were measured as a function of alloy composition. The carbon solubility in liquid alloy was significantly changed with alloy composition [3]. The specific effects of V, Ni, and Mo on sulfur was determined by considering the effect of carbon on sulfur using Wagner's formalism as the first- and second-order interaction parameters as the following relation:

$$\log f_S = \sum_{C,S}^{i} (e_S^i[\%i] + r_S^i[\%i]^2) + \sum_{V,Ni,Mo}^{i} (e_S^i[\%i] + r_S^i[\%i]^2) + \sum_{\substack{V,Ni,Mo \\ i \neq j}}^{i,j} r_S^{i,j}[\%i][\%j] \quad (1)$$

where f_S is the activity coefficient of sulfur in liquid iron, e_S^i and r_S^i are the first- and second-order interaction parameters of elements on sulfur in liquid iron, respectively, and $r_S^{i,j}$ is the second-order cross product parameter on sulfur in liquid iron which indicates the simultaneous effect of two different alloying elements (*i* and *j*) on the activity coefficient of sulfur in liquid iron.

Figure 7a shows the variation of sulfur distribution between a 40%CaO–60%Al_2O_3 slag and Fe–C_{sat}–i–S (i = V, Ni, Mo) alloy melts under CO atmosphere at 1600 °C. The sulfur distribution increased linearly with increasing Mo and V in the melt, but it decreased with Ni addition. The carbon solubility also increased linearly with increasing Mo and V in the melt and decreased with Ni addition as shown in Fig. 8 [3]. Carbon is known to increase the activity coefficient of sulfur in liquid iron significantly [4]. The specific effect of each alloying element of V, Ni, and Mo on sulfur can be obtained from the experimental data using Eq. (1) and available thermodynamic data [4–6] as shown in Fig. 7b.

In order to determine the simultaneous effect of different alloying elements on sulfur in multicomponent Fe–C_{sat}–Ni–Mo–V alloy melts, the sulfur distributions between a 40%CaO–60%Al_2O_3 slag and Fe-C_{sat}–V–Ni, Fe–C_{sat}–Mo–Ni and Fe–C_{sat}–V–Ni–Mo melts were also measured at 1600 °C. Using the relation of Eq. (1) together with e_S^i and r_S^i values determined from Fig. 7b, the second-order cross prod-

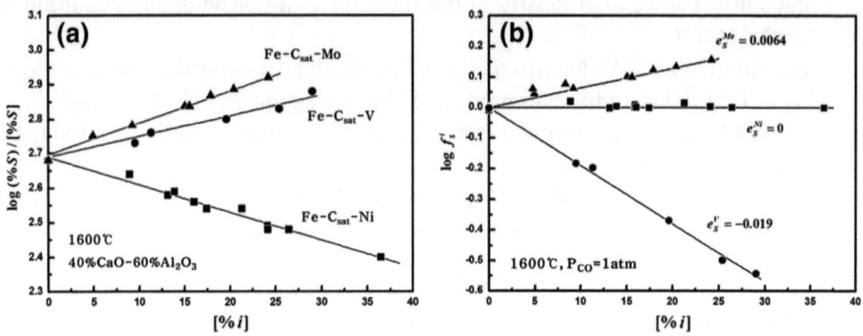

Fig. 7 **a** Sulfur distribution between a 40%CaO–60%Al_2O_3 slag and Fe–C_{sat}–i melts(i = V, Ni, Mo), and **b** the relation of log f_S^i *versus* [%*i*] in Fe–C_{sat}–i–S melts(i = V, Ni, Mo) at 1600 °C

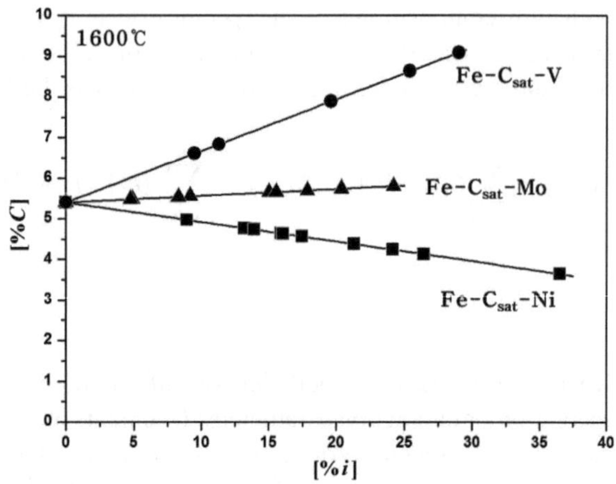

Fig. 8 Carbon solubility in Fe–V, Fe–Ni, and Fe–Mo melts at 1600 °C

uct parameters on sulfur in liquid iron, $r_S^{i,j}$ were determined from the experimental results. It was noted that the cross product effect of pair alloying elements of V-Ni, Mo-Ni, and V-Mo was not significant. The values of $r_S^{V,Ni}$, $r_S^{Ni,Mo}$ and $r_S^{V,Mo}$ were determined as 0, 0, and 0.0003, respectively, at 1600 °C. Using these thermodynamic parameters, the sulfur distribution between a slag and multicomponent iron alloy melts of Fe–C_{sat}–V–Ni–Mo could be predicted as a function of melt composition.

Conclusions

(1) A reductive smelting process has been developed to recover Mo, Ni, and V in the form of ferroalloys by carbothermic reduction of spent catalysts utilizing an EAF process.
(2) The recovery ratio of valued elements in the metal phase was higher than 95% and sulfur can be kept mostly in the slag under carbon saturated condition at 1500–1600 °C.
(3) Vanadium in a Fe–V–Ni–Mo melt can be selectively oxidized by oxygen blowing or iron oxide addition to produce a V-rich slag and Fe–Ni–Mo alloy.
(4) Fe–60%V alloy could be manufactured directly from the vanadium-rich slag by an aluminothermic process.
(5) Thermodynamics of sulfur in liquid Fe–V–Ni–Mo alloys were determined to predict the desulfurization limit during the reductive smelting of high sulfur containing spent catalyst in EAF.

Acknowledgements This study was supported by the R&D Center for Valuable Recycling (Global-Top Environmental Technology Development Program) funded by the Ministry of Environment (Project No.: 2016002230003).

References

1. Marafi M, Stanislaus A (2008) Spent hydroprocessing catalyst management: a review Part II. advances in metal recovery and safe disposal methods. Resour Conserv Recycl 53:1–26
2. Pak JJ, Kim DH, Park CH, Paek MK, Jo JO, Lee KI (2012) Recovery method of valuable metals from spent catalysts. KR Patent 10-1189183
3. Kim DH, Paek MK, Kim TJ, Won SY, Pak JJ (2014) Carbon solubility in liquid iron containing V, Mo and Ni. Mater Trans 55(3):610–615
4. Banya S, Chipman J (1969) Sulfur in liquid iron alloys: effects of alloying elements. Trans Metall Soc AIME 245:133–143
5. Ishii F, Fuwa T (1981) Activity of sulphur in liquid iron alloys. Tetsu-to-Hagane 67(6):98–106
6. Kim DH, Won SY, Han JS, Nam HS, Pak JJ (2018) Thermodynamics of sulfur in carbon saturated liquid ferro-alloys containing Ni, Mo and V at 1873 K. ISIJ Int 58(3):408–414

References

Reactivity of Crystalline Slags in Alkaline Solution

Brian Traynor, Hugo Uvegi, Piyush Chaunsali and Elsa Olivetti

Abstract Slags with varied amorphous and crystalline content, typical of iron and steel production, are generally underutilized. One promising reuse pathway for these wastes is chemical activation, producing alternatives to conventional building materials with lower embodied energy. The formation of a hardened binder is dependent on the slag mineralogy and, specifically, the reactivity of relevant phases. Reactivity can be understood by monitoring elemental dissolution rates through inductively coupled plasma (ICP-OES) analysis. Post-dissolution ICP analysis of activating solution and spectroscopic analysis of remaining solids was performed on several highly crystalline slags and on relevant synthetic minerals to track changes in chemical and phase composition. Amorphous and ionic phases have been observed as more reactive than other crystalline phases. This work aims to inform future studies on waste blending in alkali-activated systems, a promising avenue for valorization of industrial wastes with varied physicochemical properties. To this end, dissolution tests with varied initial Si, Al, and Ca concentrations in activating solution were also performed.

Keywords Slag · Alkali activation · Γ-C_2S

Introduction

Concerns over the CO_2 emissions associated with the cement industry have led to greater research efforts being devoted to the development of alternative, lower energy-embodied binders. One such family of alternative binders is alkali-activated

B. Traynor · H. Uvegi · E. Olivetti (✉)
Department of Materials Science and Engineering, MIT, Cambridge, USA
e-mail: elsao@mit.edu

P. Chaunsali
Department of Civil Engineering, IIT Madras, Chennai, India

G. Gaustad et al. (eds.), *REWAS 2019*, The Minerals, Metals & Materials Series,
https://doi.org/10.1007/978-3-030-10386-6_21

materials. Alkali-activated materials use precursors rich in silica and alumina with variable amounts of calcium in an aqueous alkaline environment to form hardened aluminosilicate networks with comparable physical properties to Portland cement-based binders. While traditional cements are hydraulic in nature, relying on the reaction of calcium silicates in the presence of water to form calcium silicate hydrate, alkali-activated materials require an alkali activating solution to promote the reaction of an aluminosilicate source to form a hardened material. One of the strengths of alkali-activated materials is the versatility in potential precursors. One high-volume source of potential precursors is industrial by-products from the metals and energy generation industries. Beneficiation of industrial wastes in alkali-activated materials can provide a construction material with lower environmental impact than conventional construction materials [1]. For example, a great deal of research has been devoted to the development of binders based on blast furnace slag and fly ash activated by an alkali activator, typically sodium hydroxide or sodium silicate [2, 3].

However, any attempt to reduce the environmental effect of the production of conventional construction materials cannot rely on these materials (blast furnace slag and fly ash) alone. Expansion of the range of suitable precursors to other industrial by-products represents an opportunity to develop a more adaptable pool of precursors for construction materials in which locally available wastes can substitute for existing construction materials. Mismatches between local supply of suitable wastes and demand for construction materials require a more flexible toolkit of load-bearing materials based on industrial by-products. One strategy to negate this problem is to design a load-bearing material, in which the shortcomings in a locally available precursor may be overcome through chemical activation (alkali activation) and/or by blending with another precursor. In terms of material chemistry, highly siliceous precursors can be blended with calcium-rich precursors in appropriate ratios to synthesize a calcium silicate hydrate phase, for example [4]. Blending of materials is particularly pertinent in the case of poorly reactive wastes or wastes with high variability in their composition or physical properties. One family of underutilized wastes is crystalline slags.

Crystalline slag is a byproduct of the steel, iron, and copper industries. Crystalline slags are produced when slags are air-cooled upon tapping from the blast furnace. A poorly cementitious material, crystalline slag is used in low-value applications, such as land reclamation and aggregate in construction materials. However, slags of this kind are rich in silica, alumina, and calcia—chemical constituents that form reaction products that provide strength to a load-bearing material. Globally, an estimated 220 million metric tons of steel slag and an estimated 25 million metric tons of copper slag were produced in 2016 [5, 6].

The reaction pathway for strength-giving phases in a construction material takes place in aqueous phase. Precursors release chemical species under the influence of an activator and organize in solution to form a reaction product that binds the structure together, endowing the material with strength. This study examines the dissolution kinetics of crystalline phases commonly found in crystalline slags with the goal of improving our understanding of how blending crystalline slags with other materials will impact the rate of dissolution.

Background

Thermodynamic Basis

Substituting for existing high embodied energy construction materials with slags requires an understanding of the properties of the constituent crystal phases of the slag. The dissolution of crystal phases in solution is dependent on many factors, including temperature, activities of dissolved species, ionic strength of dissolving solution, surface area of dissolving mineral, and surface structure of dissolving mineral. The basis for alkali activation rests on the fact that alkaline environment promotes rapid dissolution of a range of aluminosilicate precursors. In order to expand the pool of potential construction material precursors, the effect of alkali solution on the dissolution of the constituent phases of any potential precursor must be understood.

However, controlling all variables in a dissolution experiment is a fundamental challenge in determining solubility constants and understanding changes in dissolution rate as a function of undersaturation. Controlling for temperature and ionic strength is generally achievable, the latter being achieved by performing dissolution experiments in dilute environment, such that species dissolving from minerals do not appreciably affect the overall ionic strength of the solution, while still being measurable through ICP. Some control over mineral surface area can be achieved through limitations on particle sizes used. However, this parameter can change as a function of time if minerals dissolve rapidly. In this study, the minerals of interest are expected to dissolve slowly, suggesting that it is possible to control for surface area. The other most significant factor in mineral phase dissolution is the activities of the dissolved species. The thermodynamic driving force for dissolution of a mineral phase is the difference is Gibbs energy between the mineral phase and its constituent chemical species in the aqueous phase. For a given calcium aluminosilicate phase, the (unbalanced) dissolution reaction in basic medium may proceed as

$$\mathrm{Si}_x\mathrm{Al}_y\mathrm{Ca}_z\mathrm{O}(s) + \mathrm{OH}^-(\mathrm{aq}) \rightarrow x\mathrm{HSiO}_3^-(\mathrm{aq}) + y\mathrm{Al(OH)}_4^-(\mathrm{aq}) + z\mathrm{Ca(OH)}^+(\mathrm{aq})$$

This reaction will proceed to the right until equilibrium is reached, defined by the Gibbs free energy of the reaction. The Gibbs free energy of this reaction is given by;

$$\Delta G_{rxn} = \Delta G^0_{rxn} + RT\ \ln(IAP)$$

where ΔG_{rxn} is the Gibbs free energy of the reaction, ΔG^0_{rxn} is the Gibbs free energy for the reaction in its standard state, and IAP is the ion activity product of the aqueous species. At equilibrium, the Gibbs free energy of the reaction is zero, and IAP $= K_{SP}$. The value K_{SP} is the ion activity product for the activities at equilibrium. A new quantity, the chemical affinity can then be described, which is a measure of the distance from equilibrium of a reaction, and hence the magnitude of the driving force for the reaction to proceed. The chemical affinity can be written as

$$A = -RT \ln\left(\frac{IAP}{K_{SP}}\right)$$

Rates of dissolution are a complex function of the chemical affinity. The chemical affinity does not account for surface areas and defects on surfaces, which can influence dissolution kinetics profoundly [7]. However, there generally exists the so-called dissolution plateau at very high chemical affinities in which the rate of dissolution is high enough to be independent of surface defects and is constant over a range of ion activities [8]. As the concentration of aqueous species increases over time, other phases are liable to precipitate from solution. Maintaining a highly dilute solution can also prevent the precipitation of these phases.

Through carefully designed experiments, the rate of dissolution of a given crystal phase can be investigated. Additionally, the initial concentration of other elements can be varied to examine the effects of blending of crystalline slags with other materials. The most important elements in this regard are Al, Si, and Ca. Concentrations of these elements in solution model the effects of simultaneous dissolution of other cementitious and pozzolanic materials, such as Portland cement, fly ash, biomass ash, blast furnace slag, and hydrated lime. The presence of these elements may change the ion activity product of the solution, changing the chemical affinity for dissolution of the mineral of interest. However, in highly dilute systems, the more likely effect is on the rate of dissolution is expected to come from inhibition or catalysis of dissolution of the mineral phase by aqueous Al, Si, or Ca. For example, the mechanism of aluminosilicate glasses is known to be inhibited by the presence of aqueous Al [9]. Experiments such as these take a macroscopic approach to understand the mechanism of material dissolution through analysis of dissolution data.

γ-C_2S

Gamma dicalcium silicate (γ-C_2S) is a crystalline phase commonly found in steel slags and is identified as an understudied abundant mineral phase. Dicalcium silicate has five main polymorphs; α, α', α'_H, β, and γ. The γ polymorph is the only phase stable at room temperature, forming below 500 °C. The β polymorph when stabilized in the presence of other impurities is known as belite in cement chemistry and reacts rapidly with water. Studies have assessed the cementitious properties of γ-C_2S and concluded that is generally non-hydraulic phase [10–12]. Other work has sought to develop ways in which γ-C_2S can be hydrated, both by means of chemical activation and mechanical activation [13, 14]. However, little work has been done to determine the solubility constant, the relationship between undersaturation and dissolution rate, or the effect of aqueous species on the mechanism of dissolution. Additionally, little work has been done to explore the effects of blending of γ-C_2S containing materials with other, more reactive phases. This study investigates the effects of undersaturation and aqueous concentration of Al, Si, and Ca on dissolution rate in sodium hydroxide solution to lay the groundwork for exploration of blending of γ-C_2S bearing slags with other materials.

Experimental Procedures

Materials

Three crystalline slags were used in this study. Two slags were sourced from Haryana, India. These slags are identified as Bansal slag and DA slag. Both slags are cupola slags, a byproduct of pig iron production in the so-called cupola furnaces. The third slag is a steel ladle slag sourced from the US (Arcelor Mittal), identified as "ladle slag". Ladle slag is a by-product of steel refinement following the production of steel from a basic oxygen furnace or electric arc furnace.

γ-C_2S can be synthesized via a modified Pechini synthesis or sintering of precursor oxides. Sunde et al. [15] define the modified Pechini process for the synthesis of oxides as starting with a homogeneous aqueous solution containing the desired cation precursors in stoichiometric ratio and selected additives, which by evaporation and reactions is converted to a rigid cross-linked polymer hindering segregation of cation. The polymer is further converted to a homogenous powder by heat treatment. Nettleship et al. used this process to synthesize high purity γ-C_2S from calcium nitrate and colloidal silica precursors [11]. The additives used to produce the host gel are ethylene glycol and citric acid monohydrate. γ-C_2S may also be synthesized by sintering of a pellet of the comprised of the constituent oxides or carbonates—silica and calcium carbonate in this case—at high temperatures, as described by Kriskova et al. [16].

Mineral Content Analysis

Identification and quantification of the mineral phases present in slags were achieved through X-ray diffraction and Rietveld analysis. XRD data was collected using high-speed Bragg–Brentano optics on a PANalytical X'Pert Pro MPD operated at 45 kV and 40 mA. Data was obtained between 5° and 70° (2θ) using a step size of 0.0167° with each sample scan lasting 52 min. The diffractometer was configured with a 1/2° divergent slit, a 0.04 rad soller slit, and a 1° anti-scatter slit. The powdered sample was packed into a 27 mm diameter sample holder. HighScore Plus software was used for quantitative X-ray diffraction.

Dissolution Experiments

Dissolution experiments were performed in batch reactors at 20 °C; 500 mL pre-cleaned HDPE bottles containing 250 mL of solution. Containers were cleaned using nitric acid and Millipore water. Solutions were gently agitated initially, but not stirred to prevent particle abrasion. Sodium hydroxide was used as an alkali activator at concentrations of 0.1 M and 1 M. Sodium hydroxide solutions were prepared using Millipore water (18.2 MΩ cm) and analytical grade NaOH. The pH of these solutions was measured before the introduction of mineral phase. Solutions with initial concentrations of Al, Si, and Ca were prepared through additions of reagent grade $Al(NO_3)_3 \cdot 9H_2O$, $Ca(OH)_2$ and SiO_2. Appropriate quantities of the synthetic mineral of interest were added to the batch reactor to initiate dissolution. 5 mL aliquots of solution were removed at regular intervals, with the removed volume of solution being replaced by fresh NaOH solution of the appropriate concentration. Aliquots were filtered (<0.2 μm) and diluted in 2% nitric acid matrix for ICP-OES analysis on an Agilent 5100 Vertical Dual View ICP-OES. Al, Si, and Ca concentrations were measured directly by ICP-OES analysis at concentrations between 0.1 and 200 mg/L, with the dilution in 2% HNO^{-3} reflecting this desired concentration. Remnant solids were immersed in isopropyl alcohol to halt hydration before being dried and stored under nitrogen in a desiccator. Hydroxide ion concentrations were calculated using the extended Debye–Hückel equation [17] and pH measurements taken using an Orion Star A111 pH meter. The pH meter was calibrated against a series of NaOH solutions of known concentrations. The solution–mineral mass ratio was chosen to ensure no precipitation of reaction products, and to measure the rate of dissolution far from equilibrium in the so-called dissolution plateau.

Ion Activities

Ion activity products were calculated using Gibbs free energy minimization software, GEM-Selektor v3.3 with PSI-Nagra and CEMDATA18 thermodynamic databases. The activities of charged aqueous species were calculated using the Helgeson form of the (Truesdell–Jones) extended Debye–Hückel equation with ion size parameters and extended term parameters for NaOH [17]. The activity of water was calculated using the osmotic coefficient and the extended term used to calculate the activity coefficients for neutral aqueous species.

Results

Crystalline Slags

Chemical and Mineral Composition

Chemical composition from X-ray fluorescence (XRF) measurements is shown in Table 1. Bansal and DA slags have similar chemical compositions as a consequence of the similar feedstocks used at both plants. The ladle slag had relatively high calcia and alumina content. High calcium and aluminum content is typical for ladle slags and thus is expected [18, 19]. The composition of the Bansal and DA slags is typical of slags from pig iron production [20].

The XRD patterns of the three slags of interest are shown in Fig. 1. The phase distribution of these slags is markedly different. Only the major phases are identified in each case. The ladle slag is primarily composed of calcium aluminate phases and periclase. Despite having similar chemical compositions, the Bansal and DA slags have contrasting mineralogy. The Bansal slag is primarily composed of the pyroxene mineral diopside. The closely related mineral hedenbergite and augite are also detected in the Bansal slag due to the substitution of Fe, Na, and Ti. In contrast, the DA slag is semicrystalline and shows the presence of anorthite and quartz as major phases as well as an amorphous phase.

Table 1 Chemical composition of crystalline slags as determined by XRF

	Bansal slag	DA slag	Ladle slag
SiO_2	50.8	46.49	3.84
Al_2O_3	11.3	18.79	28.63
Fe_2O_3	10.8	6.11	4.66
CaO	17.7	17.06	51.38
MgO	5	6.15	6.88
Na_2O	0.01	0.42	0.07
K_2O	0.5	0.93	0.07
TiO_2	1.3	1.48	0.14
MnO_2	2.98	1.96	0.78
P_2O_5	0.16	0.15	0.06
SrO	0	0.05	0.03
BaO	0.18	0.12	0.01
SO_3	0.12	0.29	1.51
LOI	0	0.02	1.94

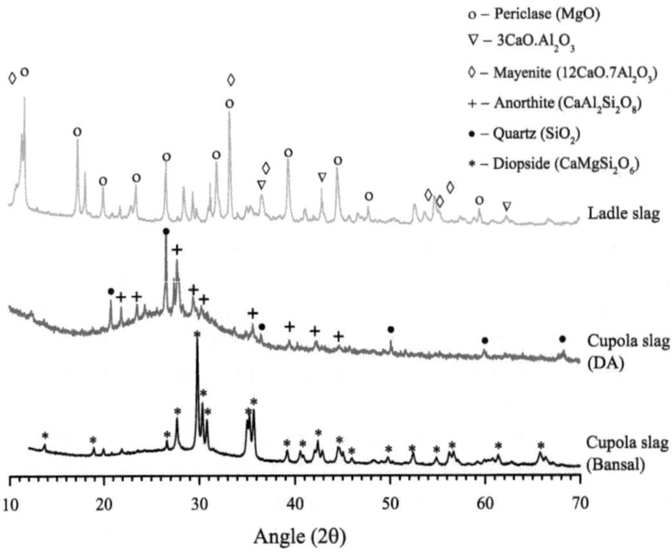

Fig. 1 Phase distribution of crystalline slags. Only major crystalline phases are identified here

Dissolution of Slags

Each of the slags was exposed to NaOH solutions of 0.1 M and 1 M for up to 48 h to assess the reactivity of the slags upon exposure to an alkali activator. The phase distribution of each slag was determined after prolonged dissolution (1 week) to those phases which are reactive in the presence of alkali activator (Fig. 2). In the case of the Bansal and DA slag, no major differences in phase distribution were observed and are thus not shown here. However, upon exposure to NaOH solution, calcium aluminate phases of the ladle slag are observed to be consumed and portlandite is observed to precipitate. Mayenite and C3A ($3CaO \cdot Al_2O_3$) are observed to react rapidly, liberating calcium into solution. The high concentration of OH^- ions in solution drives formation $Ca(OH)_2$.

Dissolution of major elements from slags is recorded using ICP-OES analysis, shown in Figs. 3 and 4 at NaOH concentrations of 1 M (pH $\approx$ 13.6) and 0.1 M (pH $\approx$ 13), respectively. Dissolution data for Al, Ca, and Si is shown. Dissolution data for Fe, Mg, Na, and S were also recorded but the negligible concentration of these elements was observed and thus was not included here.

After 1 h, the dissolution Al and Ca from the ladle slag is approximately 1 order of magnitude larger than concentrations of elements from the other slags. This very high initial concentration is indicative of the presence of highly reactive phases in the ladle slag, already identified in Fig. 2 as being due to C3A and mayenite. The concentration of silicon from the ladle slag remains low, as expected from the low silicon content of the slag. The evolution of the Al and Ca concentrations over time sees a gradual decrease in concentration in the case of the 1 M NaOH solution. This

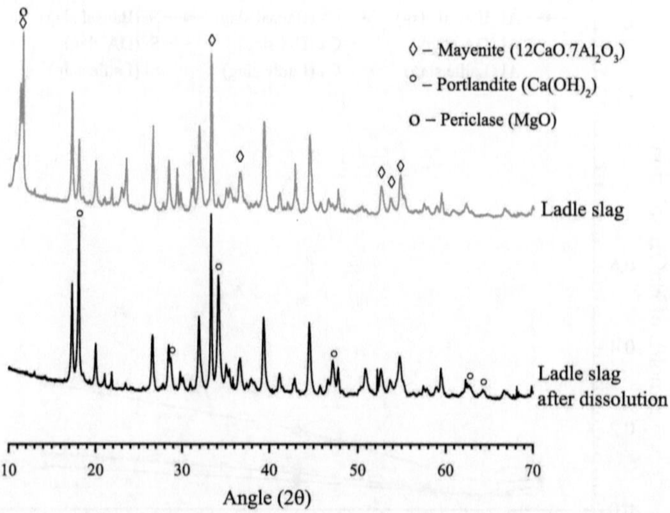

Fig. 2 Change in phase distribution of ladle slag after exposure to NaOH activating solution

decrease is suggestive of the formation and precipitation of reaction products due to supersaturation of reaction products. In the case of the 0.1 M NaOH solution, the Al concentration gradually increases, while the Ca concentration decreases over time.

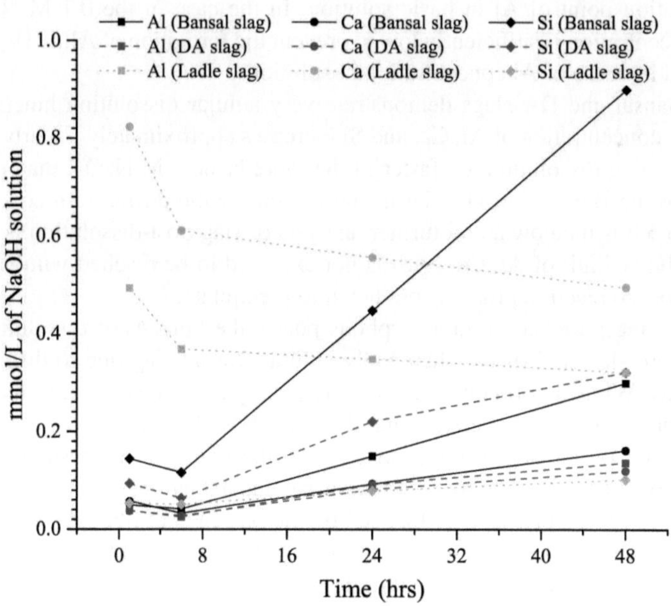

Fig. 3 Dissolution kinetics of Bansal, DA, and ladle slag in 1 M NaOH solution at L/S of 1000

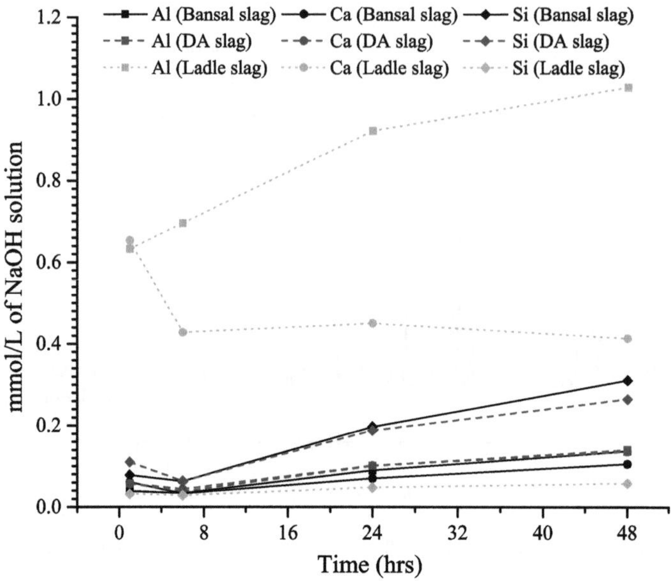

Fig. 4 Dissolution kinetics of Bansal, DA, and ladle slag in 0.1 M NaOH solution at L/S of 1000

The decrease in Ca concentration is understood to be due to precipitation of $Ca(OH)_2$. The saturation point of Ca in solution is reached at much lower concentrations than the saturation point of Al in basic solution. In the case of the 0.1 M NaOH, the OH^- concentration is sufficiently low to prevent the formation of $Al(OH)_3$, allowing continued increase of Al concentration in solution.

The Bansal and DA slags demonstrate very similar dissolution kinetics. In all cases, the concentration of Al, Ca, and Si increases approximately linearly after 6 h. The kinetics of dissolution are faster in the more basic 1 M NaOH than the 0.1 M NaOH for the Bansal and DA slags. The reason for the decrease in concentration from 1 to 6 h is unknown, and further analysis of slag post-dissolution is required. The solubility limit of Al, Ca, or Si is not expected to be reached within this time period, nor are reaction products predicted to precipitate.

Despite the presence of an amorphous phase, the kinetics of dissolution of the DA slag are slower than the chemically similar Bansal slag due to differences in mineralogy. While amorphous content is frequently a good indicator of reactivity, it is necessary to build an understanding of the kinetics of dissolution of other mineral phases found in the slag to better understand the potential of crystalline slags in construction materials. A more comprehensive understanding of how the kinetics of dissolution of these phases is influenced by the presence of other chemical species in solution can also provide insight into the potential effects of blending, whereby a more reactive material is mixed with a less reactive one to produce a load-bearing material.

Acknowledgements We would like to acknowledge the financial support for this research through the Tata Center for Technology and Design as well as the Environmental Solutions Initiative, both at Massachusetts Institute of Technology (MIT), Cambridge. We also acknowledge the support from NSF CAREER #1751925.

References

1. McLellan BC, Williams RP, Lay J, Van Riessen A, Corder GD (2011) Costs and carbon emissions for geopolymer pastes in comparison to ordinary portland cement. J Clean Prod 19(9–10):1080–1090
2. smail I, Bernal SA, Provis JL, San Nicolas R, Hamdan S, van Deventer JS (2014) Modification of phase evolution in alkali-activated blast furnace slag by the incorporation of fly ash. Cem Concr Compos 45:125–135
3. Duxson P, Provis JL (2008) Designing precursors for geopolymer cements. J Am Ceram Soc 91(12):3864–3869
4. Chaunsali P et al (2018) Mineralogical and microstructural characterization of biomass ash binder. Cem Concr Compos 89:41–51
5. van Oss HG (2017) Mineral commodity summaries
6. Brininstool M, Flanagan DM (2015) Minerals yearbook—copper
7. Luttge A (2006) Crystal dissolution kinetics and Gibbs free energy 150:248–259
8. Nagy KL, Lasaga AC (1992) Dissolution and precipitation kinetics of gibbsite at 80 °C and pH 3: the dependence on solution saturation state. Geochim Cosmochim Acta 56(8):3093–3111
9. Snellings R (2015) Surface chemistry of calcium aluminosilicate glasses. J Am Ceram Soc 98(1):303–314
10. Ghosh SN, Rao PB, Paul AK, Raina K (1979) The chemistry of dicalcium silicate mineral. J Mater Sci 14:1554–1566
11. Nettleship I, Shull JL, Kriven WM (1993) Chemical preparation and phase stability of Ca_2SiO_4 and Sr_2SiO_4 powders. J Eur Ceram Soc 11:291–298
12. Bensted J (1978) Gamma-dicalcium silicate and its hydraulicity. Cem Concr Res 8:73–76
13. Kriskova L et al (2012) Effect of mechanical activation on the hydraulic properties of stainless steel slags. Cem Concr Res 42(6):778–788
14. Kriskova L et al (2014) Influence of mechanical and chemical activation on the hydraulic properties of gamma dicalcium silicate. Cem Concr Res 55:59–68
15. Sunde TO, Grande T, Einarsrud MA (2016) Modified Pechini synthesis of oxide powders and thin films, pp 1–30
16. Kriskova L et al (2014) Hydraulic behavior of mechanically and chemically activated synthetic merwinite. J Am Ceram Soc 97(12):3973–3981
17. Helgeson HC (1981) Prediction of the thermodynamic properties of electrolytes at high pressures and temperatures. Phys Chem Earth 13–14:133–177
18. Shi C (2002) Characteristics and cementitious properties of ladle slag fines from steel production. Cem Concr Res 32(3):459–462
19. Adolfsson D, Robinson R, Engström F, Björkman B (2011) Influence of mineralogy on the hydraulic properties of ladle slag. Cem Concr Res 41(8):865–871
20. Monshi A, Asgarani MK (1999) Producing portland cement from iron and steel slags and limestone. Cem Concr Res 29(9):1373–1377

Extraction of Zinc, Silver and Indium via Vaporization from Jarosite Residue

Stefan Steinlechner and Jürgen Antrekowitsch

Abstract Based on technological reasons or in case of indium, which has gained its importance only in recent decades, metals such as zinc, indium and silver can be found in the residues or by-products of the corresponding industry like the iron precipitate—jarosite—from primary zinc production. As a result of low zinc and lead contents, targeted economic recycling of this material has to take into account also the present minor elements, such as indium and silver. In this context, a pyrometallurgical process is investigated which recovers zinc, indium and silver simultaneously as a dust product and the iron as an alloy. The reductive vaporization process is realized by the addition of a second residue, the electric arc furnace dust, acting as a chlorine carrier. Therefore, the paper describes the investigated process development and the achieved recovery rates for the main metals zinc, indium and silver.

Keywords Jarosite · Silver · Indium · Vaporization · Process development

Introduction

Technologies are getting more complex in daily life as well as industry. Hand in hand, the metals and alloys utilized are getting more specialized and unique in their composition. This in turn leads to an increase in the importance of so-called technology metals [1]. Many such valuable elements are produced as a by-product of another carrier metal, in the absence of own specific mineral mined for its primary production. These so-called hitchhiker or alternatively minor metals such as indium, gallium and germanium, or different other metals like the precious metals, contribute, if present in sufficient amounts, to the overall economy of a production, which is summarized in Table 1 [2, 3].

In general, they are of less economic value for the metallurgical plant than the carrier metal. Typically, they do not constitute the main product of the production site with regard to revenue and thus only indirectly influence the production quantity.

S. Steinlechner (✉) · J. Antrekowitsch
Montanuniversitaet Leoben, Franz-Josef-Str. 18, 8700 Leoben, Austria
e-mail: stefan.steinlechner@unileoben.ac.at

G. Gaustad et al. (eds.), *REWAS 2019*, The Minerals, Metals & Materials Series,
https://doi.org/10.1007/978-3-030-10386-6_22

Table 1 Hitchhiker metals, their sources, recovery efficiencies and contributions to refinery revenues [2]

Hitchhiker metal	Sources of production	Share of production (%)	Recovery efficiencies	Max. share of total revenues
Gallium	Alumina	90	10%	≈4%
	Zinc	10	–	–
Germanium	Zinc	70	≈12%	≈2%
	Coal	25	–	–
Indium	Zinc	≈100	25–30%	≈3%
Palladium	Platinum	60	40–60%	≈15%
	Nickel	40	–	≈15%
Platinum	Nickel	15	–	≈10%
Silver	Lead-Zinc	35	>95%	≈45%
	Primary	30	–	–
	Copper	23	>99%	≈25%
	Gold	12	–	–

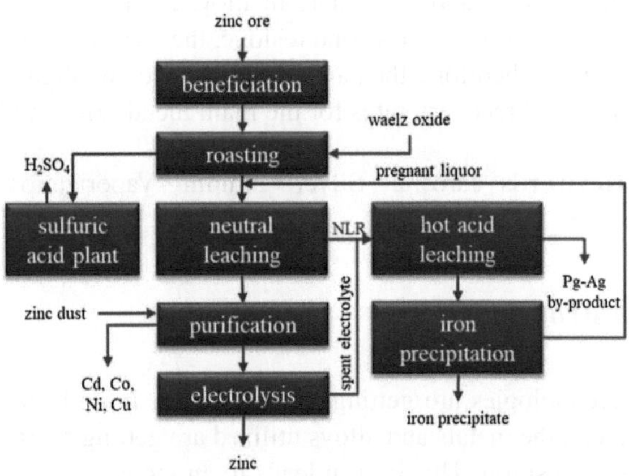

Fig. 1 Flow sheet of zinc production by roasting, neutral- and hot-acid leaching, and electrowinning [4]

Aside not every industrial site has the required infrastructure for the recovery of those potential minor elements. Consequently, they can be present in the generated residues as well [2].

Today, the hydrometallurgical zinc extraction from oxidic or sulfidic zinc concentrates dominates, with a ratio of more than 90% of primary zinc production. Even though several new technologies have been implemented in the last few decades (e.g. direct leaching of sulfidic ore and solvent extraction for solution purification), the typical flow sheet of a hydrometallurgical zinc plant still features the concentration, roasting, neutral- and hot-acid leaching of the zinc ore followed by electrowinning, as shown in Fig. 1.

Without going too deep into details concerning the various steps, the one of interest for the investigated material within this paper is the iron precipitation step. Over time, various iron removal technologies (hematite, goethite and jarosite) have evolved. Although the jarosite precipitate has the lowest iron content and, with this, the highest level of moisture, it is the most commonly used source of iron removal among those listed and offers the highest possibility of also finding silver and indium in the residue.

Materials and Methods

Jarosite is a complex basic iron sulphate with the chemical compound $R_2Fe_6(OH)_{12}(SO_4)_4$, where R is a cation [4, 5]. Typical cations are K^+, Na^+, NH_4^+, Ag^+, H_3O^+ or R_2 can be Pb^{2+}. Equations 1–3 summarize the reactions during the formation of jarosite [4]:

$$3Fe_2(SO_4)_3 + 6H_2O \rightarrow 6Fe(OH)SO_4 + 3H_2SO_4 \quad (1)$$

$$4Fe(OH)SO_4 + 4H_2O \rightarrow 2Fe_2(OH)_4SO_4 + 2H_2SO_4 \quad (2)$$

$$2Fe(OH)SO_4 + 2Fe_2(OH)_4SO_4 + 2NH_4OH \rightarrow (NH_4)_2Fe_6(SO_4)_4(OH)_{12} \quad (3)$$

Combining the above-mentioned reactions, the overall reaction can be stated as shown in Eq. 4 [4].

$$3Fe_2(SO_4)_3 + 10H_2O + 2NH_4OH \rightarrow (NH_4)_2Fe_6(SO_4)_4(e)_{12} + 5H_2SO_4 \quad (4)$$

It can be seen that sulphuric acid is formed during this reaction, leading to the necessity of the addition of calcine for neutralization. This is also the reason why the jarosite—and the same holds true for goethite—always carries a certain amount of concentrate. In some production plants, no separation of the Pb–Ag residue from the hot-acid leaching step is carried out in favour of lower investment costs, resulting in additionally increased lead and silver content in the iron precipitate [5]. Moreover, Zn, Ag, Cu and In can be assimilated into the jarosite structure [6].

The concept studied pursues the strategy of decomposing the jarosite in a first step and then capturing indium, silver and zinc in a dust product. During the heating up process, the jarosite decomposes, such as shown in Eqs. 1–4, for ammonium jarosite.

$$(NH_4)Fe_3(SO_4)_2(OH)_6 \cdot xH_2O \rightarrow (NH_4)Fe_3(SO_4)_2(OH)_6 + x\{H_2O\} \quad (\text{Temp.}: 120\,°C) \quad (5)$$

$$(NH_4)Fe_3(SO_4)_2(OH)_6 \rightarrow (NH_4)(FeO)_3(SO_4)_2 + 3\{H_2O\} \quad (\text{Temp.}: 260\,°C) \quad (6)$$

Fig. 2 Untreated jarosite (left), dried jarosite (centre) and decomposed jarosite (right) [8]

$$6(NH_4)(FeO)_3(SO_4)_2 \rightarrow 6\{NH_3\} + 3\{H_2O\} + 5Fe_2O_3 + 4Fe_2(SO_4)_3 \quad (Temp.: 350-400\,^{\circ}C) \tag{7}$$

$$2Fe_2(SO_4)_3 \rightarrow Fe_2O_3 + Fe_2(SO_4)_3 + 3\{SO_2\} \; (Temp.: >500\,^{\circ}C) \tag{8}$$

By this decomposition, the entrapped silver and indium become available for the formation of volatile compounds again. The aim was to collect indium as well as silver in a zinc oxide product to be utilizable in primary zinc industry, without the need of additional investment in any separation technology (Fig. 2).

So, after the decomposition, the Indium is present in oxidic form and requires the formation of a volatile compound to be separated with the zinc under reducing conditions. As a base thermodynamic calculations with the software HSC Chemistry 8.1.4. were carried out. The vapour pressure curves for selected indium and silver compounds as well as the metallic state can be seen exemplarily in Fig. 3 [7].

It can be observed that the indium in metallic state starts to have a relevant vapour pressure above 925 °C while its oxide, In_2O_3, is not volatile in the given area. Theoretically, the chlorine compounds have a lower temperature where a relevant vapour pressure exists, starting at ~190 °C for $InCl_{(g)}$ and ~275 °C for $InCl_{3(g)}$. The assumption for the calculation is that the underlying compounds are stable at present conditions but the diagram does not give any reference on if this is the case. Therefore, phase stability diagrams as shown in Fig. 4, also known under the label predominance diagram, were calculated, illustrating the conditions required to form a specific stable compound [7].

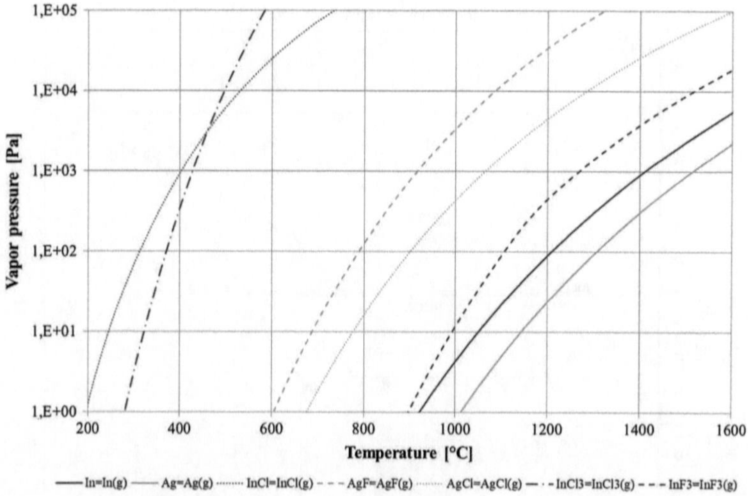

Fig. 3 Vapour pressure for selected indium (blue lines) and silver compounds (red lines) as a function of temperature [7]

It is evident that as long as the $O_{2(g)}$-partial pressure is high enough and the $Cl_{2(g)}$-partial pressure is low, In_2O_3 is stable over a wide temperature field. The transformation to metallic indium is only possible by lowering the $pO_{2(g)}$, which can be realized by adding reducing agents, for instance. Another kind of predominance diagram, shown in Fig. 4b, illustrates for four selected $O_{2(g)}$-partial pressures, starting at 0.21 bar (blue line) to 1E-15 bar (green line), the stability of metallic indium, In_2O_3, InCl and $InCl_3$. It can be recognized that over a wide temperature field, predominantly In_2O_3 is again stable and in case of high enough $Cl_{2(g)}$-partial pressure, $InCl_3$ is predominant. In the case of very low, technically more difficult to implement partial pressures of oxygen, InCl does show a stable region at a higher temperature.

The conclusion in regard to a metallurgical process, aiming to extract the indium via the vapour phase, is that based on the present thermodynamic calculation in a temperature field of up to 1200 °C, a high $Cl_{2(g)}$-partial pressure next to a low $O_{2(g)}$-partial pressure in the atmosphere is required to form a volatile chlorine compound. In the case of present non-volatile In_2O_3, as it is the case if roasted jarosite is treated, a transformation can be supported by binding the oxygen to another compound. This can be a reducing agent, such as carbon, forming $CO_{(g)}$ lowering the $pO_{2(g)}$ in parallel. In case of silver, the thermodynamic calculations showed that potentially present silver oxide decomposes at temperatures above 180 °C and easily reacts in the presence of chlorine or chlorine carriers to a volatile silver chloride under oxidizing as well as reducing conditions.

So as a result, the second step after the decomposition has to be reductive vaporization, under sufficient activity of chlorine. This can be achieved by the addition of carbon and a chlorine source. The concept developed was evaluated in a series of experiments with industrial jarosite, various chlorine carriers and operational param-

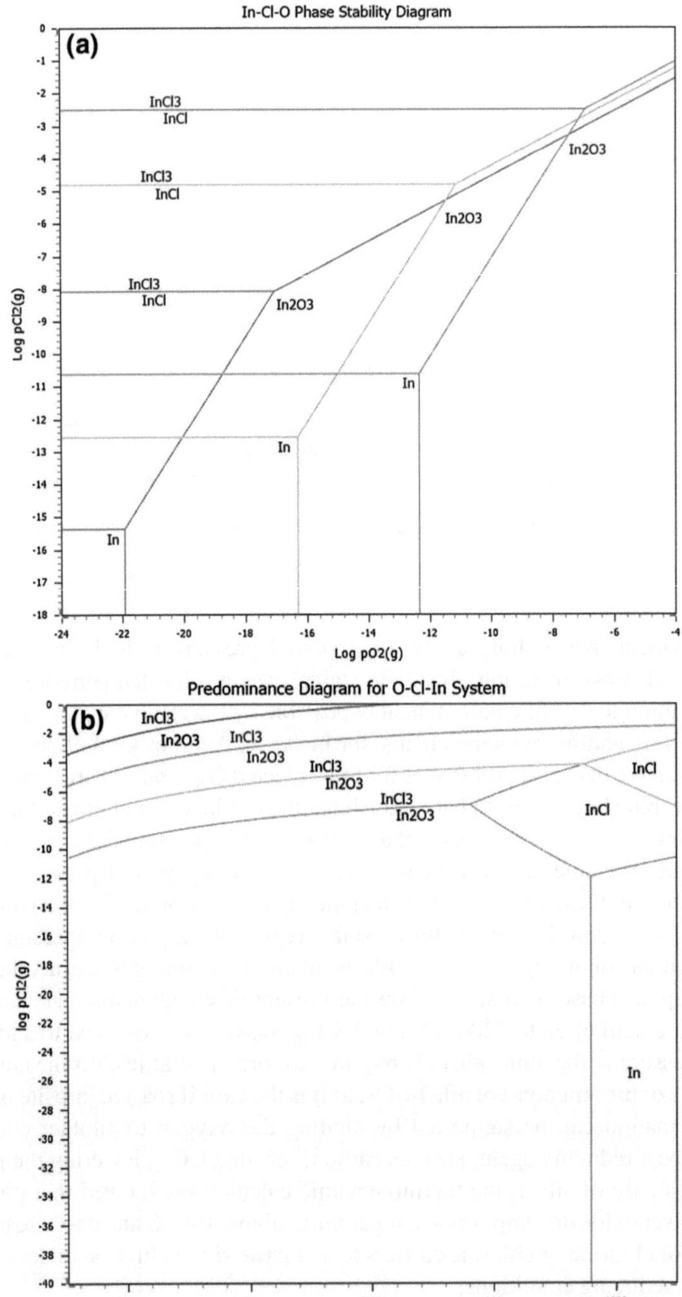

Fig. 4 **a** Predominance diagram (LPP) for the system O–Cl–In at three temperatures (green: 700 °C; yellow: 900 °C; red: 1100 °C). **b** Predominance diagram (TPP) for the system O–Cl–In at 4 $O_{2(g)}$ partial pressures (green: 1E-15 bar; yellow: 1E-10 bar; red: 1E-5 bar; blue: 0.21 bar) [7]

eters. With the aim of reduction of operational costs, also the addition of electric arc furnace dust, a residue paid for getting treated, as a further chlorine carrier was examined and lead to the satisfying extraction yields of 78–85% of silver, 76–81% of indium and almost 100% of zinc and lead [7].

Results and Discussion

For the development of an ecological and economical process, the multi-metal recovery in parallel to the formation of an added value product from jarosite was the target. As the iron precipitate typically is low in its value, elements like indium and silver have to be considered during an extraction process as a possible contributor to the revenues. Due to the low content of those minor elements, this is only possible by forming a combined product, which is easily utilizable in an industry sector with existing infrastructure for the separation of zinc, silver and indium, like selected primary zinc plants. With the reductive vaporization step, the targeted separation of Zn, In and Ag via the gas phase and the formation of such an added value product were successfully developed. Aside the iron compound optionally is reduced as well to form an iron alloy and minimizing the amount of final residue drastically. As the iron reduction is a cost-intensive step, it was also considered to leave the iron at least partly in the slag to reduce costs further.

Acknowledgements The authors thank the Austrian Research Promotion Agency (FFG) and the Federal Ministry of Science, Research and Economy (BMWFW) for the financial support of this research activity (FFG project number: 844725).

References

1. Paschen P (2007) Heavy nonferrous metals in world economy over the last 50 years. World Metallurg ERZMETALL 60:263–272
2. Chapman A, Arendorf J, Castella T, Thompson P, Willis P, Espinoza LT, Klug S, Wichmann E (2017) Study on critical raw materials at EU level. http://ec.europa.eu/DocsRoom/documents/5605/attachments/1/translations/en/renditions/native. Accessed Nov 24 2017
3. Worrell E, Reuter MA (2014) A handbook of recycling. Elsevier, Waltham, Mass
4. Metallhüttenkunde PF (1983) Walter de Gruyter: Berlin, Germany. ISBN 3-11-007458-3
5. Sinclair RJ (2005) The extractive metallurgy of zinc. AusIMM, Carlton, Vic
6. Dutrizac JE, Jambor JL (2000) Jarosites and their application in hydrometallurgy. Rev Mineral Geochem 40:405–452
7. Steinlechner S, Antrekowitsch J (2018) Thermodynamic considerations for a pyrometallurgical extraction of indium and silver from a jarosite residue. Metals 8 (Online Proceedings)
8. Wegscheider S, Steinlechner S. and M. Leuchtenmüller M (2017) Innovative concept for the recovery of Silver and Indium by a combined treatment of Jarosite and electric arc furnace dust. JOM 69(2017):388–394

Efficient Utilization of Zinc-, Lead- and Copper-Containing By-Products

Juergen Antrekowitsch and Stefan Steinlechner

Abstract Ores nowadays often show a higher complexity, containing different valuable metals as well as disturbing elements. In the state-of-the-art smelters, the majority of these elements end up as by-products. Treatment of such by-products is in many cases not done and if processes exist, they often recover only one metal, leaving a lot of value in newly generated residues. New strategies have to be employed to allow more efficient utilization of these important raw materials. The paper describes developments in bath smelting processes allowing a simultaneous recovery of different metals in one step, generating valuable products and an inert slag which in best case enables utilization in the construction industry. The proposed system is described in detail for zinc-, lead- and copper-containing by-products. Especially economic considerations are explained, showing how different metals contribute to a feasible process and what are the critical factors that have to be considered.

Keywords Multi-metal recycling · By-products · Precious metals · Assessment

Introduction

For many decades, residues from metal production, such as slags, dust and sludge have often been qualified as waste that was simply dumped. This fact was—at least at that time—accepted by industry and involved authorities. Except for metal producing companies, no market participant really recognized that metal-bearing wastes were often landfilled to an amount that was more or less as high as the metal production itself. In the eighties and nineties of the last century, first environmental concerns triggered more attention on such dumps and residues. Nevertheless, due to missing technologies and low metal prices, the interest in processing residues to recover valuables was still very low.

During the past few years, multi-metal recovery from concentrates has been developed for both, primary ores and secondary materials. In addition, minor metals con-

J. Antrekowitsch (✉) · S. Steinlechner
University of Leoben, Leoben, Austria
e-mail: antreko@unileoben.ac.at

G. Gaustad et al. (eds.), *REWAS 2019*, The Minerals, Metals & Materials Series,
https://doi.org/10.1007/978-3-030-10386-6_23

tained in such residues became important for daily life in industry. In other words, metals that did not have a real value in the past nowadays may essentially contribute to the economic viability when recovered from residues.

In combination with increased metal prices and the need for special and rare metals in various highly sophisticated technologies (e.g. the electronics industry), more attention must be paid to such residues which often represent a potential secondary resource. Currently, recycling rates are still low and huge amounts of residues are dumped annually.

Recycling is mainly based on scraps, where the elements in focus are available as pure metals.

However, scraps have a defined volumetric limit and their treatment is already optimized to a certain level. As a matter of fact, recycling rates must be improved in the near future considering continuously dropping grades in primary ores, implying that new sources need to be exploited.

As an important solution to this problem, by-products of various metallurgical processes, already dumped or continuously being produced, started to move into the focus of interest of both industry and research.

Screening the metallurgical area, residues emerging from the lead, zinc and copper industries (marked in Fig. 1) must be considered. Processed ores commonly contain a high number of critical and valuable metals at the same time. This can be demonstrated by the so-called Metal Wheel which contains the main ore metals (dark blue inner circle) with associated by-products (light blue circle) and ‘unwanted’, respectively, non-recoverable, elements (white and green outer circles); (Fig. 1).

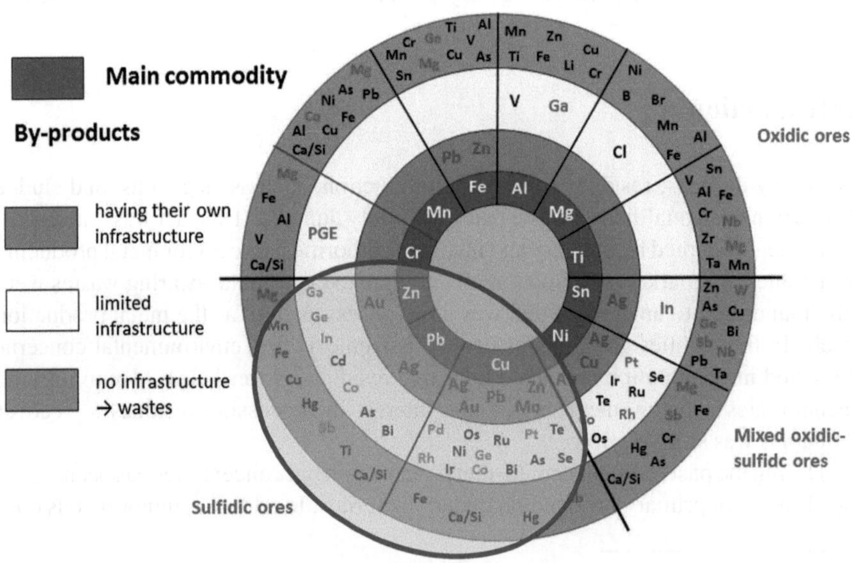

Fig. 1 Main metals in non-ferrous industry and their associated elements [1]

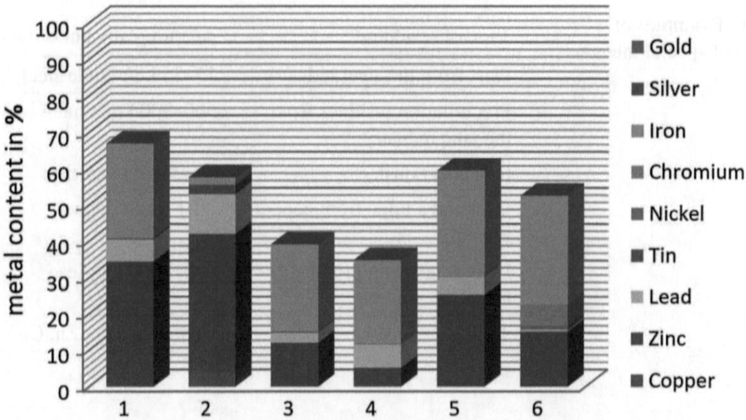

Fig. 2 Examples for different residues and their metal content (except PGM). 1: Electric arc furnace steel mill dust, 2: dust from copper recycling, 3: slags from lead industry, 4: jarosite from zinc industry, 5: dust from cupola furnaces, and 6: stainless steel production dust

Some of the metals displayed in the Metal Wheel are already recovered to a certain amount by primary metallurgical processes but the bulk ends up in different waste streams.

Very often sludge from hydrometallurgical operations and dust from pyrometallurgical operations offer a significant number of metals at concentration levels that might be of economic interest. In addition, some recycling processes of highly diversified and contaminated scraps (e.g. car bodies or waste electric and electronic equipment) generate complex by-products. Figure 2 illustrates the possible recoverable elements but also shows the high overall metal content available in currently produced residues.

Some of the residues described in Fig. 2 are already recycled to a certain extent. However, also for these secondary materials, the worldwide recycling rates are still below well 50% due to two main reasons:

- the lack of optimized processes and
- missing information and data about the materials themselves (e.g. compositional, chemical and granulometric data).

Both factors are responsible for the inefficiently utilized potential on secondary resources [2].

Another reason for the weak exploration and use of residues is the lack of any guideline or competence describing how to evaluate residual materials [3]. This is partly due to non-existent adequate databases. For primary materials like metal ores, codes do exist that serve as accepted and applied guidelines for a qualified evaluation of predefined steps, leading to a 'bankable feasibility study'. Such guidelines currently do not exist for secondary raw materials and due to several reasons, the existing codes are not suitable for the residues described above.

Table 1 Examples of residues of special interest

Type of residues	Accruing amount of residue
Dust from the steel industry	15–25 kg/t crude steel
Precipitation residues from the zinc industry	500–800 kg/t zinc
(Foundry) dust	10–20 kg/t cast iron
Dust and sludge from copper winning	20–60 kg/t copper
Lead slags from primary lead/zinc winning	600–1000 kg/t lead
Tailings	Containing Pb, Zn, Cu

As a consequence, a major task of a running project at University of Leoben is to create a comprehensive database, to define and design a similar guideline (code) for these residual materials. Such a guideline must be proposed, formulated and applied by a competent consortium in the sector of primary and secondary raw materials that has deep knowledge and experience in the fields of:

- geochemical and mineralogical characterization of raw materials,
- mineral processing of heavy metal-containing by-products,
- metallurgical processing principles and methods for various metals,
- access to dumps and companies producing these secondary raw materials,
- analytical devices to allow a specific characterization and
- devices for testing concepts in mineral processing and metallurgy.

Typical by-products are produced either in hydrometallurgical or pyrometallurgical process routes.

Some examples of typical residues from the metallurgical production routes and their accruing by-products are shown in Table 1.

These residues can be found worldwide in huge quantities either on dumps dating from the end of the last century or being currently produced and form an interesting secondary resource of metals with high availability. Due to the long European history in mining and metallurgy, such landfilling sites and producing mills are still present and have to be considered as ‘part of the market’ [2, 4].

A consequent ‘stepwise approach’ must be applied right from the beginning of the project. A practical and useful scheme complying with this approach is shown in Fig. 3. Prior to characterization, representative samples have to be taken from continuous production or from landfills. The state-of-the-art methods have to be checked and adapted to individual requirements. Assessing the most suitable sampling methods for complex and heterogeneous materials is an integral part of the work carried out. Nevertheless, it is a necessary part of the characterization. In a first step, it is necessary to obtain detailed information about the structure and the properties of the secondary material that becomes subject of investigation, followed by an assessment of a possible processing route. With these first basic definitions, it becomes possible to develop a new process concept or optimize an existing process. The last step refers

Fig. 3 Steps for the assessment of wastes and by-products

to product quality after the treatment or some optimizations to enrich the quality of existing products.

In some cases, one of the steps pointed out in Fig. 2 might already exist and sufficient data for this particular step is available.

Characterization and Assessment of Residues

Depending on the feed material used and the treatment technology applied, various residues with different properties emerge that can be discriminated by their origin or properties. Some of the above-described residues have already been investigated to a certain extent with respect to characterization and a first evaluation.

Detailed geochemical and mineralogical characterization of metallurgical residues is of the highest importance in order to develop appropriate mineral processing recipes and further on appropriate metallurgical methods to recover all metals of interest. Hence, the following questions need to be answered:

(1) Which particles contain the metals to be recovered?
(2) What is their particle size, shape and texture, i.e. zoned grains, coated grains?
(3) How intensive/complex are incorporated phases intergrown in different particle fractions?
(4) What is the concentration level of metals in the major particles?
(5) Are deleterious elements (such as As, Cd, Hg) present? If yes, what are their concentrations?

In order to separate minerals or metallurgical phases by means of mineral processing techniques with physical, physical–chemical and/or chemical methods it is necessary to know the relevant phases regarding the key aspects mentioned above. In the end, the size, shape, structure and the chemical and mineralogical composition define a particle.

Deriving from these parameters, the density, optical, electrochemical, thermal, chemical and mechanical characteristics of any particle can be determined. As the way of managing a metallurgical process and its actual parameters highly affect the residues not only in composition but also in structure, the variance of the residues' properties with regard to mineral processing characteristics is large, even for an existing plant in operation. In analogy to primary mineral deposits, processing results may vary widely if these changes are not taken into account. Thus, the residue has to be investigated and characterized properly.

Based on that, the main goal is to generate a basis for assessing possible metal or metal compound recycling methods. Additionally, the different metallurgical possibilities for a treatment—hydro- or pyrometallurgical—need to be determined. These results are mandatory for evaluating possible process steps that are required to generate products from the residues. To generate this information, the main properties such as melting behaviour, volatilization behaviour, reducibility and solubility need to be determined in advance by using special metallurgical characterization procedures.

Process Development and Product Quality

To cope with the continuously stricter environmental legislation next to a shortage of critical metals in Europe, especially the process development for treatment of secondary resources and the optimization of existing processes are two major concerns. The challenge of the years to come is to, at the same time, minimize newly generated residues and to achieve maximum product quality within the recycling process. In addition, the quality of the products influences the targeted market and achievable revenues. Therefore, the state-of-the-art processes have to be continuously adapted regarding energy consumption, mass balancing and further improvement of the product quality.

A specific geochemical, mineralogical and metallurgical characterization creates the basis for improving such existing processes. Thus, it will be possible to obtain higher product qualities and lower amounts of newly generated waste streams.

As demonstrated by several investigations carried out at Montanuniversität Leoben, the residues are not at all homogenous. Hence, it is not possible to recycle such residues as one combined single input stream. In order to increase the recoverable metal grade and to reduce the content of substances that hinder the metallurgical process, some separation by physical/mechanical techniques is required. If such a process shall be developed systematically rather than by trial and error, the residue must be characterized properly in terms of its properties relevant for a separation process. The fractional analysis is the standard tool in mineral processing for assessing the amenability of primary mineral resources to be upgraded to saleable products of defined and consistent quality. Taking into account the heterogeneity of the residues it is expected that some upgrading is possible which may pay off during recycling. The fractional analysis yields the optimum results for separating a given feed material by assuming perfect separation. It thus also serves as a benchmark for assessing the performance of an existing process. The possible benefits of assessing and processing the residues in a similar way to primary mineral resources are obvious. As some common properties of the residues (e.g. the generally fine particle size distribution of dusts, just to name one) are problematic in the fractional analysis, their methods require further adaptations.

In order to find a suitable processing chain, test work needs to be carried out. Once done for a residue, the obtained data will be collected and condensed in a database, which acts in future assessments as a decision support tool. Although similarities within the same type of material (e.g. slags, dust) can be found, each residue is generated in slightly different processing routes with varying input material. This leads to variable extraction yields, consumable consumption, energy input, obtained product quality, etc. Based on the experimental results, the database will grow, and the more materials are catalogued; the more accurate the data for the assessment becomes.

Metal Bath Processes

One major disadvantage of past attempts in recycling by-products was that often just one single metal was recovered while possible further valuable metals were distributed into newly generated wastes and were therefore lost for production. A well-known example is the Waelz kiln process for the recycling of dust where only zinc is in focus of the recovery, while iron, lead, silver and others remain generally unnoticed.

Pyrometallurgical concepts allow the recovery of valuable metals by applying different basic operations: forming a volatile phase, where, e.g. zinc or lead can be collected as oxides in the off-gas; forming a liquid metal phase, where, e.g. iron or lead act as an efficient collector for various minor metals. In combination with mineral processing steps, where separation of metal-containing fractions could be realized prior or subsequent to a metallurgical step, the targeted multi-metal recovery can be realized (Fig. 4).

For hydrometallurgical concepts, the dissolution of a larger number of elements followed by selective precipitation again combined with mineral processing would offer multi-metal recycling. However, also a combination of hydro- and pyrometallurgy could lead to appropriate solutions.

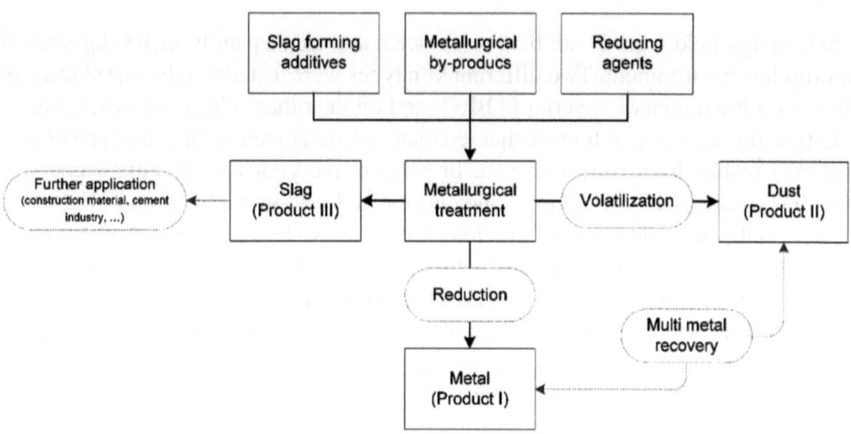

Fig. 4 Example of multi-metal recovery

Facing the situation that for some by-products recycling concepts do not exist and for those which are already partly recycled but with the majority recovering only one metal, this approach must be regarded as one of the essential novelties.

One general concept for multi-metal-containing residues was developed by the chair of non-ferrous metallurgy, Montanuniversität Leoben in Austria together with other companies. This process idea is somehow a little bit more complex follows the basic scheme in Fig. 4, but would allow at the end of the recovery of different valuable metals mainly via a liquid lead bath. Trials in lab-scale as well as technical scale (50–100 kg/batch) showed high recovery rates as well as moderate energy consumption, if the pretreatment and the metallurgical treatment are combined in a reasonable way. Lead and zinc were recovered to more than 90% and also the silver recovery reached values above 90%. The typical products out of such a recycling process are listed below:

- Zinc oxide, received by reduction and subsequent volatilization, may also contain a certain amount of indium,
- Lead bullion contains the reduced lead as well as silver and other lead soluble metals, and
- Remaining is slag, which is widely free of heavy metals.

The viability of the described concept is first of all dependent on the following major factors:

- the amount of silver and gold,
- the recovery of further elements like tin, copper, indium, germanium, etc.,
- the LME notation of the valuable metals and
- the accessibility of the dump.

Results from Various Campaigns

Trials in this field were done; first in lab-scale and subsequently in 100 kg-scale in appropriate environment. Two different kiln types were tested: on the one side a gas-fired top-blown rotary converter (TBRC) and the other side, an electrical heated DC-type furnace. Depending on charged material, parameter setting and type of kiln, different results have been received. In general, the yields were quite successful, which is summarized for the main metals zinc and lead in the diagram in Fig. 5.

Especially, the collection of precious metals as well as copper and others in the metal bath showed a successful extraction and therefore utilization of the residues.

The somewhat lower lead yields when using the TBRC may result from the higher turbulence and with this the presence of small metal droplets in the slag. However, this could be optimized by adding an appropriate settling unit or by reducing the rotation speed at the end of the process.

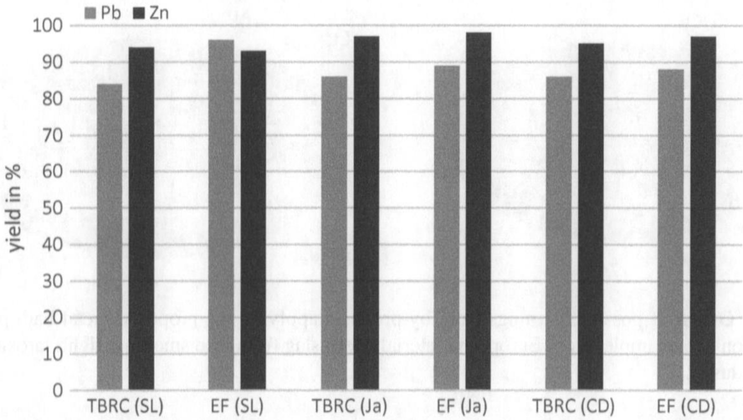

Fig. 5 Example of different reduction trials in metal bath processes (mean values). TBRC: Top-blown rotary converter); EF: electric furnace; SL: slag from lead industry; Ja: jarosite from zinc industry; and CD: dust from secondary copper industry

Economic Considerations

Within the investigations also a model for economic verification was established and different dumps worldwide have been investigated. Within this evaluation high promising dump sites as well as residues of poor quality have been detected.

Economics were done for different materials and metal prices. As already mentioned above, gold and silver contribute in a very important way to the overall revenues. The following pie charts show the splitting between the different sources of income—first of all the metal-containing products—for different by-products. Two examples for jarosite and slag from lead industry are shown. The values are based on achievable product prices and include determined mean metal yields.

In case of the slag from the lead industry (Fig. 6, left), the zinc product represents the main income. Therefore, the economy of the process is quite dependent on the zinc LME notation. For the jarosite (Fig. 6, right), the situation is different. If silver is present in a higher amount (in the present case 450 ppm), it is of main importance for the income, closely followed by lead and zinc. In case gold is present too, as it was seen for some residues, it could play an important role too.

However, the key message is that a multi-metal recovery is essential for the economic stability of a recycling process and allows higher flexibility regarding varying metal prices.

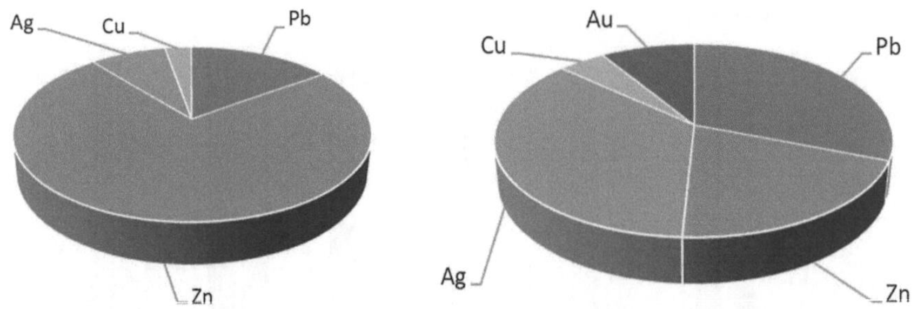

Fig. 6 Origin of possible earnings from by-products applying the proposed metal bath process. Based on two examples from European materials; left: slag from lead smelting; right: jarosite from zinc industry

Summary

By-products from metallurgical industry play an important role as secondary resource for different metals. Especially, precious metals are important to realize an economical process. The University of Leoben, Austria, together with other companies developed a process concept which would allow efficient recovery of different metals, following the concept of a multi-metal recovery via a reducing metal bath concept. Currently, investigations are ongoing which also intend a concentration of the recoverable metals by means of mineral processing. Such a concentration step would save energy and makes the overall concept more insensitive against varying metal prices. The idea of recovering more than only one metal from described by-products would allow an economic and ecologic utilization of such wastes in many cases.

References

1. Reuter MA (2013) Metal recycling. United Nations Environment Programme
2. Antrekowitsch J, Steinlechner S (2011)The recycling of heavy-metal containing wastes: mass balances and economical estimations. JOM 63(1):68–72
3. JORC—competent persons. http://www.jorc.org/competent.asp. Accessed June 2018
4. Mineral Yearbooks (2018). http://minerals.usgs.gov/minerals/pubs/country/index.html#pubs. Accessed June 2018

Production of High-Purity Mo and Fe–Mo Alloys from Recycled Mo Oxide and Mill Scale Through Hydrogen Reduction

Min-Kyu Paek, Do-Hyeong Kim, Daniel Lindberg and Jong-Jin Pak

Abstract The present study handled an economic and clean route for the synthesis of pure Mo and Fe–Mo alloys through utilization of waste materials such as spent acid containing Mo and mill scale removed from the surface of steel slabs. The Mo oxide, MoO_3, was successfully synthesized from the spent acid by ammonia gas neutralization method and reduced under hydrogen atmosphere to produce a high-purity Mo powder. The optimum condition for the spherical shape and uniform particle size of Mo powder was determined. In addition, the recycled MoO_3 from the spent acid was mixed with mill scale as an iron oxide source to produce Fe–Mo alloy by the hydrogen reduction. Based on the thermogravimetric analysis, the reduction rate of the mixed MoO_3 and mill scale was measured at 700–1100 °C. The Arrhenius equation with the approved mathematical formulations for the heterogeneous gas–solid reaction was applied to calculate the activation energy (Ea) values and determine the rate controlling mechanisms.

Keywords Spent acid · Mill scale · MoO_3 · Hydrogen reduction · Fe–Mo alloy · Reduction kinetics

Introduction

In manufacturing tungsten filament coils in lamp industries, very fine tungsten wire is coiled around molybdenum mandrel wire/rods to form the so-called coiled coil. The formed coiled coil is then heat-treated and cut to a size, and immersed in a solution of mixed nitric acid and sulfuric acid which dissolves the molybdenum mandrel leaving the coiled coil tungsten as shown in Fig. 1. Such operation results in the generation of large volumes of spent acid containing H_2SO_4, HNO_3, and dissolved

M.-K. Paek · D. Lindberg
Department of Chemical and Metallurgical Engineering, Aalto University, Kemistintie 1, 02150 Espoo, Finland

D.-H. Kim · J.-J. Pak (✉)
Department of Materials Engineering, Hanyang University, ERICA, 426-791 Ansan, Korea
e-mail: jjpak@hanyang.ac.kr

G. Gaustad et al. (eds.), *REWAS 2019*, The Minerals, Metals & Materials Series,
https://doi.org/10.1007/978-3-030-10386-6_24

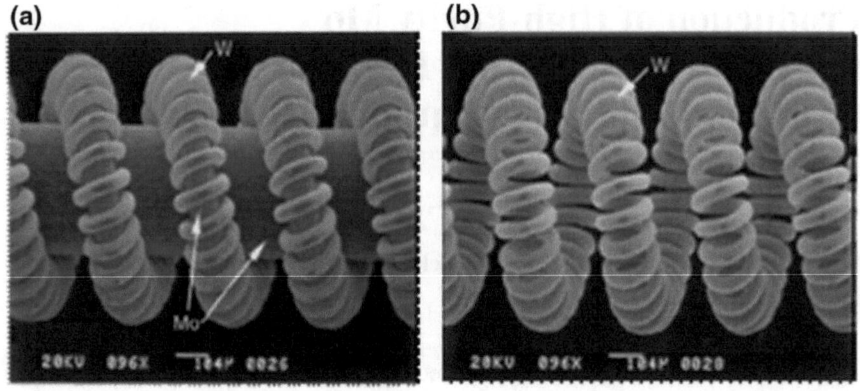

Fig. 1 Tungsten coil on Mo mandrels: **a** before Mo dissolution and **b** after Mo dissolution

molybdenum. Molybdenum is usually dissolved in the amount of 80–90 g/L of spent acid solution. The volume of spent acid generated from a typical coil manufacturing company in Korea ranges from 300 to 600 L per day. In the past, it was a common practice in these companies to neutralize the spent acid with sodium hydroxide and discharge in into wastewater system, even though there have been several methods to recover the molybdenum by sulfide precipitation, carbon absorption, ion exchange, solvent extraction, and ammonium salt precipitation. The loss of molybdenum in this manner is economically significant, and the discharge of nitrate and sulfate ions and heavy metals like molybdenum into the nation's rivers is undesirable from the environmental view.

In the present work, the molybdenum recovery process using ammonia gas neutralization was implemented [1, 2] and improved. The new process used a high acid/water ratio for high efficiency of molybdenum recovery, and also for high total nitrogen content as ammonium and nitrate ions in neutralized acid solution. The molybdic oxide, MoO_3, is successfully synthesized from the spent acid by roasting the ammonium molybdate. It was reduced by hydrogen to produce a commercial grade molybdenum metal powder, and it was also mixed with mill scale for the synthesis of Fe–Mo alloy. The present investigation provided a hydrogen reduction process to produce molybdenum metal powder of uniform spherical shape and an economical clean route for the synthesis of Fe–Mo alloys from a secondary resource such as mill scale and spent acid. Gaseous reduction of mixed iron oxide (mill scale) with MoO_3 (recycled from spent acid) was applied. Under the effect of reaction temperature, the reduction behavior and kinetics in hydrogen atmosphere with the synthesis of Fe–Mo alloys were investigated.

Experimental

Ammonia Neutralization and Precipitation

A schematic of Mo recovery process from spent acid is shown in Fig. 2. After predetermined volumes of acid and water (1:3, 1:1 and spent acid) were mixed in the reactor, anhydrous ammonia was introduced into the solution at a rate of 10 NL/min *via* nozzles immersed near the bottom of the reactor. The aimed solution pH was in the range of 2.5–3.0 in the present work. After precipitation, the solution was drained to a centrifuge hydroextractor to collect all ammonium molybdate crystals precipitated. The separated ammonium molybdate was dried at 120 °C and roasted in air at 500 °C to obtain MoO_3 powder.

Hydrogen Reduction of Recycled MoO_3

Two stages of hydrogen reduction of MoO_3 to Mo were carried out. The first stage reduction of MoO_3 to MoO_2 was carried out at 600°C in hydrogen atmosphere for 2 h. The second stage reduction of MoO_2 to Mo was carried out at 930–1000 °C for 8 to 14 h depending on Mo powder sizes to be obtained. The molybdenum products obtained were analyzed by X-ray diffraction (XRD) analysis for the phase identification, scanning electron microscopy(SEM) for the particle sizes and morphologies. Thermogravimetric analysis was also carried out to check the condition of roasting of ammonium molybdate and hydrogen reduction of MoO_3 to produce Mo powder.

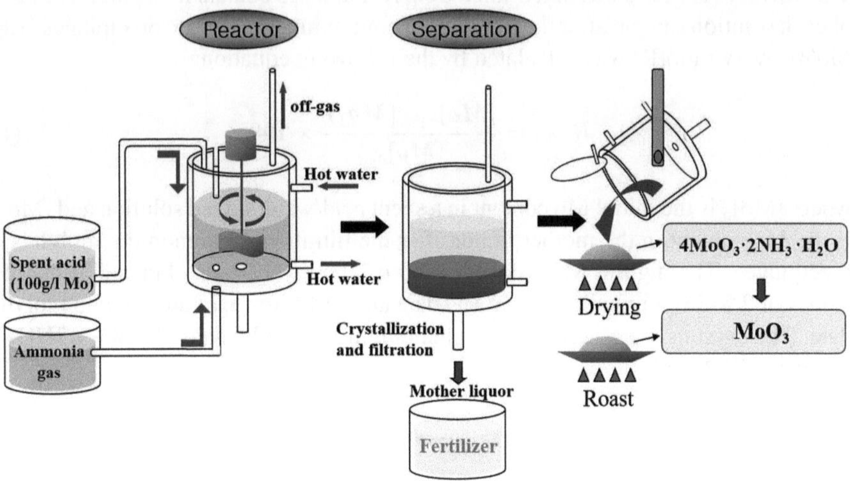

Fig. 2 Mo recovery process from spent acid

Synthesis of Fe–Mo Alloys

Mill scale samples with 74.1% total iron, 0.057%TiO_2, 0.957%MgO, 0.6%SiO_2, 0.18%CaO, 0.357%P_2O_5, 0.873%Al_2O_3, 0.42%MnO, 0.047%S, 0.246%Cr_2O_3, and 0.028%C were first ground to a very fine powder and then thoroughly mixed in 1:1 molar ratio with prepared MoO_3 using agate mortar and dry ball milling for 6 h. The well-mixed oxides powder was dried at 105 °C for 24 h. Equal weights of dried powder were compressed into compacts of about 1.5 g weight, 7 mm diameter and 3 mm thickness. The prepared compacts were reduced in continued two stages of hydrogen reduction. The first stage reduction was carried out at 600 °C for 15 min to convert MoO_3 to MoO_2. The second stage reduction was carried out at 700, 800, 900, 1000, and 1100 °C in hydrogen gas atmosphere.

In each experiment, after the furnace was heated up to the required temperature the mixed oxides compact was put inside a basket to be hanged in the balance and adjusted in the middle zone of the tube furnace in flow of purified Ar gas. Then, the gas was switched to hydrogen. The oxides compact was reduced with hydrogen gas at 600 °C for 15 min, and the temperature was increased to the required value (700, 800, 900, 1000, or 1100 °C). The reacted compact was kept in the reducing atmosphere till a constant weight was achieved.

Results and Discussion

Ammonia Neutralization and Precipitation

The Mo recovery ratio was determined from residual Mo content in the mother liquor of each solution sample after the filtration of ammonium molybdate precipitates. The Mo recovery ratio(R) was calculated by the following equation:

$$\boldsymbol{R}(\%) = \frac{[\boldsymbol{Mo}]_o - [\boldsymbol{Mo}]_f}{[\boldsymbol{Mo}]_o} \times \mathbf{100} \tag{1}$$

where $[Mo]_o$ is the initial Mo content in a spent acid/water mixed solution and $[Mo]_f$ is the Mo content in the mother liquor after the filtration of ammonium molybdate precipitates. The highest Mo recovery ratio of above 99% was obtained at the pH values of 2.5–3.0. Figure 3 shows XRD data and SEM image of ammonium molybdate. The crystals were identified as ammonium tetramolybdate ($4MoO_3 \cdot 2NH_3 \cdot H_2O$) by XRD analysis.

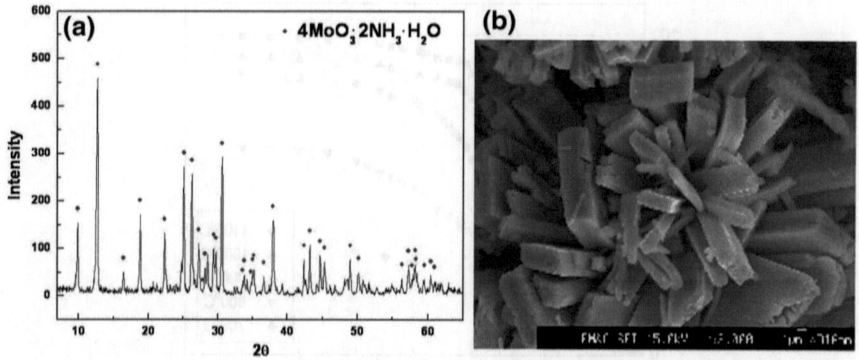

Fig. 3 Identification of ammonium molybdate: **a** XRD data and **b** SEM image

Hydrogen Reduction of Recycled MoO_3

Ammonium tetramolybdate decomposed NH_3 and H_2O to form MoO_3 at above 450 °C. The powder obtained after roasting was confirmed as MoO_3 by XRD analysis. In the commercial process for the reduction of MoO_3 to Mo, two-stage reduction method of MoO_3 to MoO_2 and MoO_2 to Mo is generally used because MoO_3 shows a relatively high vapor pressure at a temperature above 650 °C. The temperature for second stage reduction was varied from 930 to 1000 °C depending on the Mo powder size to be obtained. As the reduction temperature decreased, the size of reduced Mo powder decreased. However, at a lower reduction temperature, a longer reduction time was required for the completion of Mo reduction. Figure 4 shows XRD data and SEM image for the morphology of Mo powder obtained by hydrogen reduction at 950 °C. The reduced Mo grains had the spherical shape and uniform particle size range under 5 μm.

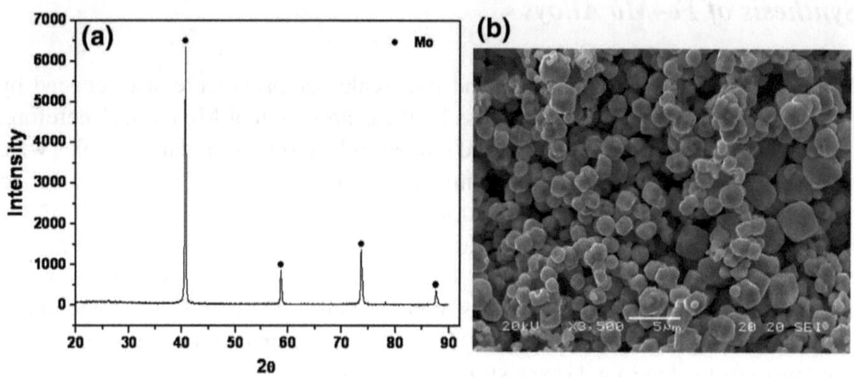

Fig. 4 **a** XRD and **b** SEM image of Mo powder obtained by H_2 reduction of MoO_2

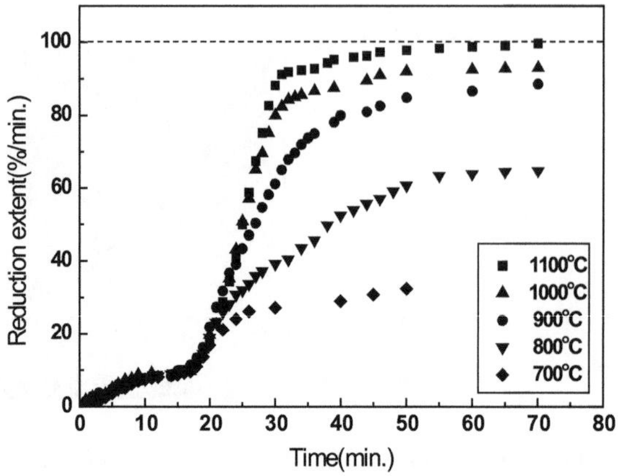

Fig. 5 Reduction curves of MoO_3/mill scale mixture in the temperature range from 700 to 1100 °C

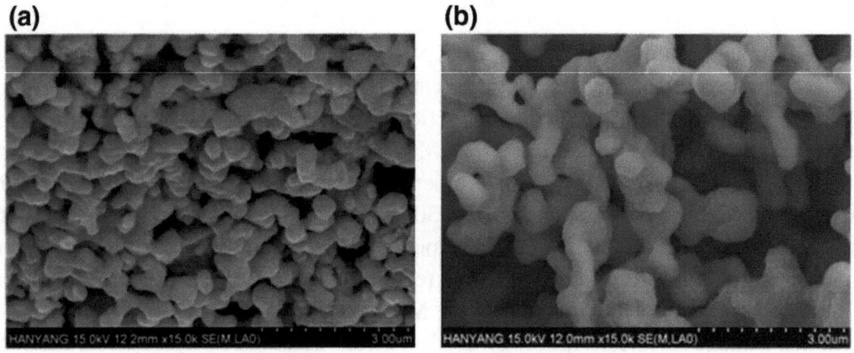

Fig. 6 SEM images of synthesized Fe–Mo alloys at **a** 900 and **b** 1100 °C

Synthesis of Fe–Mo Alloys

The mixed oxides of recycled Mo and mill scale compacts were also reduced by the two-stage method to avoid Mo loss by the evaporation of Mo oxide. Therefore, the reduction experiments were carried out partially at 600 °C and the samples were isothermally reduced at the various temperatures from 700 to 1100 °C. The reduction curves are given in Fig. 5. Each reduction curve after partial reduction at 600 °C, the rate of reduction increased as the reduction temperature increased. An incomplete reduction was detected from 700 to 1000 °C while the complete reduction was achieved at 1100 °C with the synthesis of Fe–Mo alloy. Figure 6 shows SEM image of reduced compacts at 900 and 1100 °C. It is clearly seen that the grains of Fe–Mo alloy were formed in round shape structure with size about 500 nm that are partially agglomerated all over the sample.

The activation energy of the reduction was calculated at overall reduction stage to illustrate the rate controlling mechanism depending on the Arrhenius equation. To confirm the reduction mechanism concluded from apparent activation energy values, different mathematical models for heterogeneous gas–solid reactions derived by Szekely et al. [3] were applied. The following equations are the mathematical formulation for gaseous diffusion, interfacial chemical reaction, and mixed control reaction, respectively.

$$t* = X + (1 - X)\ln(1 - X) \tag{2}$$

$$t* = 1 - (1 - X)^{1/2} \tag{3}$$

$$t* = \left[1 - (1 - X)^{1/2}\right] + \sigma^2[X + (1 - X)\ln(1 - X)] \tag{4}$$

where t* is dimensionless time, X is the fractional reduction degree at a given reduction time, and σ^2 is shrinking core reduction modulus. The present experimental results show a straight line on the application of Eq. (4) at the overall reduction stage. This indicates that the reaction is controlled by the combined gas diffusion and interfacial chemical reaction mechanisms.

Conclusions

In the present study, the molybdenum recovery process using ammonia gas neutralization of spent acid was tested. The ammonium molybdate obtained from spent acid was roasted and reduced by hydrogen to produce molybdenum metal powder of spherical shape and uniform sizes under 5 μm. And the mixed MoO_3 with mill scale was reduced in hydrogen atmosphere in the temperature range from 700 to 1100 °C. The apparent activation energy was calculated using the Arrhenius equation and indicated that the reduction rate of the mixture was controlled by combined interfacial chemical reaction and gas diffusion mechanisms. The concluded mechanisms were confirmed by applying heterogeneous gas–solid reaction model.

References

1. Kulkarni AD (1976) Molybdenum recovery from spent acid solution. U.S. Patent 3,963,823, 15 June 1976
2. Heytmeijer HR (1981) Rapid and efficient recovery of molybdenum from Spent Mandrel Dissolving Acid Solution. U.S. Patent 4,307,065, 22 December 1981
3. Szekely J, Evans J, Sohn HY (1976) Gas solid reactions. Academic Press, New York

Alkali Elution of Various Mineralogical Phases in Steelmaking Slag

Zuoqiao Zhu, Xu Gao, Shigeru Ueda and Shin-ya Kitamura

Abstract To suppress alkaline elution from steelmaking slag, the dissolution behaviors of mineralogical phases in slag need to be studied. In this paper, primary crystal phases ($CaO{\cdot}SiO_2$, $CaO{\cdot}SiO_2{\cdot}FeO$, and $CaO{\cdot}SiO_2{\cdot}2MgO$) of $CaO–SiO_2–FeOx$ and $CaO–SiO_2–MgO$ systems were synthesized, and their dissolution behaviors and the change in pH were investigated. The results showed that $CaO{\cdot}SiO_2$ caused the increase in pH by the dissolution of alkali. Therefore, the alkali elution from steelmaking slag is caused by not only free CaO and $2CaO{\cdot}SiO_2$ but also $CaO{\cdot}SiO_2$. Then, slags corresponding to actual steelmaking composition in the system $CaO–SiO_2–FeOx$ and $CaO–SiO_2–MgO$ systems were synthesized and their dissolution behaviors were observed. The result showed that the pH values after the dissolution of every slag were lower than that of $CaCO_3$, even though water soluble phases were contained in slag. Therefore, alkali elution from slag containing soluble phases can be suppressed when the mass fractions of these phases in slag are low enough.

Keywords Steelmaking slag · Mineralogical phase · Alkaline elution · Primary crystal phase · pH

Introduction

As a by-product of steelmaking process, steelmaking slag has a huge annual production and is often utilized as a raw material for road construction and civil works. According to statistics, in Japan, about 14 million tons of steelmaking slag is generated every year, and more than 60% is used for the above two purposes [1]. Due to the high basicity (CaO/SiO_2) of steelmaking slag, the alkaline elution, which leads to increase in pH, causes environmental problem when steelmaking slag is used near

Z. Zhu (✉)
Graduate School of Engineering, Tohoku University, Sendai, Japan
e-mail: zhu.zuoqiao.t3@dc.tohoku.ac.jp

X. Gao · S. Ueda · S. Kitamura
Institute of Multidisciplinary Research for Advanced Materials, Tohoku University, Sendai, Japan

G. Gaustad et al. (eds.), *REWAS 2019*, The Minerals, Metals & Materials Series,
https://doi.org/10.1007/978-3-030-10386-6_25

coastal area or river side. About the slag treatment, the slag aging process can suppress the volume expansion of slag [2], but cannot prevent the alkali elution, as this treatment change free CaO to $Ca(OH)_2$. The carbonation of slag, which changes free CaO to $CaCO_3$, have been conducted [3, 4], but the industrial operation is still limited. Hot stage modification of slag is another method to suppress alkali elution and volume expansion [5]. In this process, silica sand is added to the molten slag through oxygen injection to lower basicity. The heat, generated by the oxidation reaction of metallic Fe and FeO, has a role to keep slag at molten state and the dissolution of silica is achieved. This process has been industrialized [6] and operating in some steelmaking shops [7, 8].

In our previous study, mineralogical phases that been identified in industrial steelmaking slag were synthesized and their dissolution behaviors and the change in pH were investigated [9]. As a result, in addition to free CaO, the solid solution of $2CaO{\cdot}SiO_2$ and $3CaO{\cdot}P_2O_5$ (C_2S–C_3P) caused the increase in pH by the alkali elution. This meant the amount of both free CaO and C_2S–C_3P should be restrained. After that, a slag of CaO–SiO_2–Fe_2O_3 system with low basicity (CaO/SiO_2) was synthesized and its dissolution behavior was investigated. Although this synthetic slag contained neither free CaO nor $2CaO{\cdot}SiO_2$, the solid solution of $CaO{\cdot}SiO_2$, which formed during slag solidification, dissolved in water and increased pH. This indicated that, when the slag composition is modified, the newly formed primary crystal phases should also be paid attention. In other words, to properly modify the slag composition, the primary crystal phases which may cause alkali elution should be specified in advance.

Therefore, in this study, first, several primary crystal phases that form in the slag with low basicity of the CaO–SiO_2–FeOx (C–S–F) [10] and the CaO–SiO_2–MgO (C–S–M) [10] systems were synthesized, and their dissolution behaviors and the change in pH were studied. Based on the above results, several synthetic slags of CaO–SiO_2–FeOx and CaO–SiO_2–MgO systems which consisted of the above investigated primary crystal phases were prepared, and their dissolution behaviors were observed and discussed.

Dissolution Behavior of Primary Crystal Phases in CaO–SiO_2–FeOx and CaO–SiO_2–MgO Systems

Experiment Method

Crystal phases of $CaO{\cdot}SiO_2$ (CS), $CaO{\cdot}SiO_2{\cdot}FeO$ (CSF^{2+}), and $CaO{\cdot}MgO{\cdot}2SiO_2$ (CMS_2) were synthesized by sintering chemical reagents at high temperatures for 48 h. The sintering temperatures were determined to be lower than the solidus temperature of each phase based on phase diagrams. CSF^{2+} was sintered under the protection of argon gas in order to prevent the oxidation of FeO. After sintering, the samples were quickly removed from furnace and quenched. The initial composition

Table 1 Initial composition of synthetic mineralogical phases, mass%

	CaO	SiO_2	MgO	FeO
CS	48.3	51.7	–	–
CSF^{2+}	29.8	31.9	–	38.3
CMS_2	25.9	55.6	18.9	–

of CS, CSF^{2+}, CMS_2 are shown in Table 1. After synthesis, all mineral phases were identified by XRD, and their chemical compositions were confirmed using ICP-AES. Afterwards, each synthetic phase was crushed into fine powders with particle sizes smaller than 53 μm. To study the dissolution behavior of each phase, 1 g of the synthesized powder was added into 400 ml of deionized water and then stirring started. The temperature of the deionized water was previously adjusted to 298 K using a water bath. Right after the stirring started, measurements on both pH and ORP were conducted, and the solution samples were taken at certain time intervals until the pH values reached to a maximum. The sampled solution was analyzed by ICP-AES. For comparison, the dissolution behavior of $CaCO_3$ powder (<53 μm) which is the main component of natural stone was measured using the same method.

Results and Discussion

The measured maximum pH values and the amount of dissolved alkaline substances (Ca and Mg) after dissolving various primary crystal phases are summarized in Table 2. Comparing to the maximum pH of $CaCO_3$, CS showed a higher value and CMS showed a similar value. Moreover, CSF^{2+} showed lower value than $CaCO_3$. Corresponding to the results of maximum pH, the sum in the dissolved Ca and Mg from CS was larger than that of $CaCO_3$, and the CSF^{2+} and CMS_2 showed smaller values.

Based on these results, to suppress the alkali elution from steelmaking slag, after the hot stage modification process, the amount of CS in steelmaking slag should be lowered.

Table 2 Comparison on the maximum pH and the dissolved Ca and Mg between synthesized primary crystal phases and $CaCO_3$

	Maximum pH	Content of Ca in solution (mg/L)	Content of Mg in solution (mg/L)
CS	10.52	18.98	–
CSF^{2+}	8.13	2.15	–
CMS_2	9.22	2.08	0.62
$CaCO_3$	9.98	8.84	–

Dissolution Behavior of Synthetic Slag

Experimental Method

The initial composition of slag is shown in Table 3. To prepare these slags of CaO–SiO_2–FeOx system, chemical reagents were mixed and then melted at 1623 K using a Fe crucible under Ar atmosphere, and then the slag was slowly cooled inside furnace. To prepare the slag of CaO–SiO_2–MgO system, chemical reagents were mixed and melted at 1673 K using a Pt crucible in air, and then the slag was slowly cooled inside furnace. After synthesis, generated phases in each slag were identified by XRD and the slag composition was confirmed by chemical analysis using ICP-AES. Then, each synthetic slag was crushed into fine powders (<53 μm), and the same dissolution test, as described in Sect. 2.1, was adopted to study the dissolution behavior.

Results and Discussion

Figure 1 showed generated phases in synthesized slag observed by XRD. Slag A showed mineralogical phases of CS, 2CaO·SiO_2 (C_2S) and CSF^{2+}. Slag B showed mineralogical phases of C_2S, CSF^{2+} and FeO. Slag C showed mineralogical phases of CMS and MgO·SiO_2 (MS). Based on initial composition of slags, the composition of slag A was located in the composition triangle of CS–C_2S–CSF^{2+}, that of slag B was located in composition triangle of C_2S–CSF^{2+}–FeO, and that of slag C was located in boundary line of composition triangles of CMS_2–MS–SiO_2 and CMS_2–MS–M_2S. Therefore, these results showed that slags have been synthesized successfully.

The measured maximum pH values and the amount of dissolved alkaline substance (Ca and Mg) after the dissolving test of synthetic slag are summarized in Table 4. Comparing to the maximum pH of $CaCO_3$, all of the slag synthesized in this study showed lower value. In slag C, this result is reasonable because the maximum pH of all phase is lower than that of $CaCO_3$. For slag A and B, although they contained soluble phases, which showed higher values of pH according to the dissolution test of each phase, the maximum pH values were still lower than that of $CaCO_3$.

Table 3 Initial composition of synthetic slag and mass fraction of precipitated crystal phase (mass%)

	CaO	SiO_2	FeO	MgO	Precipitated phases
A(C-S-F^{2+})	33.6	34.8	31.6	–	CS(8.5), C_2S(2.1), CSF^{2+}(89.6)
B(C-S-F^{2+})	29.0	31.0	40.0	–	C_2S(3.3), CSF^{2+}(91.7), FeO(5)
C(C-S-M)	20.0	56.3	–	23.7	CMS_2 (69), MS (31)

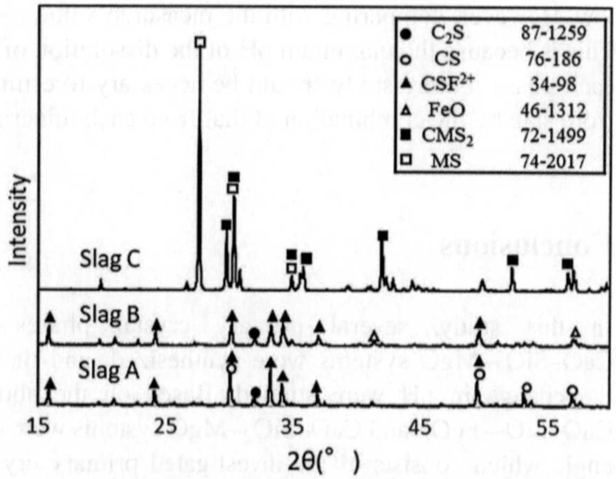

Fig. 1 XRD results of synthesized slag

Table 4 Comparison of maximum pH and dissolved Ca and Mg between synthesized slag and $CaCO_3$

	Maximum pH	Content of Ca in solution (mg/L)	Content of Mg in solution (mg/L)
A(C–S–F)	8.78	0.95	–
B(C–S–F)	7.32	0.93	–
C(C–S–M)	9.66	0.96	0.75
$CaCO_3$	9.98	8.84	–

Table 5 Measured and estimated concentration of dissolved Ca from slag (mg/L)

	Estimated value	Observed value
Slag A	3.76	0.95
Slag B	3.38	0.93

The dissolved of Ca from slag A and B was estimated by Eq. (1) and shown in Table 5 with measured values. In this equation, it was assumed that the dissolution of Ca from each mineralogical phase took place independently.

$$D_{Ca} = m_S \times V^{C_2S} \times C_{C_2S} + m_S \times V^{CSF^{2+}} \times C_{CSF^{2+}} + m_S \times V^{CS} \times C_{CS} \quad (1)$$

Here, D_{Ca} is the concentration of dissolved Ca (mg/L), m_S is the mass of slag (g), V is the mass fractions of phases in slag, and C is the concentration of dissolved Ca of each phase observed by the dissolution test using 1 g of each phase at its maximum pH, as shown in Table 2 ((mg/L)/g). The concentration of dissolved Ca of C_2S was measured in our previous study [9]. Estimated concentration of dissolved Ca in slag A was 3.76 ppm and that of value in slag B was 3.38 ppm. These values are lower than that of $CaCO_3$. Therefore, alkali elution from the slag containing water solvable phases can be suppressed when the mass fractions of these phases in slag are enough low. However, comparing with the measured values, estimated values were larger. This is because the maximum pH of the dissolution of slag is different from that of each phase. Future study should be necessary to estimate the dissolution behavior from slag by the combination of that from each mineralogical phase.

Conclusions

In this study, several primary crystal phases of $CaO–SiO_2–FeOx$ and $CaO–SiO_2–MgO$ systems were synthesized, and their dissolution behaviors and the change in pH were studied. Based on the above results, several slags in $CaO–SiO_2–FeOx$ and $CaO–SiO_2–MgO$ systems were chosen from composition triangle which consists of the investigated primary crystal phases and synthesized. Their dissolution behaviors were observed and discussed. Results were shown as follows:

1. For the *primary crystal phases* synthesized in this study, $CaO{\cdot}SiO_2$ showed a higher maximum pH than that of $CaCO_3$ (main component of natural stone), but other phases showed lower value. This result indicates that the alkali elution from steelmaking slag is caused not only by free CaO and $2CaO{\cdot}SiO_2$, but also by $CaO{\cdot}SiO_2$.
2. All *synthesized slags* showed lower values of maximum pH than that of $CaCO_3$, even though some of them contained soluble phases. Therefore, alkali elution from slag containing soluble phases can be suppressed when the mass fractions of these phases in slag are low enough.

References

1. Nippon Slag Association: Iron and Steel Slag Statistics. http://www.slg.jp/statistics/jyukyu/index_nendo.html. Accessed 23 June 2018
2. Sasaki T, Hamazaki T (2015) Nippon Steel Sumotomo Metal Technical Report 109:23–26
3. Tsutsumi N, Mukawa S, Tasaki T, Kameyama E, Amada K, Kudo K, Imado H, Miki O (2015) Nippon Steel & Sumitomo Metal Technical Report. 109:90–95
4. Ko MS, Chen YL, Jiang JH (2015) Constr Build Mater 98:286–293
5. Wada K, Yamaguchi K, Fukuda M, Takahashi H, Naito S, Murakami Y (1980) Seitetsu Kenkyu 301:13403–13414
6. Kuhn M, Drissen P, Geiseler J Schrey H (1997) Proceedings of 2nd European oxygen steelmaking congress, AIM, Italy, pp 445–453
7. Kuhn M, Drissen P, Schrey H (2000) Proceedings of 2nd European slag conference, Düsseldorf, Germany, pp 123–135
8. Tseng Y, Lee Y, Sheu B Proceedings of 8th European slag conference, Linz, Austria, USB
9. Zhu et al (2017) CAMP-ISIJ 30(2017), p 755
10. Verein Deutscher Eisenhuttenleute: Slag Atlas, 2nd ed., Verlag Stahleisen GmbH, Dusseldorf, Germany, (1995), 126, p 134

Feasibility Assessment for Recycling Copper Slag as Ferrous By-Products in FINEX®: An Alternative Ironmaking Process

Moo Eob Choi and Taehyeok Kim

Abstract The oversupply of steel product has led the global steel industry to a significant decline in revenue and profit. To make things harder, increasing demand to reduce greenhouse gas emissions has intensified pressure on steelmakers. As an effort to make progress from economical and ecological perspectives, the ironmaking field is expanding the recycling of ferrous by-products. Copper slag has been widely recycled in the abrasive media, the pavement, and the cement industries. However, there was no successful activity of recycling copper slag in the steel industry. The FINEX process is known for the flexibility of recycling ferrous by-products. There are chances that the ferrous by-products that cannot be recycled in the blast furnace due to inferior mechanical, thermal, and chemical properties can be recycled in the FINEX process. In the present study, the feasibility assessment of recycling copper slag in the FINEX process has been performed on an industrial pilot-scale.

Keywords Copper slag · FINEX® · Ironmaking · Melter-Gasifier · Recycling

Introduction

POSCO launched in 1992 the FINEX® development program in cooperation with Primetals Technologies (formerly VAI) in order to create an alternative ironmaking process to compete with the blast furnace process and operating on low-grade, low-cost raw materials [1–3]. As indicated in Fig. 1, POSCO has been successful in commercialization of the process through a series of scale-up development from a 15 t/day model plant started in 1996 to a 150 t/day pilot plant in 1999 followed by a 0.6 MTPA FINEX® demonstration plant in 2003. Upon successful demonstration of the process, the first commercial FINEX® plant with a capacity of 1.5 MTPA was constructed at Pohang Works and commenced operations in April 2007. Based on the successful results of the first commercial plant, the second commercial plant with a capacity of 2.0 MTPA was built at Pohang Works and the plant has been put into

M. E. Choi (✉) · T. Kim
POSCO, Pohang 37859, South Korea
e-mail: pneuma@posco.com

G. Gaustad et al. (eds.), *REWAS 2019*, The Minerals, Metals & Materials Series,
https://doi.org/10.1007/978-3-030-10386-6_26

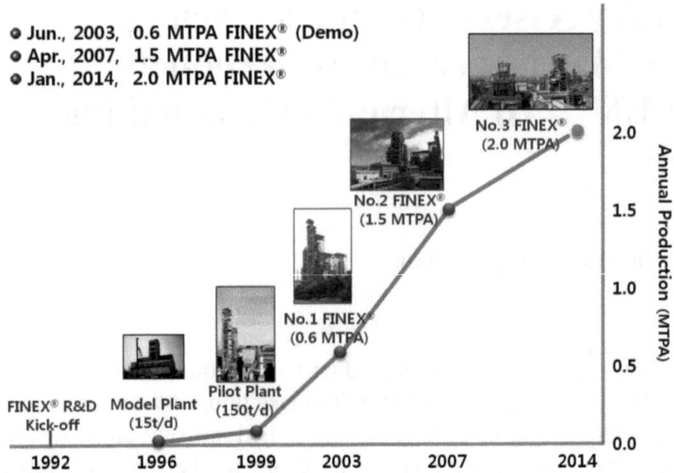

Fig. 1 Milestones of FINEX® process development

operation since January 2014. Up to now, it has shown noticeable operational results in the aspect of performance, raw material, environment, and energy. They indicate that the process is not completed but can be continuously evolved to be suited to various local conditions. POSCO is poised to make a significant contribution to the global steel industry with an innovative ironmaking technology, FINEX®.

The FINEX® process consists of a cascade of fluidized-bed reactors, a coal briquetter, a hot DRI compactor, and a melter-gasifier. Figure 2 shows the process flow of FINEX®. Ore fines are charged into the fluidized-bed reactors together with additives such as limestone or dolomite. As the solid materials pass through the reactors, they are heated and reduced/calcined. The partially reduced iron ore (DRI) is compacted at an elevated temperature and then the hot compacted iron (HCI) is charged into the melter-gasifier. Either the briquetter or the pulverized coal injection (PCI) facilities process coal fines. Coal briquettes are put into the dome of the melter-gasifier, while pulverized coal is injected through the tuyeres. The reducing gas generated by coal combustion with pure oxygen in the melter-gasifier is transported to the fluidized-bed reactors to reduce the iron ore. A portion of outlet gas of the reactors goes through CO_2 removal utilities for recycling. It leads to higher gas utilization efficiency.

Feasibility Assessment for Recycling Copper Slag in FINEX®

Using ferrous by-products or ferrous waste is very attractive way to *reduce production cost* in ironmaking process. FINEX process also strives to reduce production cost compared to that of conventional ironmaking processes by feeding cheap ferrous by-products or ferrous wastes instead of iron ores. Figure 2 shows detail routes to

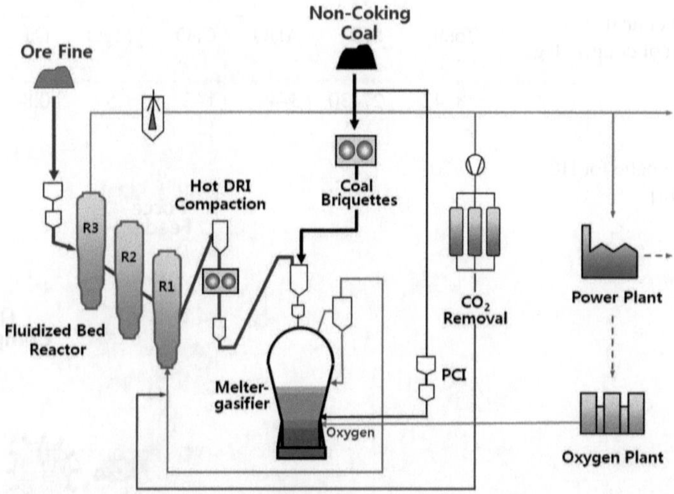

Fig. 2 Schematic diagram of the FINEX® Process

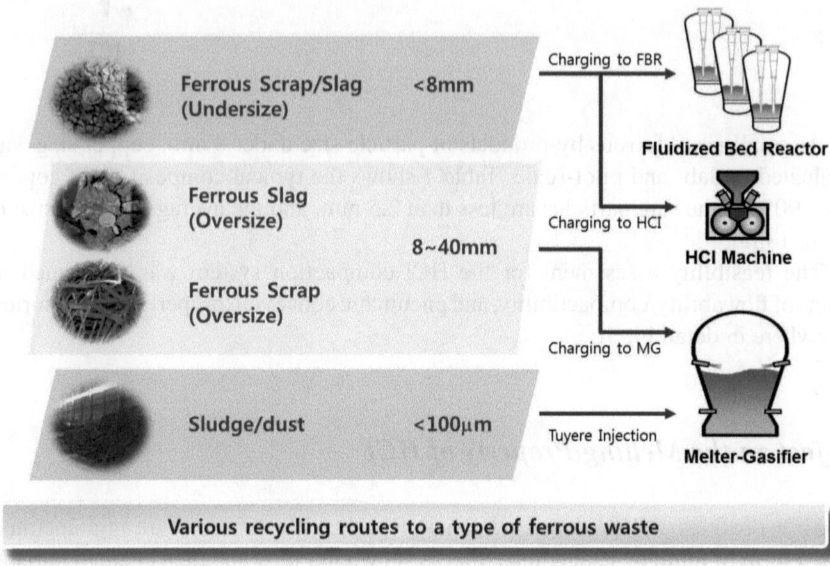

Fig. 3 Current status of ferrous byproduct/waste recycling in FINEX

recycle ferrous by-products in FINEX. Large-sized ferrous by-products of particle size over 8 mm but less than 40 mm are directly charged to melter-gasifier. Relatively small-sized ferrous by-products of particle size under 8 mm are charged to fluidized bed or to HCI process via additional recycling facility. Sludge and dust are also recycled mixing it with PCI (Fig. 3).

Table 1 Chemical composition of copper slag (unit in %)

Total Fe	SiO_2	Al_2O_3	CaO	MgO	Cu	Zn
38–42	27–30	3–4	1–2	1.5	0.8–1	1–1.5

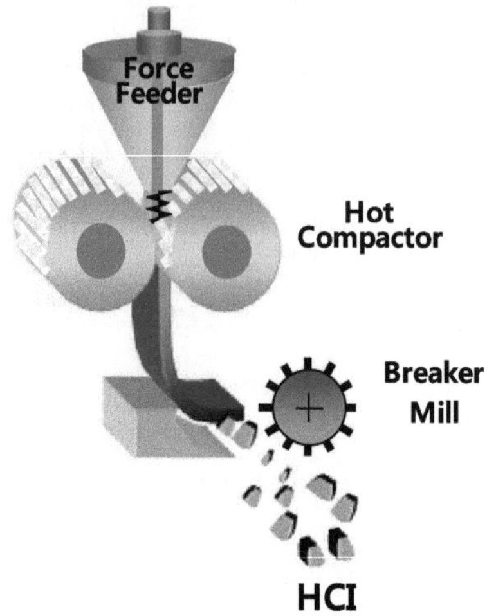

Fig. 4 Schematic for HCI compactor [4]

As small-sized ferrous by-products of particle size under 8 mm, copper slag was evaluated on lab- and pilot-scale. Table 1 shows the typical composition of copper slag. 90% of the slag particles are less than 2.5 mm, and the average particle size is about 1 mm.

The feasibility assessment for the HCI compaction system was performed in terms of flowability, compactibility, and pneumatic conveying properties and reported elsewhere in detail [4, 5].

Effect on the Melting Property of HCI

The iron ore fines (DRI) reduced in a series of fluidized-bed reactors are agglomerated into a form of lump by HCI compactors as shown in Fig. 4 and then charged into the melter-gasifier for smelting. HCI consists of 60–70% Fe, 5–8% gangue materials, 8–11% flux, and the rest is oxygen attached to Fe in the form of iron oxide. The flux material is mainly the mixture of limestone and dolomite.

The melting property of HCI is of great concern. At around 1300 °C, the flux and gangue materials start to react and produce molten species while leaving pores. Flux having relatively large size consumes all of gangue materials around and then starts to form dicalcium silicate (C2S) layer on the boundary because locally the ratio of CaO over SiO_2 is high. C2S has a high melting point of 2,130 °C. At 1400–1500 °C, the molten species moves through the pores inside HCI and makes contact with C2S

Fig. 5 Optical micrograph of copper slag added HCI

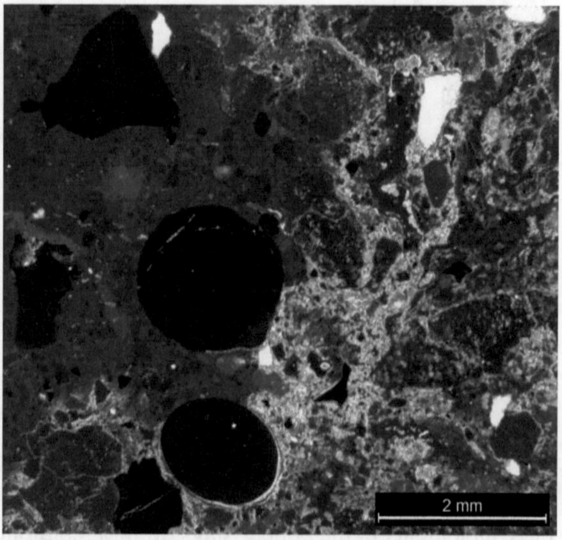

layer on the surface of large-sized flux material and start to melt the C2S layer first and the flux next. Molten slag is formed while all of the flux and gangue materials react with each other. At the same time, carburization and melting of Fe component occur. When everything is molten, metal slag separation and dripping occurs. At the bottom of the melter-gasifier, two distinct layers of metal and slag are formed and tapped. When HCI was compacted by adding copper slag shown as spherical dark particles in Fig. 5, the melting property of HCI changes. Thus, it is important to set up appropriate criteria in the selection of recycling ferrous by-products. First, the melting point of the recycling material itself should be lower than the constituents of HCI. Second, when it mixed with HCI, it should not increase the melting point of HCI.

A series of lab-scale tests were carried out using heating microscope to evaluate the melting property of copper slag and compare it with other ferrous by-products. Copper slag containing fayalite (m.p. 1200 °C) represents excellent melting property itself (Fig. 6).

In situ X-ray observation during HCI melting represented that copper slag decreased the melting point of the HCI and microscopic analysis indicated that fayalite facilitates FeO-based slag, which leads to the accelerated melting of C2S and flux.

Effect on the Composition of Molten Iron

In addition to the major component, copper slag contains impurities such as Cu, S, Zn, Cr, V, and P. These impurities affect the composition of molten iron and

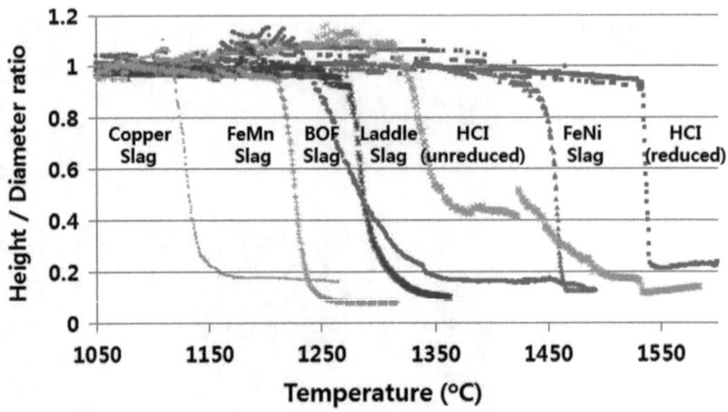

Fig. 6 Heating microscope results of various ferrous by-products

melter-gasifier operation. According to the industrial-scale pilot test results, most of the Cu elements in the copper slags go into the molten iron. Evaporation of Cu by the formation of CuS(g) was suggested under reducing atmosphere in the melter-gasifier [6]. However, it seems negligible to form CuS(g) and evaporate in the real operation. Considering the Cu limit required by downstream process to produce steel product, the maximum amount of copper slag input was limited up to 3%. For some specialty product, more strict requirement was imposed and thus the input amount was allowable up to 1%. When per 1% copper slag input, the S amount in the molten iron was increased by 0.001% and thus it was verified that copper slag can be fed up to 3% in this regards. The typical Zn input limit for the blast furnace process is from 500 to 800 g per ton of pig iron. Implying the same criterion, the plausible input amount of copper slag to melter-gasifier ranges from 2 to 4%. Other impurities like Cr, V, and P did not have any significant impact on the composition of molten iron.

Commercial-Scale Facility for Recycling Copper Slag

Having considered the physicochemical properties of copper slag itself and HCI containing copper slag, it was determined to inject copper slag directly to the HCI compaction process. Figure shows two possible options for the commercial-scale facility. The first option is to feed copper slag to "DRI Feed bin" which is a buffer hopper receiving DRI from the final stage of fluidized-bed reactors and distributing it to each of HCI machines. The second option is to feed copper slag directly to individual HCI machine having an intermediate hopper in-between. Although the second option gives us more flexibility to control the input amount of copper slag to each of HCI machines, the intermediate receiving and feeding system was not reliable and durable. Thus, the commercial-scale facility was constructed taking the first option and long-term usage test is currently in progress (Fig. 7).

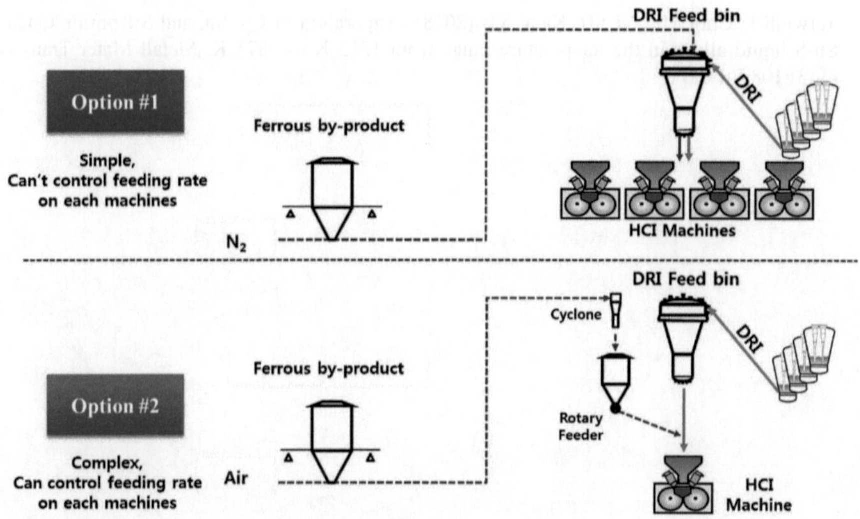

Fig. 7 Schematic of candidates of recycling process [4]

Conclusion

Feasibility assessment of charging copper slag in the FINEX process was performed on a lab- and pilot-scale. The evaluation was performed for each unit process such as fluidized-bed reactors, HCI compaction system, and melter-gasifier. For the industrial pilot-scale test, copper slag was directly charged to the HCI process via additional recycling facility. The mixing of copper slag was improved the melting property of HCI. The input amount of copper slag was suggested from 2 to 4% in consideration of the effect of impurities on the composition of molten iron.

References

1. Lee HG, Yi SH, Kang CO (2012) Recent development of FINEX process. In: ICSTI proceedings
2. Yi S-H, Choi M-E (2014) Recent update of BF & FINEX route ironmaking technologies in Korea. In: JSPS International Symposium Proceedings
3. Yi S-H, Lee H-G (2015) Ironmaking in South Korea—Beyond the Blast Furnace, AISTech Proceedings (2015)
4. Yoon S-K, Kim H-S, Choi M-E, Kim S-Y (2017) Development of ferrous by-product recycling process in FINEX®. In: 35th conference proceedings on Institute for Briquetting and Agglomeration (IBA)
5. Kim H-S, Yoon S-K, Cho MY (2015) Influence of composition of partially reduced iron ores on consolidation and strength of hot compacted iron. In: 34th conference proceedings of Institute for Briquetting and Agglomeration (IBA)

6. Tafwidli F, Choi ME, Yi SH, Kang YB (2018) Evaporation of Cu, Sn, and S from Fe-C-Cu-Sn-S liquid alloys in the temperature range from 1513 K to 1873 K. Metall Mater Trans B 49(3):1089–1100

Development of Electromagnetic Interference Materials from Metallurgical Wastes

Yong Fan

Abstract In view of developing a sustainable metallurgical production, the potential of its slag and/or dusts as a secondary resource must be considered to avoid landfilling. Many products have been innovated such as cement, concrete, ceramics tiles, land reclaim, etc. High iron content industrial solid wastes from metallurgical process may be utilized in the making of electromagnetic shielding and absorbing materials due to the metal droplets and ferrite it contains. This article provides a proof-of-principle of a new high-value-added application for the high iron content industrial solid wastes as an admixture for enhancing electromagnetic interference (EMI) shielding with a potential for future improvement. These special materials are particularly needed for applications in underground vaults, electric power devices, telecommunication, anti-spying, etc.

Keywords Copper slag · Electromagnetic interference · Functional material

Introduction

Electromagnetic interference (EMI) shielding is an area gaining positive attention due to the increasing abundance of electronics. Accordingly, EMI shielding material plays a paramount role at the present age in electromagnetic pollution controlling, communication security, and the ground/underground substation containing central processing electronics, which are crucial to power transmission and telecommunication. Consequently, the development of EMI shielding materials has captured considerable attention [1]. Many researchers share a favorable attitude on cement-based EMI materials; these are not only a structural material but also have some

Y. Fan (✉)
Institut Für Eisen- Und Stahltechnologie, TU Bergakademie Freiberg,
09599 Freiberg, Sachsen, Germany
e-mail: Yong.Fan@extern.tu-freiberg.de

Y. Fan
Department of Materials Engineering, KU Leuven, Kasteelpark Arenberg 44,
bus 2450, B-3001 Heverlee, Belgium

G. Gaustad et al. (eds.), *REWAS 2019*, The Minerals, Metals & Materials Series,
https://doi.org/10.1007/978-3-030-10386-6_27

EMI shielding properties through introducing fillers like carbon, metal, and ferrite materials [2]. The primary mechanism of EMI shielding is usually reflection of the radiation through mobile charge carriers such as electrons, which interact with the electromagnetic fields. There is no doubt that high conductivity materials such as carbon or metal are suitable to be fillers. The second mechanism of EMI shielding is usually absorption of radiation through electric and/or magnetic inlets that interact with the electromagnetic fields [3].

During the whole copper smelting process, nearly 30 million tons of copper smelter slag is produced every year in the world [4]. This slag is outflowed for recycling, production of the high-value-added products, and for disposal in slag dumps or stockpiles. The recent overcapacity of the cement industry (dominating consumer of copper slag) in the developed markets has been causing increasing concern, thereby urgently necessitating other utilization of slag. There are some applications using copper slags such as concrete aggregates, abrasives, ceramic tiles, land reclamation, etc. [5]. In the present work, it has been attempted to seek an innovative application, i.e., electromagnetic interference shielding material, aiming at high-value-added product.

The development of cement matrix materials with EMI shielding effectiveness has been urgently required with the current unfavorable electromagnetic radiation environment. These emerging researches provide a new direction for the application of solid wastes and cement materials in electromagnetic field. This paper was implemented to utilize the copper slag as an admixture for the synthesis of cement-based EMI shielding composites.

Experimental

The copper slag obtained from a stainless steelmaker was used in the present experiments. The chemical composition of the dust was as follows: TFe = 41.5 mass%, SiO_2 = 32.1 mass%, CaO = 0.7 mass%, Al_2O_3 = 4.5 mass%, MgO = 0.5 mass%, Cr = 0.3 mass%, Zn = 6.8 mass%, and Cu = 0.2 mass%. The cement used was typical Portland cement. Because the present work is fundamental research, aggregate was not used, whether fine or coarse, to enhance the mechanical property.

A total mixture of 200 g of the cementitious materials were used to prepare each specimen in 5 mm thickness. After mixed and homogenized of cement and copper slag, water was added to achieve a moisture content of approximately 30 mass%. Shaping was performed with a Teflon mold with dimensions of 140 mm in side length and 5, 10, and 15 mm in depth. Then specimens were dried for 48 h, demolded, and cured finally at room temperature.

The EMI shielding effectiveness was measured using the coaxial cable method or the transmission line method in the present study. The measurement setup consists of two coaxial adapters, two 10 dB attenuators, and a network analyzer. The sample placed in the center of the tester with the input and output of the tester on the two sides of the sample.

Results and Discussion

The compositional analysis confirms the presence of iron, silicon, aluminum, and zinc as the significant metallic elements in the copper slag, and that of numerous other elements in lower concentrations, such as calcium, magnesium, copper, and chromium. The presence of heavy metals can pose threats to the environment and human health. Therefore, this waste is considered hazardous in many countries. The reusing and recycling of this waste should be carefully handled and studied. Although the environmental impacts are beyond the scope of the present study, future attention would have to be paid to these issues. The iron silicate was the main component of the water-granulated copper slag, which was the base of slag matrix. Iron is mostly crystallized as fayalite (Fe_2SiO_4) and magnetite (Fe_3O_4) during the matte smelting process or water cooling.

EMI shielding effectiveness (SE) of a material is measured in decibels (dB) and defined as the ratio of transmitted power to incident power. Total SE is the summation of shielding by absorption, reflection, and multiple reflections in the interior of the material. Coaxial transmission line equipment in the present study is performed in the frequency range of 500 MHz–1.5 GHz which corresponds to ranges for commercial applications, such as TV signal, FM radio, Mobile phone, wireless LAN, radar image, etc.

The measured shielding effectiveness of copper slag filled samples with 5 mm thickness is shown in Table 1. Generally, the shielding effectiveness of material is closely correlated with the electric conductivity and the electromagnetic parameters of its composite. The cement matrix is inert in conductivity, showing its relatively low shielding effectiveness. The copper slag filled sample with different weight mass percentages into the cement, namely, 15 mass%, 30 mass%, and 45 mass%, indicated an intensification of the shielding effectiveness. Notably, the addition of a 45 mass% of copper slag alone to the cement matrix boosts the shielding to approximately 7–8 dB as tested in the frequency of 0.6, 0.9, 1.1, and 1.4 GHz, and highlights the incident electromagnetic wave that has been reduced by approximately 60%. This phenomenon is attributed to the fayalite and magnetite embedded in the sample mixture serve as magnetic and dielectric loss absorbent, which derives from the copper slag.

Ferrites are the prevalent fillers used in cement-based composites. Reflection cannot lessen or depress radiation, and the reflected wave may interact with the incident wave, which might impact other units or devices. With the service of electromagnetic-absorbing materials transferring the energy to other forms, the EMI radiation could be vitiated to the furthest. The ferrites contained in the copper slag are exceptional choices of wave absorbing component for cement-based EMI shielding materials.

It is clear that with the increase of thickness, the SE level of the samples is reinforced. The thickness of EMI shielding material is a crucial economic factor to be considered in the future practical application. Copper slag with low-value application shows its competitive advantages in some cases, for instance, massive

Table 1 The shielding effectiveness of reference sample and dust filler sample

Fillers of cement-based materials	Thickness/mm	Shielding effectiveness/dB			
		600 (MHz)	900 (MHz)	1.1 (GHz)	1.4 (GHz)
Pure cement	5	3.7	4.2	4.1	4.3
15 mass% Slag	5	5.0	5.8	6.0	5.9
30 mass% Slag	5	5.7	6.5	6.7	6.7
45 mass% Slag	5	7.0	7.6	7.7	7.3
30 mass% Slag	10	9.2	9.7	9.6	10.8
30 mass% Slag	15	11.1	11.0	11.2	12.1

underground or remote information base construction. Moreover, the reutilization of copper slag has its social significance and benefit.

Copper slag is usually valorized by extracting metals or used to produce some construction materials (low valuable added). The present study has attempted to apply it to the cement matrix for obtaining EMI shielding function, which provides practical and high-value-added reutilization, furthermore, not only by adding the slag directly to cement, but also through metallurgical modification, such as carbothermal reduction, magnetization roast, or controlled molten oxidation.

Conclusions

The present study attempts to apply copper slag to the cement matrix for obtaining EMI shielding function, which provides a practical and high-value-added approach for its reutilization. Notably, the addition of a 45 mass% of copper slag to the cement matrix boosts the shielding value to approximately 7–8 dB as tested in the frequency of 0.6, 0.9, 1.1, and 1.4 GHz, and highlights the incident electromagnetic wave that has been reduced by approximately 60%. This phenomenon is attributed to the fayalite and magnetite deriving from copper slag embedded in the sample mixture, which serves as magnetic and dielectric loss absorbent. Moreover, with the increase of thickness, the SE level of sample is reinforced. Copper slag with low-value application shows its competitive advantages in some cases, for instance, massive underground or remote information base construction. Moreover, the reutilization of copper slag has its social significance and benefit.

Acknowledgements The author Yong Fan thanks the Japan Society for the Promotion of Science (JSPS) and Alexander von Humboldt (AvH) Foundation for their financial supports.

References

1. Tong XC (2009) Advanced materials and design for electromagnetic interference shielding. CRC Press, Boca Raton, pp 252–253
2. Khushnood RA et al (2015) Improvement in electromagnetic interference shielding effectiveness of cement composites using carbonaceous nano/micro inerts. Constr Build Mater 85:208–216
3. Fan Y et al (2017) Utilization of Stainless-steel furnace dust as an admixture for synthesis of cement-based electromagnetic interference shielding composites. Sci Rep 7:15368
4. Schilesinger ME (2011) Extractive metallurgy of copper. Elsevier B.V, Great Britain
5. Fan Y et al (2013) Review on iron recovery and cleaning of copper slag. J MMIJ 129:177–184

Part VI
REWAS 2019: Plenary Session

Recycling of Critical Metals

Toru H. Okabe and Takanari Ouchi

Abstract In modern high-tech products, rare metals play an increasingly pivotal role. To support the development of a highly sustainable society, where valuable natural resources are not wasted and most materials are recycled, new efficient, environment-friendly recycling technologies for rare metals are required. The authors have developed environment-friendly recycling technologies that efficiently extract rare metals and precious metals from scrap without generating harmful waste solutions (waste liquids) and exhaust gases. These technologies include (1) a technique for converting contaminated titanium scrap, which is expected to increase in the future, into high-quality titanium feed material; (2) a technology for extracting and separating rare metals, such as rhenium, directly from end-of-life turbine blades used in aircraft and power stations without generating any waste aqueous solutions; and (3) a method for efficiently concentrating and separating platinum group metals in automobile catalytic converters without using harmful acids or other chemicals. These technologies will help establish an environment-friendly rare metal recycling system.

Keywords Recycling · Critical metals · Rare metals · Titanium · Rhenium · Platinum group metals · Refining

Introduction

A wide variety of rare and precious metals are encountered in daily life. These metals and their alloys and compounds are used to produce industrial products such as smartphones, televisions, and hard drives (Fig. 1), which support the development and functioning of a rich, modern society. Indeed, modern lifestyles cannot be sustained without these metals. Moreover, rare metals and precious metals are also indispensable for energy generation and storage devices, and energy-saving products

T. H. Okabe (✉) · T. Ouchi
Integrated Research Center for Sustainable Energy and Materials, Institute of Industrial Science, The University of Tokyo, 4-6-1 Komaba, Meguro-Ku, Tokyo 153-8505, Japan
e-mail: okabe@iis.u-tokyo.ac.jp

G. Gaustad et al. (eds.), *REWAS 2019*, The Minerals, Metals & Materials Series,
https://doi.org/10.1007/978-3-030-10386-6_28

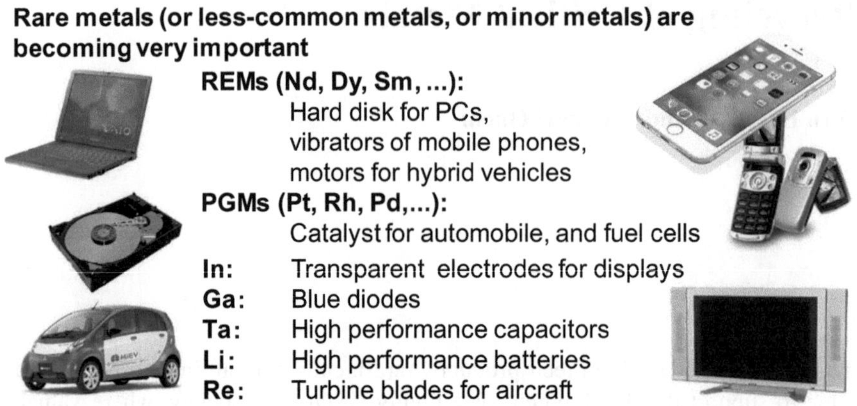

Fig. 1 Rare metals (or less-common metals or minor metals) are increasingly important materials supporting a high-quality lifestyle in advanced countries. These metals are essential for producing high-tech industrial products, which will become "scrap" after use

such as high-performance motors, batteries of hybrid and electric vehicles, and the panels and controllers used for photovoltaic power generation.

The demand for rare metals and precious metals is growing steadily owing to the worldwide improvement in industrial products, the economic growth of emerging countries, and the strengthening of environmental regulations. In this talk, we discuss current trends in the recycling of critical metals (or rare metals) and related processing technologies. Specifically, recent research on the recycling of titanium, rhenium, and some other precious metals will be introduced. The possibility of developing next-generation recycling technologies for critical metals will also be discussed from a multilateral perspective [1–3]. Furthermore, recent progress in the refining and recycling processes of titanium and other critical metals will be discussed.

Recycling of Titanium Scrap

There is an increasing demand for Ti and its alloys in various fields, particularly in the aerospace industry, because of their properties, such as high specific mechanical strength and corrosion resistance. During the fabrication of Ti and its alloys from their ingots, a large amount of scrap is generated. Currently, Ti scrap is used as an additive in steel production. Owing to the high production cost of the virgin metal (Ti sponge), it is desirable to recycle Ti scrap by remelting it with the virgin metal to produce primary ingots of Ti. However, oxygen accumulates in the Ti ingot when using the scrap as a feed material in the remelting process. The oxygen concentration in typical Ti scrap is 2000–4000 mass ppm O (0.2–0.4 mass pct O), which is higher than that in virgin metals (300–2000 mass ppm O) such as Ti sponges produced

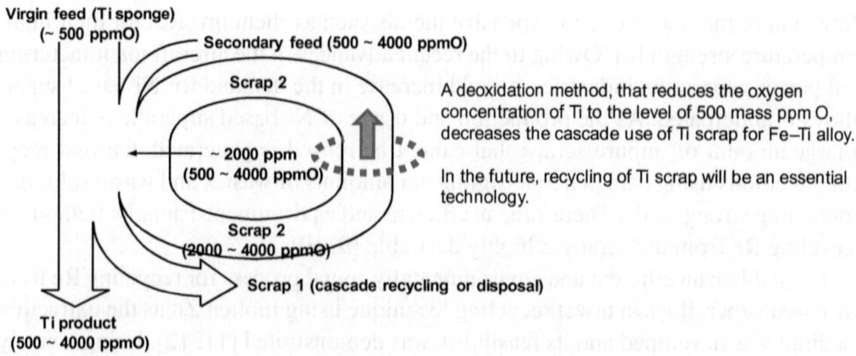

Fig. 2 Material flow of Ti and its alloys with respect to the oxygen concentration [6]

using the Kroll process (Fig. 2). Therefore, prior to recycling Ti scraps, it is critical that their oxygen content be reduced to the level of Ti sponge, ~500 ppm, via a deoxidation process. There are several deoxidation methods available for Ti. One of the most efficient processes to remove oxygen from Ti is electrochemical deoxidation [4]. The deoxidant material deposits on the surface of Ti scraps and reduces the oxygen in Ti into a molten salt containing an oxide ion (O^{2-}). The oxide ion is subsequently removed in the form of carbon oxides (CO_x) via an oxidation reaction on a carbon anode. The deoxidation limit of the electrochemical method is extremely low as O^{2-} does not accumulate in the molten salt. There are studies demonstrating an electrochemical deoxidation process in molten magnesium chloride ($MgCl_2$) at 1173 K [1, 5, 6]. This process yields oxygen concentrations lower than 100 ppm. Another noteworthy advantage of using $MgCl_2$ is that the Mg (deoxidant) and $MgCl_2$ attached onto the surface of Ti scraps can easily be removed by vacuum distillation owing to their high vapor pressures.

Recently, the possibility of direct removal of oxygen from Ti using Y and some light rare-earth metals (e.g., La, Ce, Pr, and Nd) as deoxidants was discussed from a thermodynamics perspective [2, 7]. The authors assessed the deoxidation limit of Ti using the equilibrium of M/M_2O_3 and $M/MOCl/MCl_3$ (M: Y, La, Ce, Pr, or Nd.). Furthermore, a deoxidation method of Ti that utilizes Mg as the deoxidant in the $MgCl_2$–YCl_3 flux was demonstrated; this decreases and maintains the activity of the oxide ion a_{O}^{2-} at a low level in the system via the formation of yttrium oxychloride (YOCl), which will be discussed in the presentation.

Recycling of Rhenium from Ni-Based Superalloys

Nickel (Ni)-based superalloys, which are mainly used in high-pressure turbine blades in aerospace and power generation industries, exhibit excellent corrosion resistance and mechanical strength over a wide range of temperatures. However, they require rare (scarce) and expensive metals such as rhenium (Re) for their high-temperature strength [8]. Owing to the recent advances in the aircraft manufacturing and power generation industries, a rapid increase in the demand for Ni-based superalloys is anticipated. As the production and usage of Ni-based superalloys increase, a large amount of impure scraps that cannot be reused is generated. Current recycling methods using acid generate significant amounts of wastes and waste solutions containing strong acids. Therefore, an efficient and environment-friendly method for recycling Re from the scraps is highly desirable [9, 10].

To establish an efficient and environmentally sound process for recycling Re from Ni-based superalloys, a novel recycling technique using molten Zn as the extraction medium was developed and its feasibility was demonstrated [11, 12]. Using a newly designed reaction chamber with circulating molten Zn, continuous Ni extraction and Re concentration directly from superalloys could be simultaneously accomplished. The conceptual schematic illustration of the apparatus for recycling Ni from superalloy scraps is shown in Fig. 3. It was also demonstrated that the Zn used in this process could be reused in the process by distilling it from the obtained Ni–Zn alloy.

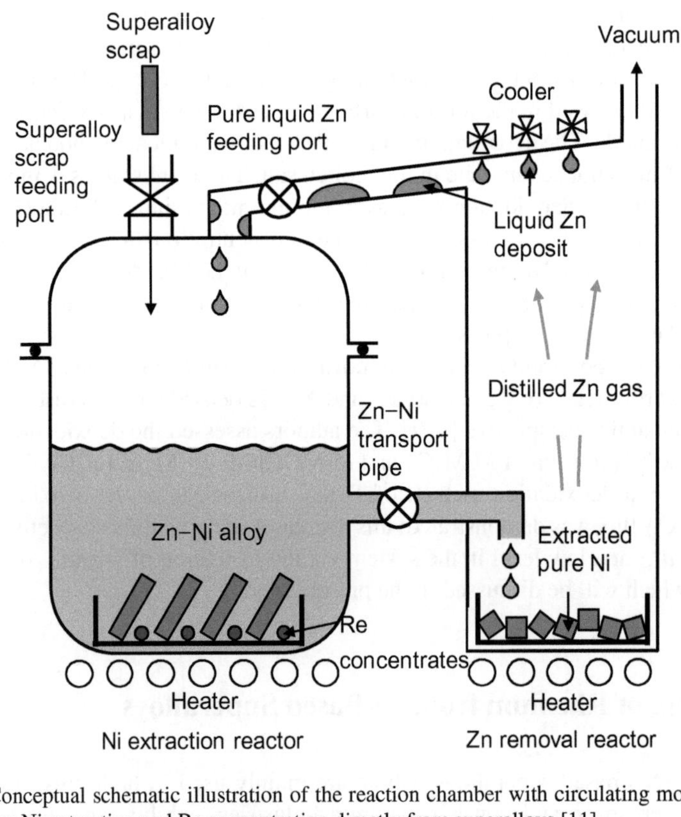

Fig. 3 Conceptual schematic illustration of the reaction chamber with circulating molten Zn for continuous Ni extraction and Re concentration directly from superalloys [11]

Therefore, the proposed technique is an environment-friendly recycling process to recover Re from superalloy scraps without the generation of any waste and waste solution.

Recycling of Platinum Group Metals

Platinum group metals (PGMs) are widely used in many applications such as for manufacturing automobile catalysts and electronic devices. In general, PGMs are recovered via pyrometallurgical and hydrometallurgical processes. In pyrometallurgical processes, PGMs are enriched using a collector metal such as copper or iron. The enriched PGMs are dissolved in acid and alkali solutions in hydrometallurgical processes, followed by separation using solvent extraction, ion exchange, etc. These processes are highly efficient and fast; however, they consume a considerable amount of energy and require large-scale equipment. Another method is a hydrometallurgical process that employs an acid solution with strong oxidants (e.g., aqua regia). Direct dissolution of PGMs from catalyst scrap can be carried out in small-scale plants with low energy consumption. However, the low recovery rate of direct dissolution processes and the generation of large volumes of toxic solutions in these methods are challenging.

There are several new processes for recycling PGMs. In one such process, it was realized that Pt dissolves in aqueous solutions without toxic chemicals (Fig. 4). The process involves a reaction between PGMs and Mg vapor to form PGM–Mg alloys and chlorination of the alloys using metal chlorides (e.g., copper chloride) [13]. The compounds containing platinum can be dissolved in a solution of hydrochloric acid without any oxidants, as well as in a simple NaCl solution. Another new process for recycling PGMs involves magnetically concentrating PGMs directly from spent automobile catalysts (Fig. 5). In this process, ferromagnetic Fe is deposited on the PGM particles via electroless plating or iron chloride vapor treatment through a disproportionation reaction (i.e., $FeCl_2$ (*g*) = Fe (*s*) + $FeCl_3$ (*g*)). These processes efficiently magnetize the PGMs, which can then be magnetically separated from nonmagnetic materials such as ceramics catalyst carriers [14]. The concentrated PGMs are highly valuable and have a significantly lower volume than untreated scrap. Therefore, they can be transported by air to refineries for further treatment. Refined PGMs can be obtained quickly using this new recycling method. These processes can be carried out in relatively small-scale plants with low energy consumption. Furthermore, these processes do not produce toxic waste solutions and are thus environmentally sound [15–19].

If time permits, possible applications of the developed recycling techniques, especially for titanium recycling, in practical industrial processes will also be discussed.

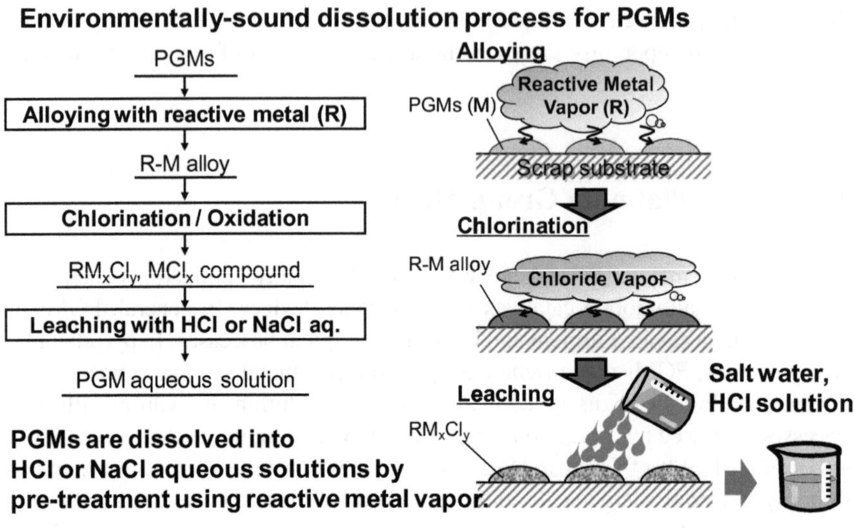

Fig. 4 Basic concept of an environmentally sound recovery process for platinum group metals (PGMs). PGMs in scrap are dissolved in aqueous HCl or NaCl solutions after a pretreatment using a reactive metal vapor

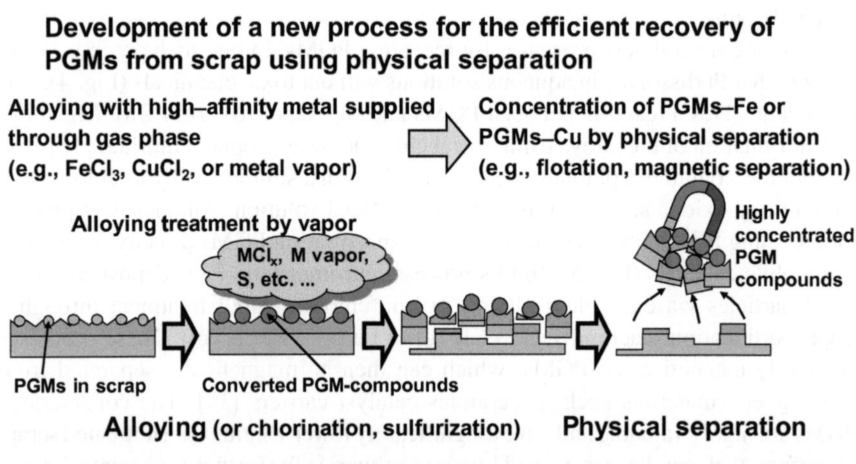

Fig. 5 Basic concept of an environmentally sound recovery process for precious metals including PGMs using physical separation

References

1. Okabe TH, Hamanaka Y, Taninouchi Y (2016) Direct oxygen removal technique for recycling titanium using molten $MgCl_2$ Salt. Faraday Discuss 190:109–126. https://doi.org/10.1039/C5FD00229J
2. Okabe TH, Zheng C, Taninouchi Y (2018) Thermodynamic considerations of direct oxygen removal from titanium by utilizing the deoxidation capability of rare-earth metals. Metall Mater Trans B 49(3):1056–1066. https://doi.org/10.1007/s11663-018-1172-4
3. Okabe TH (2017) Bottlenecks in rare metal supply and the importance of recycling—a Japanese perspective. Miner Process Extr Metall 126(1–2):22–32. https://doi.org/10.1080/03719553.2016.1268855
4. Okabe TH, Nakamura M, Oishi T, Ono K (1993) Electrochemical deoxidation of titanium. Metall Trans B 24B:449–455. https://doi.org/10.1007/BF02666427
5. Nakamura M, Okabe TH, Oishi T, and Ono K (1993) Electrochemical deoxidation of titanium. In: Proceedings of international symposium molten salt chemical technology, pp 529–540
6. Taninouchi Y, Hamanaka Y, Okabe TH (2016) Electrochemical deoxidation of titanium and its alloy using molten magnesium chloride. Metall Mater Trans B 47(6):3394–3404. https://doi.org/10.1007/s11663-016-0792-9
7. Okabe TH, Taninouchi Y, Zheng C (2018) Thermodynamic analysis of deoxidation of titanium through the formation of rare-earth oxyfluorides. Metall Mater Trans B. https://doi.org/10.1007/s11663-018-1386-5
8. Yagi R, Okabe TH (2016) Current status and smelting technologies of rhenium. J Japan Inst Met 80(6):341–349 (in Japanese). https://doi.org/10.2320/jinstmet.j2016022
9. Yagi R, Okabe TH (2016) Current status of recycling of rhenium and process technologies. J MMIJ 132(7):114–122. (in Japanese). https://doi.org/10.2473/journalofmmij.132.114
10. Yagi R, Okabe TH (2017) Recycling technologies for rare metals contained in Ni-based superalloy turbine blade. Kinzoku 87(9):39–46 (in Japanese). https://doi.org/10.2473/journalofmmij.132.114
11. Yagi R, Okabe TH (2017) Recovery of nickel from nickel-based superalloy scraps by utilizing molten zinc. Metall Mater Trans B 48(1):335–345. https://doi.org/10.1007/s11663-016-0854-z
12. Yagi R, Okabe TH (2017) Continuous extraction of nickel from superalloy scraps using zinc circulation. Metall Mater Trans B 48(3):1494–1501. https://doi.org/10.1007/s11663-017-0941-9
13. Horike C, Morita K, Okabe TH (2012) Effective dissolution of platinum by using chloride salts in recovery process. Metall Mater Trans B 43B(6):1300–1307. https://doi.org/10.1007/s11663-012-9746-z
14. Okabe TH, Mitsui J (2016) Japan Patent, P5946034
15. Taninouchi Y, Watanabe T, Okabe TH (2017) Recovery of platinum group metals from spent catalysts using electroless nickel plating and magnetic separation. Mater Trans (JIM) 58(3):410–419. https://doi.org/10.2320/matertrans.M-M2017801
16. Taninouchi Y, Watanabe T, Okabe TH (2017) magnetic concentration of platinum group metals from catalyst scraps using iron deposition pretreatment. Metall Mater Trans B 48(4):2027–2036. https://doi.org/10.1007/s11663-017-0999-4
17. Taninouchi Y, Okabe TH (2017) Enhanced dissolution of platinum group metals by utilizing electroless iron deposition. Metall Mater Trans B 48(6):2866–2872. https://doi.org/10.1007/s11663-017-1087-5
18. Taninouchi Y, Okabe TH (2018) Effective alloying treatment for platinum using iron chloride vapor. Mater Trans (JIM) 59(1):88–97. https://doi.org/10.2320/matertrans.m-m2017844
19. Taninouchi Y, Okabe TH (2018) Recovery of platinum group metals from spent catalysts using iron chloride vapor treatment. Metall Mater Trans B 49(4):1781–1793. https://doi.org/10.1007/s11663-018-1269-9

Part VII
REWAS 2019: Secondary and Byproduct Sources of Materials, Minerals, and Metals: Electronics and Battery Recycling

Li-Cycle—A Case Study in Integrated Process Development

Boyd Davis, Kevin Watson, Alain Roy, Ajay Kochhar and Darcy Tait

Abstract The resource recovery market has significant interest in new process development. One reason is that valuable materials for hi-tech applications are now present in sufficient quantities to warrant recycling efforts. Another is due to broad efforts on environment such as green chemistry and urban mining. However, many of these projects end in early development due to a lack of up-front integration on a broad range of issues such as experimental work, techno-economics, supply/demand, safety, regulatory landscape, and product quality. Using Li-Cycle, a Canadian lithium battery recycling company now engaged in process piloting, and as an example, this paper discusses the key barriers that companies developing new chemical or metallurgical processes face and how they can be overcome through an integrated approach. In the integrated approach, economic and market analyses commence as soon as possible in the project's life. These are used to establish a clear process/project scope, define specifications for the process products, identify the key cost drivers to appropriately focus technical work, and ultimately provide an objective, effective, and efficient method to evaluate the merits of the project.

Keywords Lithium · Battery recycling · Process development

B. Davis (✉) · K. Watson · A. Roy
Kingston Process Metallurgy Inc., 759 Progress Avenue, Kingston, ON K7M 6N6, Canada
e-mail: bdavis@kpm.ca

K. Watson
e-mail: kevin.watson@kpm.ca

A. Roy
e-mail: alainroy@kpm.ca

A. Kochhar · D. Tait
Li-Cycle, 2351 Royal Windsor Drive, Unit 10, Mississauga, ON L5J 4S7, Canada
e-mail: ajay.kochhar@li-cycle.com

D. Tait
e-mail: darcy.tait@li-cycle.com

G. Gaustad et al. (eds.), *REWAS 2019*, The Minerals, Metals & Materials Series,
https://doi.org/10.1007/978-3-030-10386-6_29

Introduction

There are two significant factors driving the metals industry to increasingly focus on new secondary sources and recycling to supply their resource needs. The first factor is the combination of a historical trend of decreasing grades in primary metal resources and the rapidly increasing availability of new secondary sources such as waste electrical and electronic equipment (WEEE). Gold provides a good example with its recovery from WEEE developed over the last 20 years compensating, in part, for the significant historical decrease in gold resource grades illustrated in Fig. 1 [1]. In addition, an increase in complexity of resource mineralogy for some metals has multiplied the negative impact of overall decreasing grades. Examples include arsenic-containing copper deposits, polymetallic deposits (e.g., Voisey Bay, sea nodules), refractory minerals (e.g., gold in sulfides), and deposits with nonstandard minerals (e.g., zinc carbonate).

The second factor is a growing environmental and societal context around waste. A tightening of general environmental regulations in the first world, combined with an awakening of developing nations to the problem of taking in other countries' waste means that individual countries will have to start dealing with waste within their own borders. At the same time, social and consumer pressures are forcing manufacturers to set strong environmental targets for their processes and products.

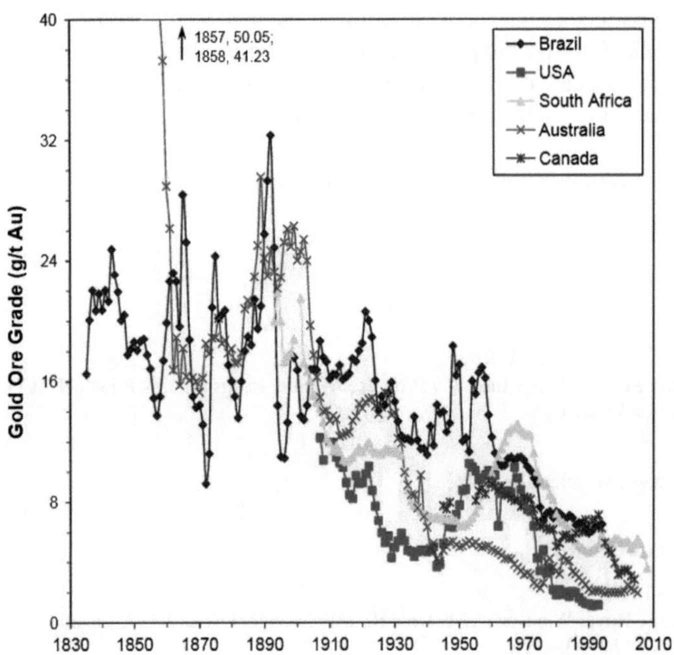

Fig. 1 Historical average gold resource grade. Reproduced from Calvo et al. (2016)

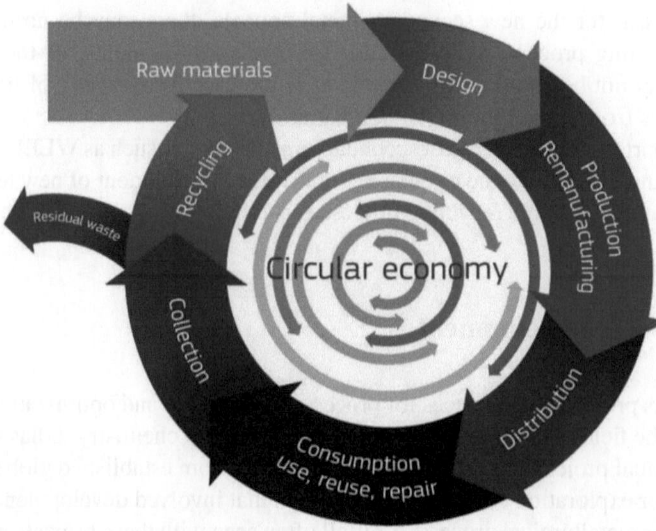

Fig. 2 Schematic illustration of the circular economy. Reproduced from European Commission (2018)

In particular, there is a push to make manufacturers internalize costs which have traditionally been borne by society, as is occurring globally with carbon dioxide emissions. This shift in cost bearing is expected to a have significant positive impact on the future economic viability of recycling as it will force manufacturers to design for remanufacturing and recycling rather than consider the end-of-life product as somebody else's problem. The ultimate goal is a circular economy, as illustrated in Fig. 2, where the value of products, materials, and resources is maintained in the economy for as long as possible, and the generation of waste is minimized [2].

The processing of secondary materials tends to follow a common evolution. At, first, used material is dumped or abandoned, then it is recovered due to pressure from governments and sent offshore to low-cost recyclers, then it is recycled in the country of origin with government financial support, and finally it becomes more economically favorable to recycle and the subsidies decrease to become a minor or no incentive. WEEE is a classic example of this, but paper, plastics, and cathode ray tube televisions have also followed the same pattern.

The integration and position of secondary metals processing in the overall metal supply chain depend on the availability of a technically viable recycling process and the economics of the recycling process taking into account government intervention and the comingling of other materials and metals. For example, aluminum and steel recycling is well integrated into the overall supply chain. These are major commodities enabling economies of scale for recycling infrastructure, the processes for recycling are well established (the first US aluminum recycling plant was built in 1904) and there is a plentiful supply of near pure form of secondary sources.

In contrast, for the new secondary metal sources, there may be no technically viable recycling process, as is the case for rare earth elements, or the available process may not be commercially viable as is the case for recovery of lithium and other values from spent lithium (li)–ion batteries.

To support the increased use of secondary metal sources, such as WEEE and li–ion batteries, and to facilitate the circular economy, the development of new technically and commercially viable recycling processes is required.

New Process Development

KPM offers professional services for process development and optimization to companies in the fields of mining, metallurgy, and industrial chemistry. It has done over 200 individual projects with a wide range of clients from established global producers to junior exploration companies to start-ups that involved development of a new chemical or metallurgical process. A pitfall often seen with these projects is a lack of integration of the technical work with the financial and market drivers of the project at an early stage. The common features of these projects are discussed below.

Figure 3 illustrates the typical time line of projects that are based on new chemical or metallurgical processes. Following the generation of the new idea, the first step for a project is a concept or scoping level study that typically involves proof-of-concept testing and the formalization of a process flow sheet. The laboratory test work required to provide the data to underpin the concept study is usually conducted at the bench scale in a research and development laboratory. This test work is often based around the chemistry of the process and is driven by the desire to show that a reaction is favorable and that a specific product can be made. This initial part of the project is usually a success as it is possible to make virtually anything in a laboratory when unconstrained by commercial cost considerations. While proof-of-concept testing does have its place, it does not necessarily provide the objective data required to properly evaluate the project and efficiently direct project activities.

At this early stage of the project, the economics of the process or the wider business case are rarely analyzed in any detail. If an economic analysis has been conducted, it usually lacks a clear definition of the process or project scope. In most cases, a specific end user is not yet involved in the project and details such as product specification, volumes, and prices are usually very poorly defined. This typically results in overly optimistic forecasts of potential product pricing, especially if the product is not traded on an open exchange like the London Metal Exchange. In many cases, secondary process by-products are overvalued to make up for the fundamentally weak economics of the main product. In the rare cases, where an effective financial model has been developed, a sensitivity analysis on the main cost drivers is either poorly done or not done at all.

Projects that include new chemical or metallurgical processes typically stall once they reach the so-called "valley of death" funding gap [3] illustrated in Fig. 3. Prior to this point, the relative project investment is small and funding is predominantly

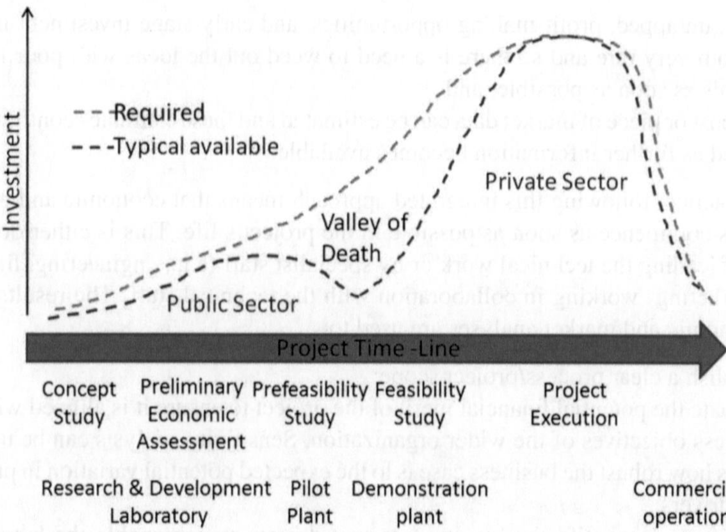

Fig. 3 Project investment timeline illustrating the valley of death funding gap

provided by the public sector in the form of government grants and university subsidies leveraging corporate research money to encourage innovation and seed new concept. It is also the time in the project when there is the maximum ability to affect change in the project or the underlying process for minimum cost. The investment required in money and human capital increases significantly as the project reaches the critical transition from a smaller scale research project to a larger scale commercialization project during the pre-feasibility and feasibility studies. These studies will typically require the operation of a pilot and/or demonstration plant to generate the data to support process engineering and evaluation of the project business case. The availability of public sector funding and support tends to drop off at this point and the project must increasingly turn to private sector funding. This private sector funding is comparatively scarce and competition for it is fierce. It is at this point, where clear quantified answers are required to questions such as: What exactly is the product? Who is buying it and for what price? How much to build and operate the plant? The projects that have been developed as described above without integration of the technical with the financial and market drivers are typically unprepared to answer these questions and lack the necessary objective data to clearly demonstrate a compelling business case.

Integrated process development as practiced by KPM is based on the following precepts:

- Paper studies (technical, economic, market, safety, and environmental) are less expensive than physical testing;
- A team approach is required, linking all aspects of the project and its business case;

- Good, untapped, profit-making opportunities, and early-stage investment money are both very rare and so there is a need to weed out the ideas with poor fundamentals as soon as possible; and
- Any cost or piece of market data can be estimated and these estimates continuously refined as further information becomes available.

In practice, following this integrated approach means that economic and market analyses commence as soon as possible in the project's life. This is either done by the staff leading the technical work or by specialist staff (e.g., engineering, finance, and marketing) working in collaboration with the technical staff. The results from the economic and market analyses are used to:

- Establish a clear process/project scope;
- Evaluate the potential financial merit of the project to ensure it is aligned with the business objectives of the wider organization. Sensitivity analysis can be used to assess how robust the business case is to the expected potential variation in process cost drivers;
- Identify and quantify the key process cost drivers, which enable the laboratory testing to be appropriately focused to ensure project costs and time frames are minimized;
- Provide clear detailed specifications for the main process product(s) and any by-products, which define the performance targets for the laboratory testing and for preparation of product qualification samples; and
- Develop the preliminary process commercialization and product marketing plans.

The economic, market, and technical data can be captured in a techno-economic model which can be continuously updated and refined as the project develops and new information becomes available. Ultimately, an integrated approach provides the necessary technical, economic, and market data to best equip the project to tackle the hurdles to funding and cross the valley of death.

Li–Ion Battery Recycling

About Li-Cycle Corp

Li-Cycle Corp. (Li-Cycle) is on a mission to leverage its innovative solution to address an emerging and urgent global challenge. Lithium–ion (li–ion) rechargeable batteries are increasingly powering our world in automotive, industrial/utility/residential energy storage, and consumer electronic applications. The world needs improved technology and supply chain innovations to better recycle these batteries, and to meet the rapidly growing demand for critical and scarce battery-grade materials.

Li-Cycle is based out of North America and has a global focus. Li-Cycle has retained and partnered with KPM as a contract piloting and laboratory service provider to scale up Li-Cycle's resource recovery technology. Li-Cycle, KPM, and other strategic/scale-up partners work together as one integrated team.

Li-Cycle Technology™ is a low-cost, safe, and environmentally friendly process that can recycle all types of li–ion batteries. It can do so with an unparalleled recovery rate of up to 100% of all materials. Li-Cycle's industry-leading innovations make it uniquely positioned to support a key element of the growing international movement toward zero carbon technologies as part of the circular economy. Li-Cycle Technology™ is a mechanical and hydrometallurgical resource recovery technology tailored to li–ion battery recycling. The flow sheet consists of three key stages that are as follows:

1. Reduce the size of all li–ion batteries in a safe, scalable, and proprietary manner;
2. Separate and recover intermediate materials; and
3. Refine the recovered materials for sale to end markets or reuse in the process.

Li-Cycle has completed two scale-up programs to date and is currently executing an integrated pilot plant, as a final step prior to commercial plant execution. Li-Cycle's physical validation work streams have been premised on a "scale-down" focus, i.e., scaled down relative to a first commercial plant. Each stage of the scale-down roadmap is focused on the validation of specific key performance indicators. Li-Cycle's integrated pilot plant is designed to process 1 ton/day of spent li–ion batteries, at a scale of 1:17 with the company's first commercial plant.

Li-Cycle's core business model is to build, own, and operate li–ion battery recycling plants tailored to regional needs. Through the integrated development of Li-Cycle Technology™, it has been identified that the best and fit-for-market commercialization strategy is one of the decentralization—i.e., with tailored size reduction plants that are close to the highest volume points of consolidation of li–ion batteries, and a centralized hydrometallurgical plant that is decoupled from the size reduction facilities. The core benefit of this approach is the generation of a nonhazardous product close to the point of consolidation/generation, minimizing transportation liability, and cost to the greatest extent. This decentralized strategy is reflected in the commercial rollout plan for Li-Cycle's first commercial plant, with a planned resource recovery capacity of 5,000 tons of spent li–ion batteries/year.

Li–Ion Recycling Industry Challenges

Feed Sourcing

Unlike mining, secondary resource recovery has a set of unique challenges that need to be addressed concurrent to process development. When sourcing feed material, the mining sector employs exploration techniques to identify high concentrations of a target element. For secondary resources, feed materials are typically inherently distributed, making it more difficult to collect and feed to a processing plant. Although the collection supply chains for some analogous industries such as lead–acid battery recycling are well established and matured by comparison, the li–ion battery recycling supply chain continues to be fluid.

Since Li-Cycle's inception, the company has been deeply focused on the li–ion battery supply chain and key stakeholders that are central to spent li–ion battery sourcing. Spent li–ion battery sources can be broadly segmented into portable/"small format" and "large format" sources, which corresponds to the relative voltage of li–ion batteries (i.e., low voltage and intermediate to high voltage, respectively). Each of these types of batteries have a diverse group of stakeholders—from manufacturers, to the dealer network, recycling programs, electronics, and vehicle recyclers.

The profile of li–ion battery feed must be central to process development. The inherently heterogenous nature of li–ion batteries must be closely considered as part of any li–ion battery recycling process. There are at least 14 different types of li–ion battery cathode chemistries currently existing in the market [4], each of which have even further permutations when considering specific constituents. Additionally, many researchers and companies are focusing on improving li–ion technology which is leading to new battery chemistries, binders, and electrolytes.

Logistics and Regulations

Li–ion batteries are currently classified as Class 9 dangerous goods due to their dual chemical and electrical hazard. Li–ion batteries can possibly undergo thermal runaway, typically resulting from internal shorting, leading to fire or explosion. There are numerous factors that can cause thermal runaway, including but not limited to overcharging, environmental conditions (e.g., extreme external temperatures), and manufacturing defects. At the onset of thermal runaway, the battery heats in seconds from room temperature to above 700 °C. As part of this complex set of chemical reactions, the electrolyte solvent in lithium–ion batteries—typically alkyl carbonate based—acts as a "fuel" source for combustion.

Added care must also be taken when handling critical or damaged/defective batteries, as there is an increased risk of thermal runaway. Specialized systems (e.g., Genius Technologie's LionGuard container [5]) are typically used in tandem with nonflammable packing material to safely transport these batteries. As time progresses, the number of critical or damaged/defective batteries is expected to increase with the increased use of li–ion battery technology across a broad swath of applications.

As the li–ion battery resource recovery industry is still maturing, regulations vary significantly around the world. These regulations can also change significantly from year to year, as new industry and research reports are released. As a result, it is important to keep close track of regulatory (including logistics) considerations concurrent to process development.

Safety and Storage

The challenges of logistics and changing regulations typically revolve around one key factor—safety. Safety is paramount for those who handle, transport, store, and process li–ion batteries, as there is a risk of thermal runaway. This raises another unique challenge for processors and consolidators, relative to the primary production of commodities and specialties. Specifically, the safest approach is to have the lowest amount of spent li–ion batteries on site as possible, in order to mitigate the risk of a thermal runaway event occurring. However, this is contradictory to the requirement to secure significant amounts of feed for processing purposes. The development of safe storage is further complicated by the currently prominent format factor of spent li–ion batteries, i.e., portable/small format batteries (e.g., from mobile phones, laptops, and other consumer products). Portable li–ion batteries are typical consolidated in drums and could be mixed with other battery types. Upon an initial inspection, the state of all collected batteries within a single drum is not always clear (i.e., whether undamaged or damaged) and often only becomes apparent when the drums are tipped for sorting or processing.

Secondary Resource Processing Challenge

From a process development standpoint, the recovery of constituents from li–ion batteries presents a unique challenge compared to traditional primary metal resources due to the highly heterogeneous nature of the feed material. With traditional metal resources, the primary concentrate stream might have 1–4 elements to be recovered (e.g., copper, gold, silver, and platinum). Li–ion batteries may, however, contain over 20 elements that demand consideration for recycling as illustrated by the example composition in Fig. 4 [6]. In addition, the metal values are typically contaminated with inorganic materials, organic materials, and plastics further complicating the recycling process. To be able to separate out the valuable constituents typically requires complex process flow sheets with many individual unit operations. Under this scenario, it is critical that the physical test work required to develop the process flow sheet is well focused and driven by techno-economic analysis.

Integrated Approach

In order to address each of these challenges, Li-Cycle has applied an integrated development approach. Specifically, the company has focused on four three-core work streams in parallel: battery supplier, end product, and process development, as well as techno-economic modeling.

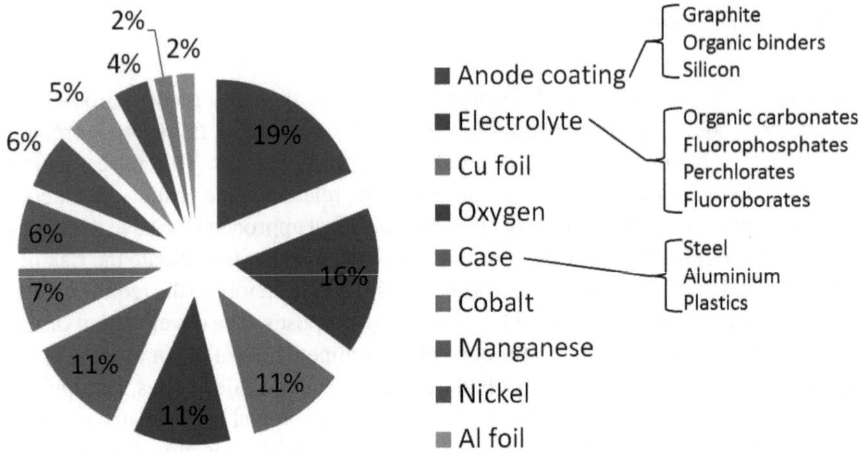

Fig. 4 An example li–ion battery composition. *Source* Diekmann et al.

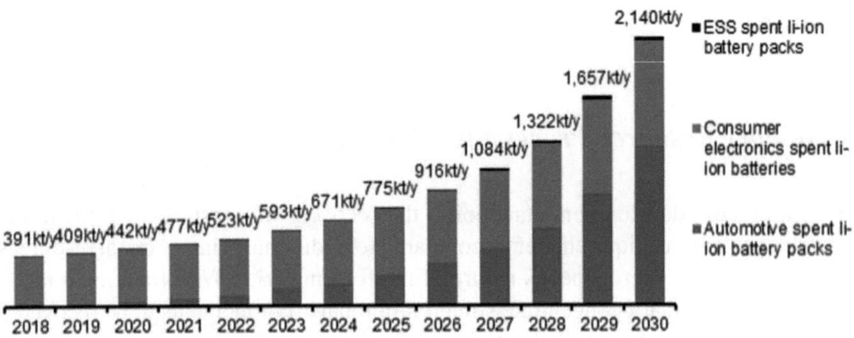

Fig. 5 Global forecasted tonnage of li–ion batteries reaching the end of their useful life per year, segmented by key prior use/application

Battery Supplier and End-Product Development

Li-Cycle's battery supplier development began with a mapping of the li–ion battery supply chain. Through this focus, Li-Cycle has directly interacted with original equipment manufacturers (OEMs) and battery consolidators for potential near-term and long-term supply of li–ion batteries. For example, Li-Cycle engages with electric vehicle (EV) producers to service the increasing volume of spent high voltage li–ion batteries requiring resource recovery. The diversity of the li–ion supply chain evidences the need to forecast a possible feed profile into the future, and to integrate this directly into process development and techno-economic analysis. Li-Cycle's forecast for the global forecasted tonnage of li–ion batteries reaching the end of their useful life is provided in Fig. 5, segmented by key prior use/application (i.e., automotive, consumer electronics/portable, and stationary energy storage systems/ESS).

During the initial stage of process development, based on battery supply development efforts, it quickly became apparent that any process technology focused on resource recovery from spent li–ion batteries must be robust to handle a broad range of chemistries. For example, a first batch of li–ion batteries received was labeled with a lithium cobalt oxide (LCO) cathode. However, based on the subsequent chemical analysis, the specific batteries were shown to have nickel manganese cobalt (NMC) chemistry. Additionally, further engagement with battery suppliers revealed that consolidators and recyclers often do not take the time to identify the chemistry of each consumer electronics li–ion battery as it would be too costly for them and/or would significantly increase Li-Cycle's costs. Moreover, battery supplier development also revealed that OEMs have proprietary cathode compositions which affect the concentration of elements in the feed stream. Therefore, an integrated approach to project development has been critical to ensuring long-term process viability.

From a commercialization standpoint and the segmentation of the Li-Cycle Technology™ flow sheet, through prioritizing the concerns surrounding the transport and handling of spent lithium–ion batteries, Li-Cycle's decentralized commercialization approach (i.e., with satellite safe size reduction sites) minimizes the risk of an incident occurring during transportation. Alongside, packaging and transport costs are minimized. Once the feed li–ion batteries have been size reduced, the resulting product is no longer classified as a hazardous material or dangerous good as the electrolyte has been separated from the rest of the battery components, eliminating the risk of thermal runaway. This approach has only been enabled by an integrated battery supplier development work stream that dialogues with the process development and techno-economic modeling work streams in an integrated approach.

In parallel to all other work streams, Li-Cycle has also focused on the development of end-product markets, premised on the re-introduction of critical materials in the li–ion battery supply chain. Traditionally, many of the products from the resource recovery of spent li–ion batteries have not reentered the li–ion battery supply chain. To enable this circular economy focus, end-product qualification alongside each of the stages of process development has been central to Li-Cycle's approach, with robust consideration of and integration with related drivers (i.e., battery supply feed and its direct influence on end products, the modeling of the revenue drivers as part of techno-economic analysis).

Process Development and Techno-Economic Modeling

Process development and techno-economic analysis included the development of a technical process model which could predict product outputs based on the feed/input, coupled with fundamental chemistry inputs (i.e., design criteria) from physical validation efforts. It is important to conduct this modeling during the initial stage of process development as it has a significant impact on the ultimate economics. Li-Cycle has utilized a robust multi-stage approach that progressively de-risks the project and technology in a structured fashion, as summarized in Fig. 6.

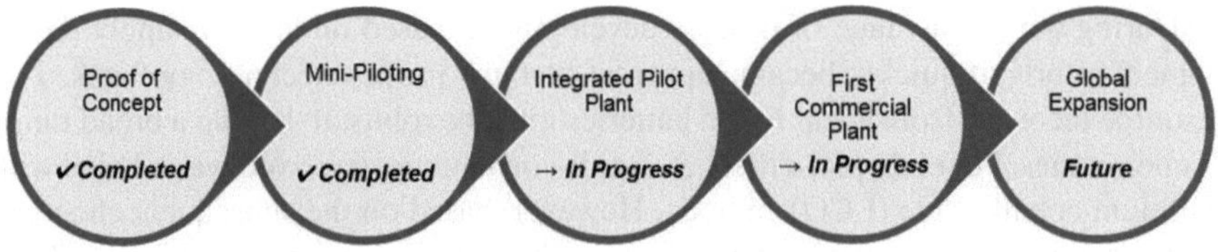

Fig. 6 Li-Cycle Technology™ commercialization roadmap

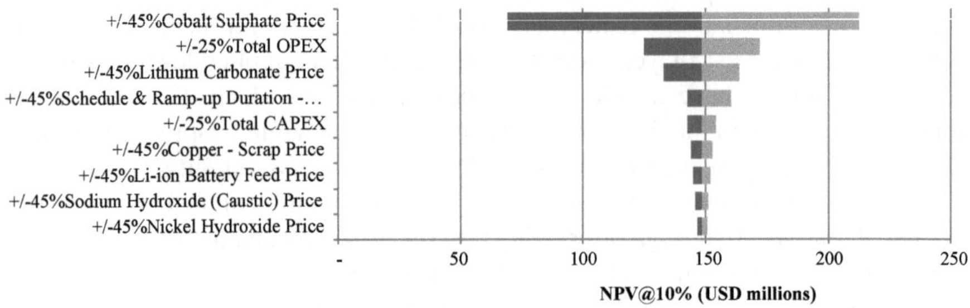

Fig. 7 Li-Cycle Technology™—first commercial plant sensitivity analysis

Techno-economic analysis has played a critical role in each stage of the multi-stage commercialization approach. Based on inputs from the battery supply and physical test work, key financial drivers are progressively iterated and refined, including but not limited to capital expenditure (CAPEX), operating expenditure (OPEX), net present value (NPV), and earnings before depreciation, interest, taxes, depreciation, and amortization (EBITDA) for the first commercial plant. Ultimately, techno-economic analysis and the associated financial model are intended to become an operational support tool, e.g., testing the financial sensitivity of specific battery feeds that could be received and processed through the combination of technical and financial drivers. Figure 7 provides a sensitivity tornado diagram for the NPV of the first commercial plant which indicates that the key financial drives are the prices for cobalt sulfate and lithium carbonate and the plant OPEX. This subsequently provided focus for product market and process development activities.

An example of the two-way feedback that has occurred between the physical test work and techno-economic analysis is the optimization of process reagent use. Specifically, acid and sodium hydroxide (NaOH) reagent consumption required for leaching and neutralization were identified as key OPEX drivers, as shown in Fig. 8. The physical test work during the mini-piloting was subsequently focused on optimizing their use. Specifically, a method was developed to separate the copper (Cu) and aluminum (Al) foils from the cathode and anode powders prior to leaching. This reduced the Al precipitated as aluminum hydroxide ($Al(OH)_3$) during leaching, resulting in an approximate 30% reduction in acid consumption and an associated reduction in sodium hydroxide (NaOH) consumption thereafter in the neutralization step. The net impact was a substantial 15% reduction in total OPEX. An additional

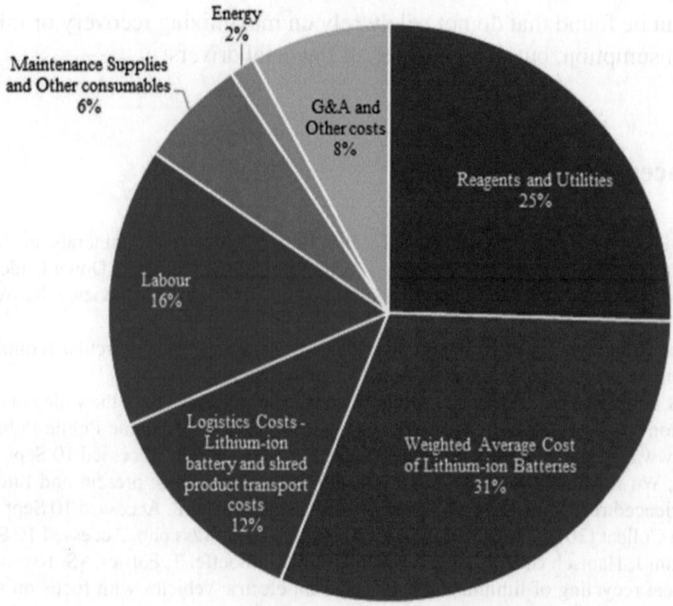

Fig. 8 Operating cost estimate breakdown, Li-Cycle first commercial plant, size reduction, and hydrometallurgical plants

benefit was a reduction in the valuable metal (i.e., Li, Ni, Co) lost as entrainment in the $Al(OH)_3$ precipitate resulting in a positive influence on key revenue drivers.

Conclusion

Li-battery recycling is an excellent example of a highly complex process that has many external drivers and cost implications. To develop a strategy for recycling, these external issues must be considered in an integrated way. Logistics, safety, automation, heterogeneity of feed, customers' needs, and market demands must all be part of the solution and must be included in the approach. This should not minimize the significant chemistry that goes into a complex process such as this, but it does mean that the focus, as it often is, should not just be on chemical reactions.

Through the effective integration of external factors along with a side-by-side development of techno-economics along with process chemistry, Li-Cycle's plan for Li-battery recycling provides a robust and effective method of handling these batteries, as well as returning as much material back to the front end of the process as is possible. Through integrated process development, unforeseen obstacles can be revealed early on and handled, and opportunities are made clear at the earliest possible stage and can be capitalized upon. As well, the best economics for the overall process can be found that do not solely rely on maximizing recovery or minimizing energy consumption, but that consider all financial drivers.

References

1. Giurco D, Prior T, Mudd G, Mason L, Behrisch J (2010) Peak minerals in Australia: a review of changing impacts and benefits. Prepared for CSIRO Minerals Down Under Flagship. https://www.uts.edu.au/sites/default/files/giurcoetal2010peakmineralsreview.pdf. Accessed 10 Sept 2018
2. European Commission (2018) Report on critical raw materials and the circular economy. https://ec.europa.eu/docsroom/documents/27327. Accessed 10 Sept 2018
3. Ford GS, Koutsky TM, Spiwak LJ (2007) An economic investigation of the valley of death in the innovation sequence, Phoenix Center for Advanced Legal & Economic Public Policy Studies. http://www.osec.doc.gov/Report-ValleyofDeathFundingGap.pdf. Accessed 10 Sept 2018
4. Nitta N, Wu F, Lee JT, Yushin G (2015) Li-ion battery materials: present and future. https://www.sciencedirect.com/science/article/pii/S1369702114004118. Accessed 10 Sept 2018
5. Lithium Collect (2018). https://ecobatgroup.com/occ-liion/en/ss.php. Accessed 10 Sept 2018
6. Diekmann J, Hanisch C, Frobose L, Schalicke G, Loellhoeffel T, Folster AS, Kwade A (2017) Ecological recycling of lithium-ion batteries from electric vehicles with focus on mechanical processes. J Electrochem Soc 168(1):A6184–A6191

Lithium-Ion Batteries, How to Generate Value Out of End of Life Mobile Units

Christer Forsgren

Abstract The use of Lithium-Ion Batteries (LIB) increase every year. Warranty for LIB in vehicles is often about 7 years. Laptop LIBs last less than half that time. Battery packs from both computers and vehicles can be used for stationary energy storage. After a second life used for energy storage, material recycling of especially Co and Li is important both from financial and resource perspective. In EU, there are about ten plants that with different technologies recycle mainly Co and Cu. Experiences from EU will be presented.

Keywords Recycling · Li-ion batteries · Cobalt · Second use

Background

The use of Lithium-Ion Batteries (LIB) in different products has increased dramatically in the last 5 years and will increase even more in the years to come. In vehicle applications, about 50% of the battery weight is active battery cells. To simplify, there are today three different types of chemistries used depending on application, which are given as follows:

A. Laptops and smartphones. Cobalt (Co) content 20–25 w%.
B. Vehicles, Co content about 5 w%.
C. Busses, no Co, often Li-Ion-Phosphate (LFP).

A lot of research is ongoing to improve capacity both related to weight and volume. Both pouch cells and cylindrical cells use aluminum foils as the backing of the cathode material that differs between chemistries as A–C above. The anode uses a copper backing and is often made out of graphite. The electrolyte is often a blend of different solvents with low flash point and a Lithium salt ($LiPF_6$). Other parts of the cell are the fluorinated binder (PVDF) and a separator (PP/PE)

C. Forsgren (✉)
Chalmers Technical University in Gothenburg/Stena Recycling International, Gothenburg, Sweden
e-mail: christer.forsgren@stenametall.se

G. Gaustad et al. (eds.), *REWAS 2019*, The Minerals, Metals & Materials Series,
https://doi.org/10.1007/978-3-030-10386-6_30

Recycling Legislation

The Battery Directive in EU requires batteries to be material recycled to at least 50 w%. The cost in EU to send a LIB to a recycling plant that recycles to at least 50% is today about 2–4 $US/kg for LFP and a similar revenue for high Co LIB. If there is a possibility for a second use the incentive to send LIB for material recycling is very low.

Reuse for Energy Storage

When a battery has 70–80% left of the original capacity they are not possible to use in most vehicle applications. To store energy from solar cells or from windmills is becoming a new market for used LIB. Fast charging stations for vehicles are another application for used LIB. Even used Laptop batteries mounted in a rack can be used as emergency power in different applications. Since LIB for vehicle applications use software for use, charging and discharging that is optimized for vehicle, not energy storage conditions, there in some cases are challenges in this area. In many cases also the use history of the battery is stored in the car, not in the battery, why the history is lost if the battery is replaced.

Material Recycling

First step in recycling a vehicle LIB is to dismantle into at least three parts (1) LIB cells, (2) Electronics and (3) Construction materials (Frame, cooling equipment, etc.)

Treatment of LIB cells in most recycling plants in EU is based on a combination of mechanical, pyrometallurgical and hydrometallurgical processes. Most plants have the first step of heat treatment at a temperature between 200 and 500 C to remove solvents that could cause fires in the next step of processing. Most cells contain 5–10 w% solvents that are very flammable. Without the solvent, the cell can be regarded as discharged and can be handled in a much safer way. Aluminum and copper are present in the elemental form and should preferably be removed and separated before any following high temperature or oxidizing process.

The material recycling value mainly depends on the Co content. Since Co is very expensive manufacturers try to reduce the use as much as possible and replace it by, e.g. nickel. The Li-content in the cell is often about 2 w% and is not recycled in EU today. There is no elemental Li in a LIB.

When elemental Al and Cu have been removed the residual fraction contain the cathode material with oxidized metals, binder, salt and graphite, a mixture called black mass. To mainly recycle Co and Ni there are two main technologies, pyrometallurgical where the material is melted or hydrometallurgical processes where often acids are used to dissolve the black mass and separately precipitate different metals. Hydrometallurgical processes can generate very pure material but are often more expensive.

Also, phosphorous and graphite are on the EU list of Critical Raw material. The cost to recycle these in comparably small volumes will be much higher than using virgin material as raw material why material recycling short term is not likely to happen.

Cobalt is the most critical element to recycle short term. Earlier, this year a Horizon 2020 project named "Crocodile" was launched with the focus of finding tools to more efficiently recycle Co from different sources. http://h2020-crocodile.eu/

Future

It is likely that vehicle batteries will last additional 10–20 years in some sort of energy storage application why the need for high capacity material recycling will take many years from now.

To avoid hazards during use of LIB there is a development towards solid electrolytes that do not contain solvents with low flash point. Also, material recycling will be easier without the content of solvents.

Advances in Lithium-Ion Battery Electrolytes: Prospects and Challenges in Recycling

Joseph Hamuyuni and Fiseha Tesfaye

Abstract Lithium-ion batteries are already playing a key role in the move from fossil fuels towards clean and renewable energy systems. This is because variabilities in renewable energy grids need to be supported by very stable storage mechanisms (batteries). In electric vehicles, lithium-ion batteries are also very important and determine applicability on the target vehicle. At their end-of-life, substances contained in them make it impossible to be discarded in an uncontrolled way. Moreover, the batteries contain critical metals that need to be recovered. While there are currently numerous studies on recycling of the cathode materials, there is very scarce research on how electrolytes impact recycling. This research reviews and enumerates on the advances in lithium-ion battery electrolytes in the context of recycling.

Keywords Electrolyte · Lithium-ion battery · Metals · Recycling

Introduction

Although the first lithium-ion battery for commercial use was produced only two decades ago, its technological advancement has moved very rapidly. It has a wide range of applications from small portable electronic devices to very large-scale applications such as in automobile industry. Lithium-ion batteries now further represent the future of most energy storage mechanisms. If the goal of a cleaner planet with very low greenhouse gases plus a digital age is to be realized, the solution lies in the developments of very powerful energy storage mechanisms.

Arguably, the six features/factors that make a lithium-ion battery competitive are: specific energy, specific power, battery performance, battery lifespan, safety of the battery and the costs associated. The six most developed lithium-ion battery

J. Hamuyuni (✉)
Outotec (Finland) Oy, Research Center, Kuparitie 10, PO Box 69, 28101 Pori, Finland
e-mail: joseph.hamuyuni@outotec.com

F. Tesfaye
Laboratory of Inorganic Chemistry, Johan Gadolin Process Chemistry Centre, Åbo Akademi University, Piispankatu 8, 20500 Turku, Finland

G. Gaustad et al. (eds.), *REWAS 2019*, The Minerals, Metals & Materials Series,
https://doi.org/10.1007/978-3-030-10386-6_31

technologies include: Lithium Cobalt Oxide, Lithium Manganese Oxide, Lithium Nickel Manganese Cobalt Oxide, Lithium Iron Phosphate, Lithium Nickel Cobalt Aluminum Oxide and Lithium Titanate. Table 1 depicts how these battery technologies fare against the above-listed factors and shows the extent of diversity in the battery technologies.

At end-of-life (EoL), recycling of all these batteries is critical to keeping a closed loop of metals. In this regard, Umicore is among the leading lithium-ion batteries recycling company. The sustainable recycling process of Umicore is schematically illustrated in Fig. 1. Current lithium-ion batteries reach their EoL already after 300–500 charging cycles or 3 years of usage after which they must be recycled to recover valuable metals. For this reason, there is intensive research in the field of recycling lithium-ion batteries, with focus on the cathode contents of the lithium-ion batteries. There is, however, less consideration on lithium-ion battery electrolytes and how they affect the recycling process. This paper elaborates on lithium-ion battery electrolytes from the critical metals recovery point of view.

Table 1 A comparison of the competitiveness of the six most common lithium-ion battery technologies. Notations: L stands for low, M for moderate, H for high

	Indicators for the six most common lithium-ion batteries					
	Specific power	Specific energy	Safety	Lifespan	Cost	Performance
$LiCoO_2$	L	H	L	L	L	M
$LiMn_2O_4$	M	M	M	L	L	L
$LiNiMnCoO_2$	M	H	M	M	L	M
$LiNiCoAlO_2$	M	H	L	M	M	M
$Li_4Ti_5O_{12}$	M	L	H	H	H	H
$LiFePO_4$	H	L	H	H	L	M

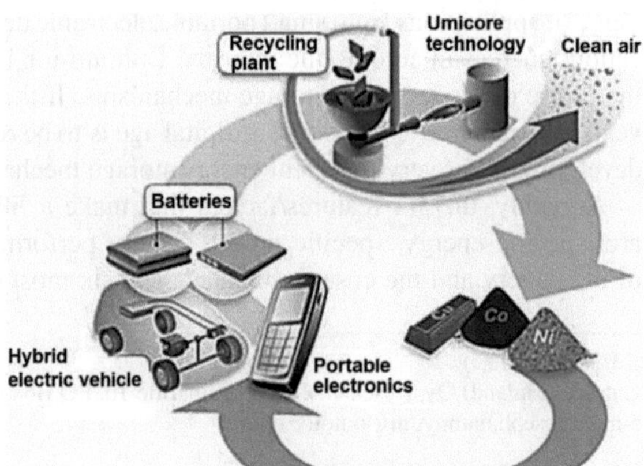

Fig. 1 Schematic diagram showing the recycling of batteries at Umicore with a closed loop of metals. Modified from [1]

Lithium-Ion Battery Electrolytes

In the operation of a lithium-ion battery, the electrolytes play a fundamental role and one which is often understated. The electrolyte is an organic liquid with dissolved substances and provides a sufficient conductive pathway for ions between electrodes during charge and discharge cycles. For this reason, the overall performance of the lithium-ion battery is a function of the performance of the electrolyte used. In fact this is why today's battery technologies have not yet been able to meet automobiles' stringent future requirements for high energy density, sufficiently long cycle life, excellent safety, and wide operating temperature range [2, 3]. Electrolytes holds the key to the success of EV batteries. State-of-the-art electrolytes primarily consist of lithium salts and organic solvents. These cause irreversible capacity losses that resulted from the formation of stable solid electrolyte interphase (SEI), which inhibit the increase in lifecycle and limit the operating temperature window, something which poses severe safety concerns on lithium batteries. Typical first-generation lithium salts and their properties are presented in Table 2.

In this respect, the replacement of currently used organic liquid electrolytes with inorganic solid electrolytes (SEs) is very appealing

Table 2 Thermal stabilities, properties, and applications of selected lithium salts

LiX	Properties of individual salt			Properties of electrolyte solutions (at 25 °C)		Application
	$T_{m.p.}$ °C	$T_{decomp.}$	Stability against hydrolysis	Passivation of Al	*E*, V	
$LiClO_4$	236 [4]	>236 [4]	Stable [4]	Yes	4.6 [7]	Primary Li cells [4]
$LiCF_3SO_3$	300–400 [4, 5]	~430 [8]	Stable [4]	No	5.0 [7]	Primary Li cells/thin film
$LiAsF_6$	340 [4]	240 [9]	Stable [4]	Yes	5.1 [7]	Primary Li cells [4]
$LiPF_6$	190–200 [4, 6]	≥50 [10, 11]	Unstable [4]	Yes	4.2–4.5	Mass-production [4]
$LiBF_4$	293–305 [4, 7]	~132–300 [7, 12, 13]	Unstable [4]	Yes	5.2	Primary cells [4]

Future Batteries Electrolytes

To increase battery-operating voltage, there is a need for electrolytes that are compatible with both anode and cathode interfaces. Development of organic solvents that are compatible with lithium salts can supply a broader electrochemical stability window whilst providing a higher operating voltage are need. In the current research, the limiting factor to achieving this is the availability of an electrolyte with a low charge transfer resistance at the solid–liquid interface. Examples of solid electrolytes pursued include: LISICON, NASICON, sulfide and poly (ethylene oxide) (PEO). The future of lithium-ion battery technology hinges on two principal drivers, increasing the cell voltage and the charge-storage capacity. It must be stated that these two tasks can only be successfully achieved with excellent electrolytes.

Chemistry and Recycling

From the contents of Table 2, it can be inferred that the variability of electrolytes for lithium-ion batteries is wide. This adds to the already existing battery chemistry associated recycling challenges of individual electrolyte battery technologies. So far, the most popular electrolyte option is the $LiPF_6$ electrolyte technology, since the voltage of a Li-ion cell (~3.6 V) is way higher than the standard potential of electrolysis of water which is only 1.23 V at 25 °C. There is a need for non-aqueous solution. Typically, solvents with high dielectric constants are desirable. Common electrolyte solvents employed are propylene carbonate (PC), ethylene carbonate (EC), and dimethyl sulfoxide. These are all able to dissolve several lithium salts. Because of very high viscosities, these solvents tend to inhibit ionic transfers. For this reason, the actual electrolytic fluids contain empirical mixtures that include low viscosity solvents, dimethyl carbonate or methyl ethyl carbonate.

Factors that Inhibit Lithium-Ion Batteries Recycling

The three potential hazards for recycling are electrical, fire or explosion, and chemical [13–15]. Recovery of the valuable metals from lithium-ion batteries is threatened by the high flammability and toxicity potential of the contents of the electrolyte on exposure to certain environments. This is also why EoL lithium-ion batteries are classified as toxic waste by many standards.

Solvents such as cyclic and linear carbonates are at present frequently used in standard commercial electrolyte formulations. Solvents such as tetramethylenesulfone, diethoxyethane and 2-methyl-tetrahydrofuran have been considered for formulations of high-voltage electrolytes due to their high anodic stability. From the analysis of the data presented in Table 1, it can be concluded that most of the solvents are liquid at room temperature and they can roughly be divided into two groups depending on their volatility:

(a) (highly) volatile solvents with low boiling temperature, high vapor pressure at room temperature and high relative evaporation rate, and (b) less volatile solvents with high boiling point, low vapor pressure at room temperature and low relative evaporation rate.

An analysis of the contents and chemistry of the components of the commercially employed electrolytes for lithium-ion batteries reveals that:

(I) Most currently used lithium-ion battery electrolytes on exposure to the environment are toxic, irritant or harmful in addition to being flammable. While flammability associated risks of electrolytes are well researched and documented such that they are well known by handlers, hazards and risks associated with their toxicity are less often addressed.

(II) Typical commercial lithium-ion battery electrolyte solvents are volatile, with volatility ranging from moderate to extremely volatile. Several studies have now shown that even small amounts of some of these solvents when exposed to the environment have irreversible serious health effects. Any recycling technology must devise a method to overcome this potential hazard.

(III) The reactions of salts including the most commercially successful–$LiPF_6$—when in contact with water, for example, is known to result in the generation of gaseous HF, which is very toxic and corrosive compound posing a serious health risk.

(IV) In addition to salts and their resultant toxic products, many of industrial available electrolytes comprise numerous additives which could also be volatile and toxic. Although the number of additives in electrolyte is generally limited, nevertheless as part of a thorough risk analysis potential risks associated with release of these additives should not be neglected.

Summary and Conclusions

Lithium-ion batteries represent the future of most energy storage mechanisms. In electric vehicles, lithium-ion batteries are becoming increasingly very important and determine applicability on the target vehicle. At end-of-life, contents of the lithium-ion batteries make it impossible to be discarded in uncontrolled way. Moreover, they contain valuable metals that need to be recovered. While there are currently numerous studies on recycling of cathode materials, there is very scarce research on how electrolytes impact the recycling process. This research highlights and enumerates the advances in lithium-ion battery electrolytes in the context of designing sustainable recycling technologies.

References

1. Toyota, Umicore Team Up on Lithium-Ion Battery Recycling, Electric Car Reports (2012) Accessed at: https://electriccarsreport.com/2012/08/toyota-umicore-team-up-on-lithium-ion-battery-recycling
2. Goodenough JB, Kim Y (2009) Challenges for rechargeable Li batteries. Chem Mater 22:587–603
3. Kotobuki M (2012) The current situation and problems of rechargeable lithium ion batteries. Open Electrochem J 4:28–35
4. Xu K (2004) Nonaqueous liquid electrolytes for lithium-based rechargeable batteries. Chem Rev 104(10):4303–4418
5. Hu L, Zhang SS, Zhang Z (2015) Electrolytes for lithium and lithium-ion batteries. In: Rechargeable batteries (pp. 231–261). Springer, Cham
6. Gnanaraj JS, Zinigrad E, Asraf L, Gottlieb HE, Sprecher M, Aurbach D, Schmidt M (2003) The use of accelerating rate calorimetry (ARC) for the study of the thermal reactions of Li-ion battery electrolyte solutions. J Power Sources 119:794–798
7. Allen JL, Han S-D, Boyle PD, Henderson WA (2011) J Power Sources 196:9737
8. Paillard E, Iojoiu C, Alloin F, Guindet J, Sanchez J-Y (2007) Electrochim Acta 52:3758
9. Dominey LA, Koch VR, Blakley TJ (1992) Electrochim Acta 37:1551
10. Sloop SE, Kerr JB, Kinoshita KJ (2003) Power Sources 119–121:330
11. Ravdel B, Abraham KM, Gitzendanner R, DiCarlo J, Lucht B, Campion C (2003) J Power Sources 119–121:805
12. Allen JL, Han S-D, Boyle PD, Henderson WA (2011) J Power Sources 196:9737
13. Lebedeva NP, Boon-Brett L (2016) Considerations on the chemical toxicity of contemporary Li-ion battery electrolytes and their components. J Electrochem Soc 163(6):A821–A830
14. Bushkova OV, Yaroslavtseva TV, Dobrovolsky YA (2017) New lithium salts in electrolytes for lithium-ion batteries. Russ J Electrochem 53(7):677–699
15. Diekmann J, Hanisch C, Froböse L, Schälicke G, Loellhoeffel T, Fölster AS, Kwade A (2017) Ecological recycling of lithium-ion batteries from electric vehicles with focus on mechanical processes. J Electrochem Soc 164:A6184–A6191

Increasing Lead Battery Performance Efficiency

Matthew Raiford, Timothy Ellis, Jagannathan Punjabkesar, Kelvin Naidoo and John Howes

Abstract An active research and development initiative in the secondary lead industry has produced a new lead alloy enabling lead batteries to function at 2× current performance standards. The need for this new lead is in response to elevated temperatures due to the new duty cycles common for rechargeable batteries. This innovation addresses the trace element impurity profile of secondary lead used in energy storage and enables improvement in cycle life, dynamic charge acceptance and capacity. Utilization of advanced structural characterization capabilities of the Advanced Photon Source at the Argonne National Laboratory has led to a fundamental reassessment of the mechanisms in Pb based batteries, driving the rate of technological improvement in this crucial and sustainable energy storage technology. The specific energy measurement of this new lead used in batteries reaches a level competitive with some lithium-ion battery materials.

Keywords Lead alloy · Lead–acid batteries · PbSO4 crystals · Dissolution · Dynamic charge acceptance · Cycle life

M. Raiford
RSR Technologies, Inc., 4828 Calvert St, Dallas, TX 75247, USA
e-mail: mraiford@rsrtechnologies.com

T. Ellis
Advanced Lead-Acid Battery Consortium, RSR Technologies, Inc.,
4828 Calvert St, Dallas, TX 75247, USA
e-mail: tellis@rsrtechnologies.com

J. Punjabkesar · K. Naidoo
Willard Battery Co., 7 Bennett St, Korsten, Port Elizabeth 6020,
South Africa
e-mail: JagannathanP@auto-x.co.za

K. Naidoo
e-mail: KelvinN@auto-x.co.za

J. Howes (✉)
Redland Energy Group, 1875 Eye Street, NW, Suite 500,
Washington, DC 20006, USA
e-mail: Jahowes@redlandenergy.com

G. Gaustad et al. (eds.), *REWAS 2019*, The Minerals, Metals & Materials Series,
https://doi.org/10.1007/978-3-030-10386-6_32

Introduction

Electrochemical energy storage, such as lead–acid batteries (LABs), serve a vital role in automotive, renewable energy, and grid regulation. LABs have been the cornerstone of energy storage for decades, and improvements have pushed battery performance to new levels [1–9]. Addition of carbons and other additives into the negative electrode, optimized production processes, and alternative electrode configurations have helped increase service life, energy density, and durability. However, more improvements are necessary for LABs to fulfill the requirements set by the Department of Energy (DOE), United States Council of Automotive Research (USCAR), and IRENA [9, 10]. Each has metrics on lowering the cost of electrochemical energy storage technology while improving energy density and lifetime. Table 1 summarizes several of the metrics with a direct comparison to a typical LAB.

Additives for the positive active material (PAM) or negative active material (NAM) are used widely, including lignosulfonates, carbons, and $BaSO_4$. Typical LAB active material is composed of primary or secondary lead with key impurities removed to decrease oxygen and hydrogen gassing. Seminal work by Lan Lam et al. demonstrated the effect of common elements on gassing reactions during float for a LAB [11]. However, little research has focused on specific additive elements in the lead on the crystal growth and dissolution behavior of the species in a LAB. Cycle life can be limited by the growth of insoluble $PbSO_4$ crystals and developing methods to control the crystal morphology of $PbSO_4$ could extend LAB life significantly. Introduction of specific alloying elements into the lead used in leady oxide production enforces a homogeneous distribution of the beneficial crystal modifiers throughout the lead oxide and resulting paste. Previous work by Masahiko Anari has shown that controlling antimony and arsenic content in lead–acid battery pastes can increase capacity and cycle life in LABs [12].

We studied these elements from cell level to 12 V battery level and correlate them to the crystal structures and properties in the alloys. Antimony in lead alloys has been used as a strengthening agent, in amounts ranging from 1.4 to 6.0% content by

Table 1 Key performance metrics set by IRENA, DOE, and USCAR

Metrics	Typical VRLA LAB	Target	Organization
DCA (A/Ah)	0.25	1.0	ALABC
Calendar life (years)	10	15	USCAR/USABC
Operating Temperature (°C)	−30 to 55 °C	−30 to 75 °C	USCAR/USABC
Energy density (Wh kg^{-1})	50	100	IRENA
Acquisition cost ($/kWh)	220–250	100–150	IRENA
ESS Cycle life (number)	2200	5000	IRENA/DOE
Operating cost ($/kWh/cycle)	0.07	0.02	IRENA/DOE
Opportunity charging time (h)	8	6	ALABC

weight. Antimony also can be utilized as Sb_2O_3 or Sb_2O_5 additives in paste mixing and production. However, this approach limits the availability of antimony ions and presents particle distribution issues. Significant enhancement was observed with micro-additions of these elements into active material lead, called SUPERSOFT-HYCYCLE®.

Performance improvement in automotive applications is further complicated by "super-hot" conditions. Consistent underhood temperatures of +70 °C are measured during operation and ~100 °C temperatures have been observed. Material compositions in LABs must be altered to perform at "super-hot" climates to decrease positive grid corrosion and positive paste shedding. Current PbSnCa grid alloys are well suited for many duty cycles and the addition of silver to these alloys has been shown to increase corrosion resistance. Added benefit is possible with barium in maintaining the grain structure at high temperatures and grid hardness. Combining the silver and barium effects are studied in a grid material lead alloy, 009™, with superior corrosion resistance and mechanical strength. The focus of this paper is demonstrating the benefit of utilizing novel micro-alloyed lead alloys, SUPERSOFT-HYCYCLE® and 009™, with As and Sb effects in active materials and Ag and Ba effects in grid alloys.

Experimental

2 V Cells were constructed from machine grade polycarbonate and leak tested over a 60 day period at 70 °C. The lid was sealed using common sealing adhesives for lead–acid battery casings from Atlas. The cells were tested with a total of nine battery electrodes (five positive and four negative or four positive and five negative), with 3 control battery cells and 21 samples tested in every battery testing regime.

Using a three-electrode cell setup, a solution of 0.1 M $Pb(NO_3)_2$ and 1.0 M HNO_3 (50 mL) was added to the electrochemical cell. A clean GC RDE was lowered into the solution until the face of the electrode was submerged under the meniscus. At this point, the disk and shaft assembly are kept stationary, or rotated at 25 or 100 rpm. Electrochemical deposition was performed galvanostatically, using the galvanostatic option in the Gamry software package. Sulfation studies were performed in a solution of 1.285 s. g. (~5 M) H_2SO_4 and the desired concentration of the ion additive were added to the electrochemical cell. The disk assembly was used in the three electrode setup, with an existing GC RDE with pre-deposited β-PbO_2 adhered to the surface as the working electrode. Cyclic voltammetry was used to simulate the charge and discharge processes in a lead–acid battery by scanning slowly between the +0.6 and +1.6 V. The deposits were then washed with ethanol and dried under air.

Results and Discussion

Initial studies of the antimony/arsenic effect were performed on model PbO_2 surfaces electrodeposited on glass carbon rotating disk electrodes (GC RDE). Pristine PbO_2 surfaces were produced in nitric acid conditions and thoroughly washed with deionized water.

Initial Electrochemical Growth/Sulfation. Glassy carbon was used as the substrate, producing a much smoother and usable deposit compared to other electrode materials [13] Visual inspection suggested that the deposition conditions produced an orange–red α-PbO_2 material on the GC electrode. The active material generated after formation in a LAB is β-PbO_2 and the α-PbO_2 material while reproducible and easily controlled is not desirable to mimic the PAM of a LAB. Various potentiostatic experiments resulted in the formation of α-PbO leading to galvanostatic experimentation to form the appropriate lead dioxide. Low current scan rate at less than or equal to 5 mV/s under galvanostatic conditions formed the desired β-PbO_2.

Figure 1 is the galvanogram for a typical lead dioxide growth experiment over a three-minute time period. The first ten seconds of the experiment are charge absorption and initial formation of the β-PbO_2 along the glassy carbon interface. The small peak in the formation curve is due to a newly conductive interfacial lead dioxide surface forming, charging, and then falling due to capacitance. This process can control the thickness of the deposit by changing the growth time, with increased time leading to an increase in thickness. Figure 2 shows SEM images (backscatter and secondary imaging) of a typical deposit of the β-PbO_2 on the GC substrate and is characterized by globular and porous particles distributed onto the electrode surface.

Cycling experiments in sulfuric acid were then performed with the reproducible β-PbO_2 PAM surface. The sulfuric acid used was a specific gravity of 1.285, typical in many lead–acid batteries. The first experiment was a slow scan rate using ten

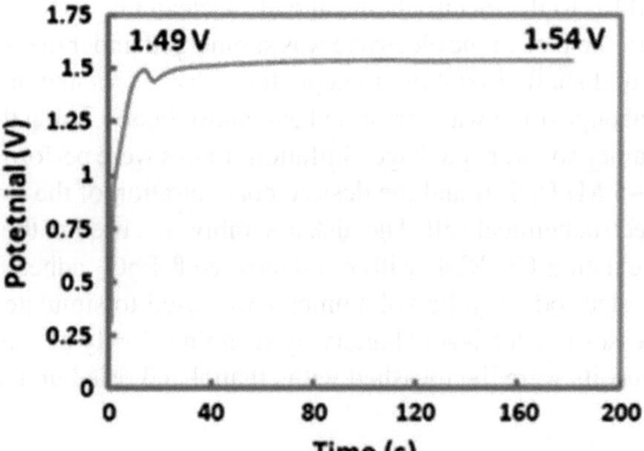

Fig. 1 The galvanogram shows the typical galvanostatic growth profile for PbO_2 formation and resulting deposition onto a GC RDE electrode

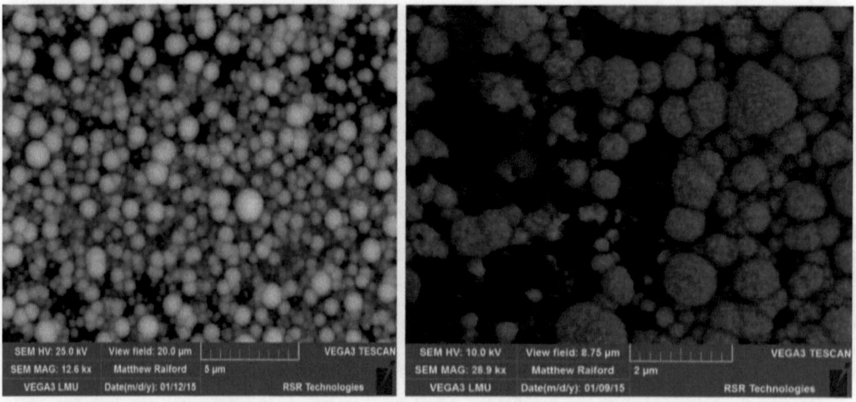

Fig. 2 SEM images of the galvanically deposited β-PbO_2 active material on the GC substrate. (**Left**) A backscatter electron image of the deposit showing the distribution of globular, spherical particles formed. (**Right**) A higher magnification image showing the porosity and fine structure of the particles

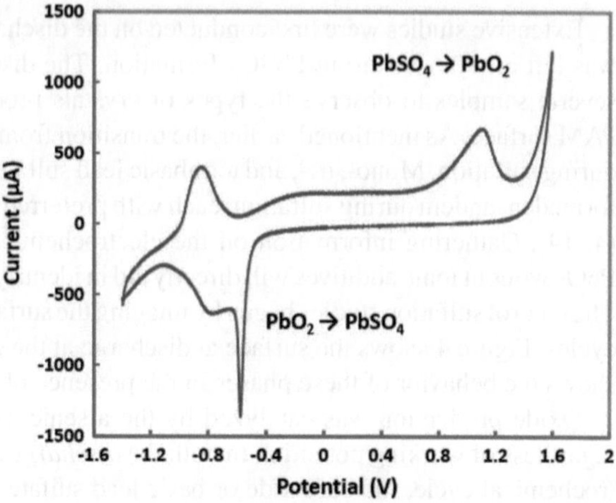

Fig. 3 A cyclic voltammogram of the sulfation process on the deposited PAM over 10 discharge/recharge cycles. The experiment was performed using an Ag/$AgSO_4$ reference electrode and graphite counter electrode

charge/discharge cycles between −1.6 V and +1.6 V. Figure 3 shows the average of the cyclic voltammagram over the ten cycles and the formation of $PbSO_4$ during discharge at −0.795 V and dissolution to PbO_2 at +1.17 V. Similar data were gathered at 50, 100, and 200 cycles. Linear dependence studies of oxidation and reduction activity indicated a slow decay of the charge potential over time. The β-PbO_2 oxidative formation peak is nonreversible which agrees strongly with previous literature. The scan rate of 5 mV/s allowed for a high electrochemical resolution and two distinct peaks are observed in the reductive region of the cyclic voltammagram, Fig. 3. The reversible peak is present at −0.795 V which correlated to other examples of sulfated lead dioxide, the sharp nonreversible peak is possibly the formation of the intermediate $Pb(OH)_2$ product formed before sulfation.

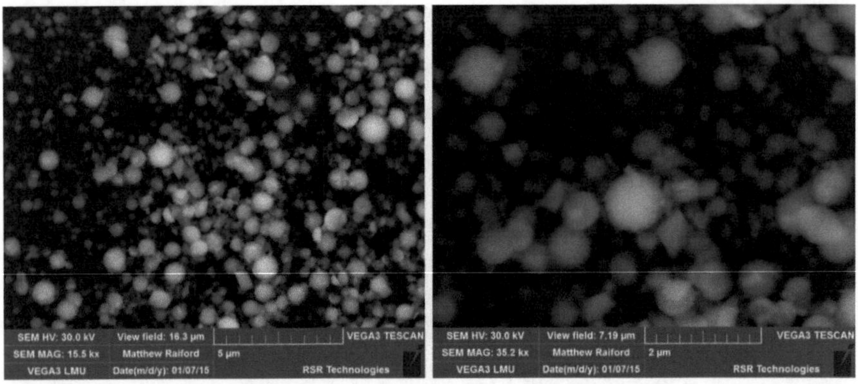

Fig. 4 (**Left**) An image of the β-PbO_2 surface after 10 discharge and recharge cycles. This particular sample was held to discharge at +0.6 V and then studied and imaged. (**Right**) An image focused on a β-PbO_2 sphere covered with small rod-like crystals located near a large prismatic lead sulfate crystal

Extensive studies were first conducted on the discharge product where the surface was left at −0.6 V during $PbSO_4$ formation. The discharge state was studied over several samples to observe the types of crystals produced during sulfation of the PAM surface. As mentioned earlier, the transition from PbO_2 to $PbSO_4$ is incomplete during sulfation. Mono-, tri-, and tetrabasic lead sulfates and different lead oxides are formed in tandem during sulfation, each with preferred sizes, shapes, and orientations [4, 14]. Gathering information on the electrochemical transformation of the pure PbO_2 without ionic additives will directly aid in identifying the effect of each element. The control sulfation studies began by imaging the surface after 10 discharge/recharge cycles. Figure 4 shows the surface at discharge at the end of the ten cycles. Figure 5 shows the behavior of these phases in the presence of Sb and As ions.

Oxide production was catalyzed by the arsenic and antimony ions in solution, regardless of working potential. In both the *discharge* and *recharge* steps of the electrochemical cycle, rods of oxide or basic lead sulfate were produced. At *discharge*, the rods measured 1.944 μm in average length with an average width of 244 nm. Similar dimensions were found at *recharge* with only a slight degree of dissolution occurring. This conductive network of oxide may be beneficial, but there is a high degree of breakdown of the PAM surface, which leaves the performance of this combination in DCA applications unclear. There is a noticeable increase in the current during voltammetric cycles, with an increase in the i_A relative to the As^{+3} and Sb^{+3} trials. The increase in current transport would suggest an increased electrochemical conductivity.

2 V Cell Testing. Cell testing in custom polycarbonate cell casings using five PAM and four NAM lead–acid plates were used to achieve similar active material to electrolyte ratios found in commercially available batteries. Initial one hour and twenty-hour capacity testing measured the benefit of antimony and arsenic on available capacity. SUPERSOFT-HYCYCLE® leady oxide was compared against control of

Fig. 5 (**Left**) *Discharge* of the PAM surface in the presence of both As(III) and Sb(III) ions formed sulfate crystals and ~2 μm in length rods of oxide. (**Right**) Similarly, *recharge* of the materials resulted in the formation of faceted rods of oxide and further oxidation of the PbO_2 globules to a new phase

a typical lead oxide with low amounts of impurities. Table 2 outlines capacity data for each leady oxide, with an increase in capacity measured for the SUPERSOFT-HYCYCLE® relative to the control. The increase in capacity is explained by the Kugelhaufen (aggregate of spheres) model, and antimony acts similarly to $SnSO_4$ [15]. There are several pathways for this improvement: (i) Sb and $Sb^{+3,5}$ ions aid in the dissolution of nonconductive or less conductive PbO_x species on the PAM (ii) Sb and As incorporate into PbO_2 as a doped oxide that is more highly conductive and more easily recharged species. Alternatively, Sb and As species may directly aid in changing the resulting discharge phase, $PbSO_4$, to a more easily dissoluted crystal form. This mechanism is supported by the benefit observed from SUPERSOFT-HYCYCLE® leady oxide used in PAM and NAM active material.

Different cycling tests were performed on the effect of varying discharge rates on the phases in LABs. Differences in capacity can be attributed to the antimony integrating into the PbO_2 and increasing electrochemical conductivity. 2 V cell cycle testing demonstrated the effectiveness of SUPERSOFT-HYCYCLE® in different DOD regimes. The highest benefit is observed at the low DOD cycling performed in Micro-hybrid testing (MHT) packets. MHT, featuring units of low DOD and small recharge times, covers conditions difficult for proper $PbSO_4$ dissolution. Increasing the conductivity and producing a more effectively dissoluted $PbSO_4$ morphology are pathways for longer cycling and lower DOD battery application. Similar principles are advantageous for geochemical dissolution of $BaSO_4$ [14]. $BaSO_4$ dissolution can be adjusted by exposure of ions (Sr^{2+}, Ca^{2+}, and others) and results in significant changes in mechanistic rates. Similarly, Sb^{3+} and As^{3+} participate in adjusting the preferred crystal surface free energies of PbO_2/$PbSO_4$ on the PAM and Pb/$PbSO_4$ on NAM (Fig. 6).

Moderate improvements were observed for three DOD cycling tests (at 17.5, 50, and 100% DOD, BCI standard). Figure 7 shows the comparison of these three tests

Table 2 Comparison of performance data from 2 V cells based on SUPERSOFT-HYCYCLE® and commercially available leady oxide (control)

	Control[a,b]	SUPERSOFT-HYCYCLE®[a,b]	Notes[a]	Improvement
Capacity (C, mAh/mg)	74	84	BCI	13.5%
Capacity (C20, mAh/mg)	113	125	BCI	10%
CA				
DCA (A/Ah)[c]	0.21	0.49	EN	×2
MHT (cycle number)[c]	8000	42,000	EN	×5
DOD 17.5 (cycle number)	3000	4170	BCI	39.0%
DOD 50 (cycle number)	810	1140	BCI	40.7%
DOD 100 (cycle number)	765	990	BCIS	40%

[a]Results were based from 2 V cells at 25 ± 5 °C
[b]Flooded configuration for cells, the cells were maintained
[c]No BCI equivalent

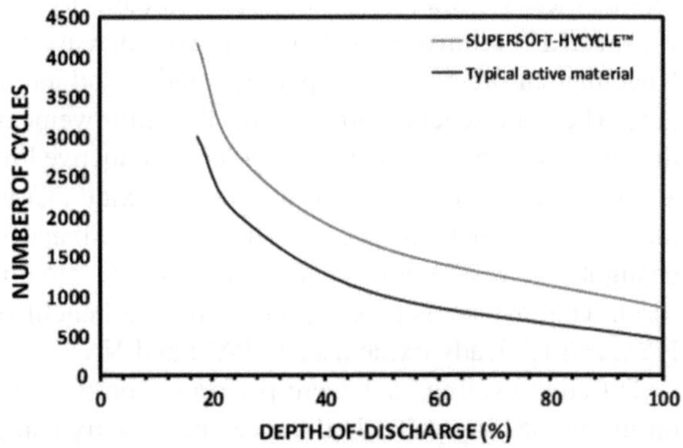

Fig. 6 2 V cell cycling performance at different depth-of-discharge duty of SUPERSOFT-HYCYCLE® compared to a typical leady oxide

compared to a typical soft lead-based leady oxide. Capacity retention is stronger when SUPERSOFT-HYCYCLE® discharges, with additional longevity offered by an increase in electrochemical conductivity over time. This effect was studied by mapping capacity retention over the length of the 2 V cells and via SEM imaging of electrodes ex situ at different points in a cycling regime. The incidence of orthorhombic $PbSO_4$ is significantly lower in samples containing SUPERSOFT-HYCYCLE® relative to the control. $PbSO_4$ particle size is 4.54 ± 0.6 μm and the control particle size is 6.79 ± 0.7 μm. SEM imaging of the battery cells after MHT confirm $PbSO_4$ dissolution is promoted by arsenic and antimony in SUPERSOFT-HYCYCLE®. Also, cured electrodes produced using SUPERSOFT-HYCYCLE® contain tribasic lead sulfate crystals of differing aspect ratio, further supporting crystal modification by the micro-alloying elements.

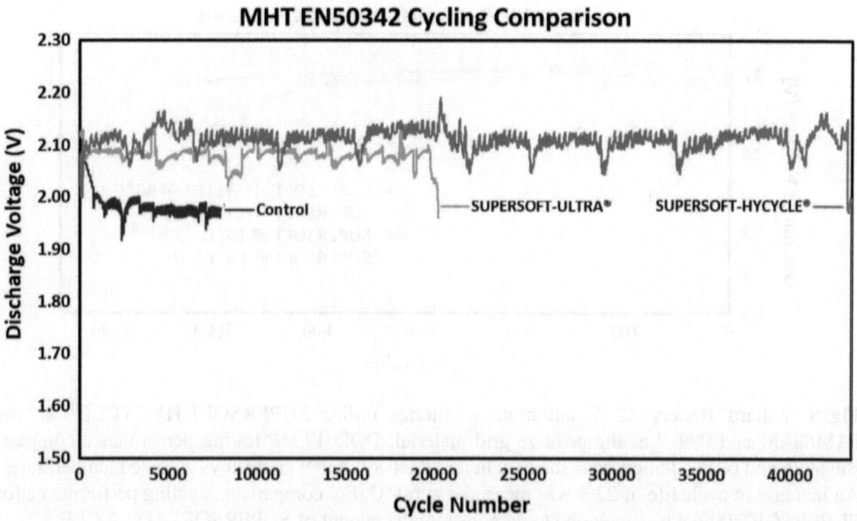

Fig. 7 2 V lead–acid cells for control leady oxide, SUPERSOFT-ULTRA® (a highly pure secondary lead) leady oxide, and SUPERSOFT-HYCYCLE® leady oxide were tested using the MHT test outlined in the EN 50432 standard. Large improvements were measured for SUPERSOFT-HYCYCLE® cells compared to other specifications

12 V Battery Testing. L3-70 Ah batteries were produced using SUPERSOFT-HYCYCLE® in the PAM/NAM and 009™ as the positive grid material. High-temperature DOD 17.5% cycling was performed at 60 °C and 25 °C to test the effectiveness of Ba, Ag at improving cycle performance. Past work has demonstrated the benefit of barium at temperatures above 50 °C. Silver offers additional corrosion resistance with a mild improvement in mechanical strength (YS). Silver and barium provide benefit at very low levels (approximately 100 ppm), providing a pathway to improving and maintaining strength and hardness. A series of batteries were produced to measure the benefit of SUPERSOFT-HYCYCLE® on cycle life at 25 and 60 °C.

Temperature (°C)	SUPERSOFT control[a,b]	SUPERSOFT-HYCYCLE®[a,b]	Notes[a]	Improvement (%)
25	412	946	BCI	130
60	783	1198	BCI	53

Figure 8 shows the cycling performance for Control/009™ and SUPERSOFT-HYCYCLE®/009™ 12 V batteries. The largest benefit is observed on the As, Sb effect OF SUPERSOFT-HYCYCLE® leady oxide-based electrodes. Cycle life increases by 130%, from 412 to 948. This can be attributed to the lower levels of lead sulfate, but also the network building effect of antimony. Antimony contributes to the formation of longer, interwoven $PbSO_4$ that is more easily dissolved. The increase in

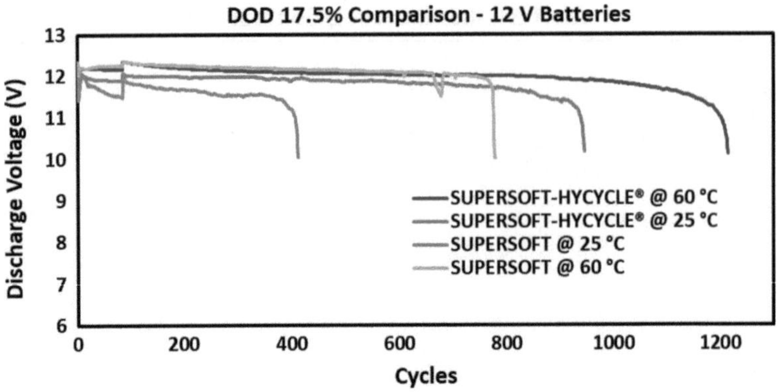

Fig. 8 Willard Battery 12 V automotive batteries utilize SUPERSOFT-HYCYCLE® in the PAM/NAM and 009™ as the positive grid material. DOD 17.5% testing performance comparison at 25 and 60 °C demonstrate the beneficial effects of 009™ grid alloys at raised temperatures. An increase in cycle life of 23% was measured at 60 °C. For comparison, cycling performance for SUPERSOFT/009™ was included to demonstrate the benefit of SUPERSOFT-HYCYCLE®

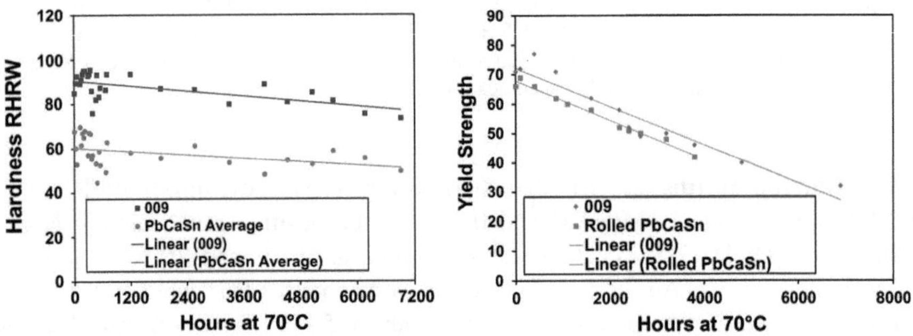

Fig. 9 Mechanical hardness and yield strength testing were performed over 7000 h at 70 °C. Tests were performed on PbCaSn and 009™ coupons in 1.285 s.g. H_2SO_4 at float voltage

conductivity from higher surface area charge pathways contributes to higher capacities and lower recharge times, but the network provides stronger cohesion in the active material paste resulting in decreased PAM paste shedding. 009™ helps inhibit positive grid corrosion. The high-temperature benefit and increased cycle life are observed with SUPERSOFT-HYCYCLE®. The 009™ decreasing PAM corrosion allows for the dissolution effects to provide greater benefit. In batteries made with SUPERSOFT-HYCYCLE®/009™, an improvement of 23% is observed at 60 °C, as shown in Figs. 8, 9.

Conclusion

Understanding the effect of metallic species and ionic species on the crystal growth and dissolution of PbO_2 and $PbSO_4$ in lead–acid battery conditions provides design criteria for more effective active material, SUPERSOFT-HYCYCLE® and grid alloys, 009™. The As and Sb effect studies in electrolyte indicated capacity and dissolution changed in the presence of these species. Further testing in 2 V cells showed improvements in capacity and a spectrum of cycling testing further demonstrating the benefit of micro-alloying and crystal growth control. Long-term mechanical testing of the hardness and yield strength of 009™ indicated the benefit of Ag and Ba in inhibiting PAM grid corrosion. 12 V battery testing strongly supports the benefit of micro-alloying by demonstrating the cycle life improvement from SUPERSOFT-HYCYCLE™ compared to normal leady oxide. High-temperature testing demonstrated the benefit of 009™ by increasing cycle life at 60 °C.

References

1. Swogger SW, Everill P, Dubey DP, Sugumaran N (2014) J Power Sources 261:55–63
2. Sugumaran N, Everill P, Swogger SW, Dubey DP (2015) J Power Sources 279:281–293
3. Pavlov D, Nikolov P (2013) J Power Sources 242:380–399
4. Ebner E, Burow D, Borger A, Wark M, Atassanova P, Valenciano J (2013) J Power Sources 239:483–499
5. Fernandez M, Valenciano J, Trinidad F, Munoz M (2010) J Power Sources 195:4458–4469
6. Zhao L, Chen B, Wu J, Wang D (2014) J Power Sources 248:1–5
7. Ponraj R, McAllister SD, Cheng F, Edwards DB (2009) J Power Sources 189:1199–1203
8. Meissner E (1997) J Power Sources 46:135–150
9. Ralon P, Taylor M, Ilas A, Diaz-Bone H, Kairies, K-P (2017) Electricity storage and renewables: costs and markets to 2030. International Renewable Energy Agency Abu Dhabi
10. Appendix A-USABC Goals for Advanced Batteries for 12 V Start-Stop Vehicle Applications, Unites States Council for Automotive Research (2016)
11. Lam LT, Ceylan H, Haigh NP, Lwin T, Rand DAJ (2010) J Power Sources 195:4494–4512
12. Anari M (2007) Lead-acid battery. J. P. Grant JP3876931B2, 2 Feb 2007
13. Egan D, Low CTJ, Walsh FC (2011) J Power Sources 196:5725–5730
14. Stack AG, Raiteri P, Gale JD (2012) J Am Chem Soc 134:11–14
15. Winsel A, Voss E, Hullmeine U (1990) J Power Sources 30:209–226

Outotec Solutions for E-Scrap Processing

Stephen Hughes, Mikael Jåfs, Hannu Johto, Jan Stål and Janne Karonen

Abstract Scrap materials generated from End-of-Life electrical and electronic goods attract a great deal of attention both as a problem for society as well as an opportunity for new value creation through efficient recovery of the metals and energy content of such materials. Whilst the arisings of such materials appear to be ever increasing with growing affluence and decreasing product life cycle, processing of these secondary materials is faced with many challenges due to their complex and evolving nature. This paper discusses processing of complex scrap materials and how a combination of pyrometallurgy and hydrometallurgy can enable recovery of a wide range of valuable metals. Special attention is needed to address the environmental aspects of processing such materials and Outotec solutions are discussed with environmental performance as a key consideration. This is borne out by the successful experience on commercial projects in countries at the forefront of environmental and operating standards.

Keywords Secondary copper · Recycling · Outotec

S. Hughes (✉)
Outotec Pty Ltd., 12 Kitchen Road, Dandenong South, VIC 3175, Australia
e-mail: stephen.hughes@outotec.com

M. Jåfs · H. Johto · J. Karonen
Outotec (Finland) Oy, Rauhalanpuisto 9, 02230 Espoo, Finland
e-mail: mikael.jafs@outotec.com

H. Johto
e-mail: hannu.johto@outotec.com

J. Karonen
e-mail: janne.karonen@outotec.com

J. Stål
Outotec (Sweden) AB, Gymnasievägen 14, 93127 Skellefteå, Sweden
e-mail: jan.stal@outotec.com

G. Gaustad et al. (eds.), *REWAS 2019*, The Minerals, Metals & Materials Series,
https://doi.org/10.1007/978-3-030-10386-6_33

The Importance of E-Scrap as a Resource

In recent years increasing attention has been drawn to the issue of the responsible processing end-of-life Waste Electrical and Electronic Equipment (so-called WEEE). The terms "Electronic Waste", "E-Waste", "Electronic Scrap" or "E-Scrap" are also widely used to describe such materials. It has been estimated that the amount of WEEE globally in 2016 was around the level of 46 Mtpa [1], hence representing a major waste disposal challenge, but simultaneously also representing a tremendous opportunity in terms of potential resource recovery. The official recovery rate of the WEEE is now only 20% [1], indicating high potential increase to available material for treatment. The rate of availability of WEEE is also growing faster than general economic growth and is estimated to be increasing by at some 3 to 5% per annum in the EU [2]. The rate of production of electrical and electronic equipment gives some leading indication of the future growth globally with increasing affluence and digitalization. As an indication, taking a consumption level of 15 kg/person as a benchmark value for a developed economy, would give a potential future global WEEE generation rate in excess of 100 Mtpa.

Material classified as WEEE can vary widely ranging from LED lights and mobile phones through to large appliances such as air conditioning systems, refrigerators and washing machines. A characteristic of such manufactured products is that the constituent materials will vary considerably between products and can be expected to change over time. Different products will also have varying life cycles.

The major constituents of WEEE, on average, are plastics and iron/steel and together these can typically account for 60–80% of the overall mass of the waste material. Whilst there are equipment and techniques available for separating and recovering plastics and iron/steel, the relatively low value of these materials can be an impediment to the recycling of WEEE. Significant amounts of aluminium and copper are often present in WEEE and copper, in particular, is subject to high levels of recycling due to its relatively high value. Copper is predominantly contained in the wiring, motors and controls of appliances and can be physically separated to a certain extent. Perhaps most attractive of all are the values of gold and precious metals that can be found in the printed circuit boards (PCB's) that are a feature of many modern appliances. The mass of PCB's, on average, is only some 2–3% of the total mass of WEEE. In addition to the materials described above a wide range of other elements are present in WEEE such as silicon, lead, tin, nickel, halides, antimony and other minor metals, Fig. 1.

Overview of Metals Production Cycle

The harmonisation of resource, technology, and environmental cycles has been described extensively by Reuter et al. [3]. Metals entering the production cycle are derived from ores mined from the earth's crust, which are typically subject to a con-

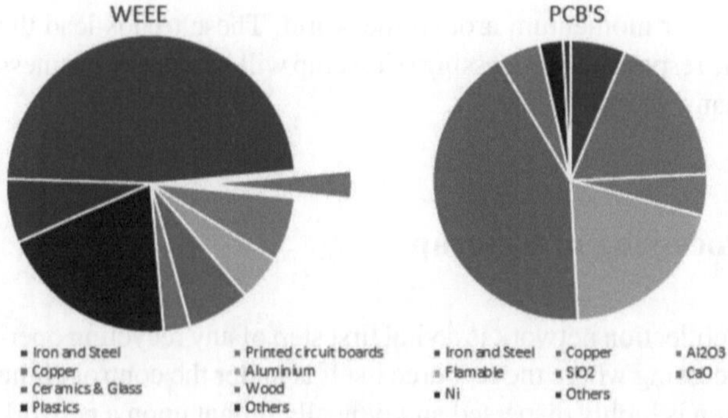

PCB's, minor elements

		min	max
Ag	ppm	10	7000
Au	ppm	0.7	600
Pd	ppm	0.5	300
Cl	%	0.01	1.9
Br	%	0.01	11
Calorific value	MJ/kg	6.5	9.2

Fig. 1 The average WEEE and PCB composition based on materials tested at Outotec

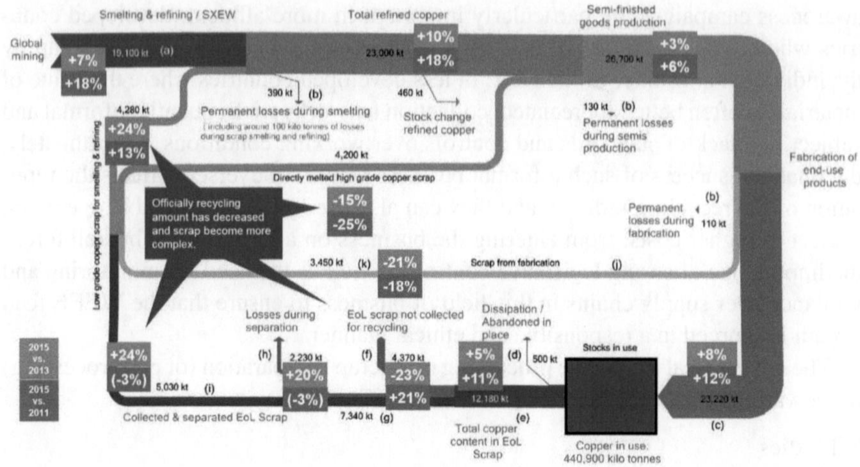

Fig. 2 Changes in global flows of copper based on data reported by the International Copper Study Group for the years 2011, 2013 and 2015

centration process, followed by smelting and refining to produce final pure forms of metal for consumption. A proportion of the metals contained in end-of-life products are then typically recovered during the secondary production cycle as illustrated in Fig. 2 [4, 5].

The recycling of metals has become increasingly important for the following reasons:

(i) Typically, lower losses and energy requirement (hence lower costs) to produce refined metals from secondary sources compared with primary sources.
(ii) Societal pressure and economic incentives to reduce wastes going to disposal, especially where hazardous materials are involved.

The primary metals production industry is faced with the dual challenge of declining ore grades and rising energy costs, whilst environmental awareness and responsibility also continue to gather momentum around the world. These trends lead the authors to conclude that the responsible processing of e-scrap will be a major business growth opportunity for many years to come.

Key Issues in the Processing of E-Scrap

An effective and efficient collection network is a vital first step of any recycling operation. Unlike primary processing where the resource itself is under the control of the processor, WEEE collection is highly dispersed and typically reliant upon a network of operators all seeking to secure some part of the value from the recycling process. Recycling schemes supported by legislation, suitable infrastructure and public awareness campaigns are particularly important in more affluent, developed countries where typically little, nil or even a negative value is ascribed to the scrap by the individual consumer of the EEE. In less developed countries where the value of materials is often better appreciated, collection networks are frequently informal and subject to a lack of standards and controls over working conditions. Unfortunately, the financial success of such informal business ventures adversely affects the reputation of the recycling industry and they can also prevent more reputable operators, subject to higher costs, from entering the business on a local level. This reinforces the importance of strong legislative controls as well as the need for monitoring and vigilance over supply chains in this field of business to ensure that the WEEE feed stream is sourced in a responsible and ethical manner.

The next critical step in the processing of e-scrap is separation (or pre-processing) of the various fractions, such as:

- Plastics
- Ferrous scrap
- Non-ferrous scrap
- Printed circuit boards and other higher value components.

Techniques and equipment can range from manual disassembly and sorting through to mechanical systems employing processes such as crushing, shredding, screening, magnetic separation, eddy current separation, electrostatic separation, x-ray and/or infrared scanning. Such equipment can be extensively automated and is available from a range of manufacturers.

Whilst separation into various intermediate product streams can be achieved using the techniques described above, there are limits to the capabilities of physical separation processes. In particular, the highest value fraction of the waste (typically the printed circuit board/complex electronics) are highly integrated, multi-metal components that cannot be easily broken down and separated physically into their basic metal components.

The analysis of such complex scrap fractions, that are subject to final processing for non-ferrous metals recovery, can be expected to evolve continually as technology evolves with factors such as miniaturisation, displacement and design for recycling.

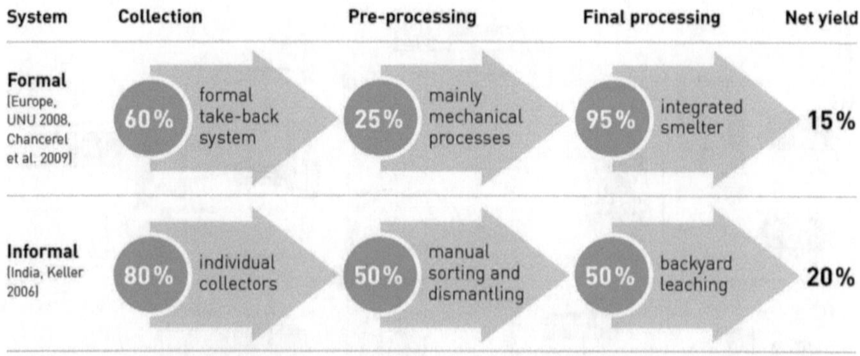

Fig. 3 Recycling efficiency between a common formal system in Europe and the informal sector in India for the gold yield from printed wire boards [6]

Typically, and as seen from the range of analyses above, such complex scrap fractions will contain a wide range of elements or components. Certain elements or components present challenges to downstream processes that require special consideration, such as:

- Plastics and other hydrocarbons
- Flame retardant materials
- Aluminium and aluminium oxide.

As seen from the data above, e-scrap also has a significant energy value that can enable pyrometallurgical processing to be performed with minimal supplementary fuel requirements as well as providing for opportunities for energy recovery in the process.

Stepping back and looking at the overall efficiency of the combined collection/pre-processing and final processing steps based on an analysis of data by UNEP [6] illustrates an interesting aspect of value realisation from WEEE, namely that the labour intensive and environmentally harmful informal sector may be achieving a higher net recovery of values compared with supposedly more highly developed systems. Whilst the final processing step that is the main subject of this paper has been developed to a high level, there would appear to be substantial room for improvement in the front-end collection and pre-processing steps applied in Western countries to maximise the overall recycling efficiency. The data would also support the goal of minimising the extent of pre-processing as far as possible in order to realise the benefit of high recovery that can be achieved in the final processing step. For example, it is significantly more efficient to direct small electronic devices to the final processing step with minimal pre-processing (e.g. removal of battery only) (Fig. 3).

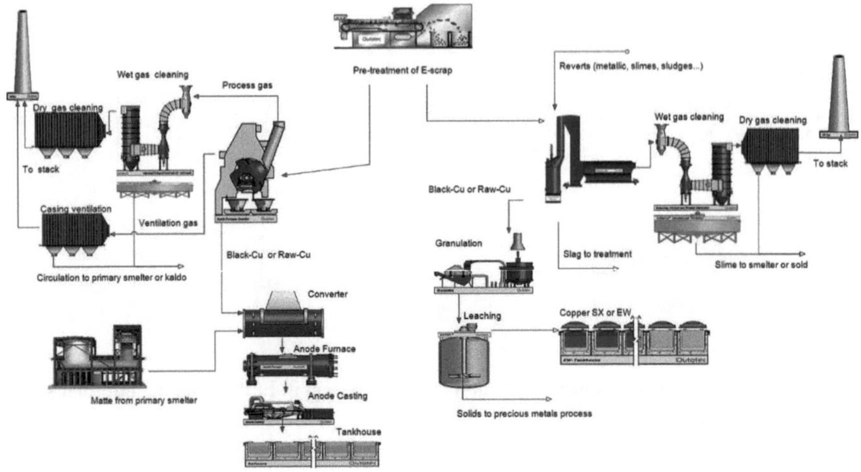

Fig. 4 Flowsheet examples of WEEE treatment with combination together with primary production (left) and on stand-alone basis (right)

Outotec Solutions

Outotec is a leading supplier of technology solutions for the metals production industries, with a long history of developing innovative equipment and processes with strong environmental credentials [7]. In the field of processing of complex secondary copper feed materials, Outotec offers a combination of proven pyrometallurgical and hydrometallurgical processes to produce high-quality refined metal products in an efficient and environmentally friendly manner. These process combinations are illustrated in Fig. 4, showing examples of overall plant schemes to treat WEEE.

Pyrometallurgy

Pyrometallurgical processes have long played a major role in the production of copper from primary and secondary sources, with most of world copper obtained through one or more pyrometallurgical processing steps.

Primary processing typically involves oxidation of sulphide concentrate/matte in a two-stage process to produce blister copper. The oxidation of sulphides liberates energy that can be used for melting of scrap, and the addition of various grades of copper scrap and reverts has been a long-standing practice in primary smelters. The addition of e-scrap as a potential feedstock, however, introduces complications in terms of impurities, energy load and environmental management. Due to the impurities and energy content of the e-scrap, addition in the first (smelting) stage of the process would be the more desirable approach as the energy liberated by combustion of plastics can be used to displace fossil fuel, whilst many impurities are typically more effectively removed across a two-step process by taking advantage of different oxygen potential conditions and the presence of a matte phase. Whilst the management of impurities can be a limiting factor, an even more significant challenge relates to the increased volumes and heat load of the smelter offgas (a limiting factor for many established smelters) and potential for dioxin and furan formation by de novo synthesis in the offgas train. These factors tend to limit the amount of e-scrap that can be added to primary smelters, but the consumption of e-scrap can be expected to increase in future as primary smelters learn to optimise their operations to manage such complex materials in the context of their overall operations.

Traditional secondary copper processing involved a combination of reductive smelting, followed by converting to produce copper of suitable quality for anode fire refining/casting followed by electrolytic refining. Typically, significant volumes of relatively clean scrap would be added to dilute impurities. More recently, several plants have been established that employ oxidative smelting/converting type processes, followed by metal granulation/leaching/electrowinning. Such a scheme can allow for treatment of less pure copper, enhancing the flexibility of a smelter.

A feature common to any smelter that introduces complex materials into its' feedstock is the need to establish plants for the treatment of by-products or link into a network of processors who can extract the full value of the valuable minor metals such as lead, zinc, tin and nickel.

Outotec offers a number of leading pyrometallurgical processes as part of its Smelting portfolio, including:

- Outotec® Flash Smelting Process, typically applied for the smelting of copper and nickel sulphide concentrates, but capable to accept a proportion of e-scrap in its feed mix.
- Outotec® Ausmelt Top Submerged Lance (TSL) Process, that can be applied to treat a wide range of feed materials, including up to 100% secondary materials in either continuous or batch method of operation.
- Outotec® Kaldo Process, that can also be applied to wide range of feed materials, including up to 100% secondary materials in a batch mode of operation.
- Outotec® Peirce Smith Converting Process, is a modern variant of this traditional technology that has been upgraded to include enhanced hooding for control of emissions and automated control features.
- Outotec® Fire Refining Process, applied for the refining of blister copper/crude copper to remove residual sulphur and certain minor elements, then de-oxidation to produce anode quality copper.
- Outotec® Anode Casting, applied for the accurate casting of anodes suitable for electrorefining.
- Outotec® Slag Cleaning Process, involving the use of an electric furnace for the recovery of values from slags prior to discard.

Integration of each pyrometallurgical process with the most appropriate furnace gas handling systems is a particular area of strength of Outotec, with a wide range of solutions available, ranging from dry/wet gas cleaning systems through to sulphuric acid production from sulphur dioxide laden gas streams. As mentioned previously, for processes treating e-scrap special consideration is required for the management of dioxin/furan, preferably by minimising the extent of formation of such compounds but, if necessary, implementing equipment to capture such compounds and avoid emission to atmosphere. The Outotec Wet Gas Cleaning technology for Electronic Scrap Treatment is a reliable solution to remove solid particles, halides and trace amounts of SO_2 in the process gas, with an efficient use of the required utilities. The system is designed to minimise dioxin formation. The process gas is quenched, scrubbed, cooled and filtered to achieve an excellent removal efficiency of pollutants as dioxin mercury and sulphur.

Hydrometallurgy

Hydrometallurgical processes have a strong role in the production of copper especially from primary sources as an anode-to-cathode electrorefining step in the pyrometallurgical route or as a direct hydrometallurgical route through the solvent extraction and electrowinning route.

Scheme 1: Anode Electrorefining

Electrorefining of copper anodes (Cu > 99%) is well-established technology to produce standard, high-grade (LME-A grade) copper cathode. Outotec technologies include anode preparation machine, permanent cathodes, electrolytic cells, busbar systems and tankhouse crane and bale including anode slime processing in PM processing plant.

Outotec digitalisation solutions enable fully automated tankhouse operations including automatic crane, robotic cathode stripping machines and process automation with CellSense® and TIMS systems.

Scheme 2: Leaching—Solvent Extraction—Electrowinning

In case high-quality copper for anodes cannot be produced economically, leaching of raw copper followed by copper solvent extraction and electrowinning is an option. This processing technology enables the same copper cathode quality and refining of precious metals than conventional electrorefining route. The main benefits are much higher tolerance towards impurities in feed material and more scalable investment.

Leaching is done atmospherically in sulphuric acid solution by introducing oxygen in the leach reactor. Copper and some other components are leached in pregnant leach solution (PLS), which is processed in solvent extraction plant to selectively extract copper and transfer into electrowinning plant. LME-A-grade copper cathodes are produced. Impurities are managed by taking a small bleed stream out of the process are precipitating impurities before recycling the solution back to the main process.

Precious metals remain undissolved in leaching residue and they are further processed in the precious-metal plant.

Precious Metals Recovery

Anode slime or leaching residues are processed in the PM processing plant. High-grade PM- and e-scrap can be fed into the smelting stage. Outotec offers solutions for complete precious metals recovery and refining and can include the following stages:

- Copper leaching with tellurium recovery
- Smelting and converting in Kaldo furnace or Trof converter and anode casting
- Silver electrorefining and silver ingot casting or granulation
- Gold refining and ingot casting
- Refining of platinum group metals
- Wet gas cleaning with selenium recovery or separate selenium roasting system
- Wastewater treatment for high recovery of valuable metals.

Case Studies

The following case examples showcase a few technologies used in the recycling of secondary copper in Asia and Europe.

Case 1: Use of Ausmelt TSL Process in Secondary Copper Plants in Asia

The Ausmelt TSL Process has been successfully applied in two important secondary copper plants in Asia, that are dedicated to the processing of secondary materials and are set up to enable multi-metal recovery.

Kosaka Smelting & Refining Co., Ltd., a subsidiary of Dowa Holdings Co., Ltd. (Dowa) operate a smelter located at Kosaka, that has been recognised for mastering the treatment of complex materials. A flash smelting furnace was implemented in 1967 [8], but due to changing market for feed materials, in 2005 Dowa elected to implement new pyrometallurgical technology that could accept up to 100% of non-sulphide feed materials. In targeting enhanced flexibility of the plant another key criterion was that the new process must conform with strict local environmental

Fig. 5 View of GRM Plant [10]

regulations. The Ausmelt TSL Process was extensively tested at pilot scale and then selected as the pyrometallurgical process for the project [9].

The new Ausmelt TSL furnace was started up in 2007 and continues to operate to treat up to 150,000 t/y of a wide range of feed materials and enables the recovery of 16 different metals by effective integration into the wider Dowa network of copper, lead and zinc smelters and refineries.

GRM Co. Ltd. was established in 2008 to develop a new secondary smelter plant at Danyang, Korea. A contract for the licensing and technology supply relating to the Ausmelt Process was signed in 2008 and the project progressed to start up in 2011. The plant has been designed with capacity of 110,000 t/y of secondary feeds and produces black copper (~75% Cu grade) as main product, with ferro silica and gypsum as by-products. The black copper acts as an effective collector for precious metals that can be recovered in further refining steps (Fig. 5).

Case 2: Use of Kaldo Process in Rönnskär Smelter, Sweden

The Boliden Rönnskär Smelter bases its operation on integrated primary smelting and secondary recovery. The smelter has been treating secondary materials and secondary copper for a long time and has been a forerunner in developing technology for treating Cu scrap with high content of organics.

The Kaldo technology was developed in the mid 70s and the first burning and smelting of copper "e-scrap" or copper with high plastic and organic content started already back in 1980. To collect mercury and dioxins a wet- and dry gas cleaning system was designed to meet the highest environmental standards. The burning of the scrap generates a lot of heat which is recycled and converted to electricity or district heating. The smelted e-scrap forms black copper, with 65–75% copper that is fed hot into the Pierce-Smith converters and integrated to the smelters main copper stream for further extraction of copper and precious metals.

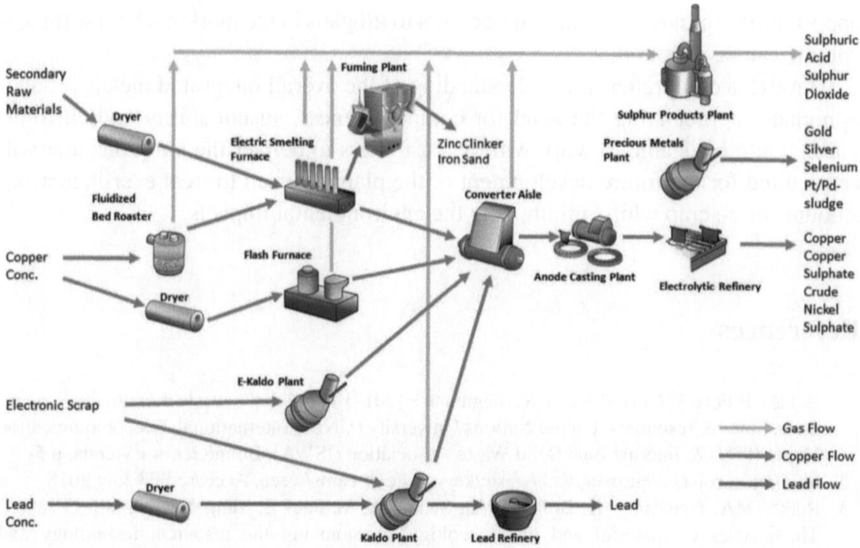

Fig. 6 Flowsheet of the Rönnskär integrated smelter [11]

In 1990, the old lead electric furnace operation was shutdown, and the lead concentrate smelting was performed using the Kaldo furnace, and the Kaldo operation was split into lead smelting- and e-scrap smelting campaigns of 1–2 months, respectively (Fig. 6).

The Rönnskär e-scrap recycling operation expanded over the years and in 2012 a new Kaldo plant and a new material sampling and processing facility was commissioned. The new plant complemented the existing Kaldo plant giving a total production capacity of 120,000 t/y of e-scrap for the Rönnskär smelter.

Concluding Remarks

The recycling of more complex scrap materials is expected to grow at a faster rate than overall metals production, due to the combined effect of metal inventory growth and great scope for improvements in collection and pre-processing efficiencies. This should present attractive business opportunities to companies with access to the necessary social license, capital, feed sourcing/product sales networks, processing plant technology and operating know-how. Companies established in the metals production chain have a significant head start, possessing many of these requisites and are likely to shape their operations in future to incorporate additional secondary feeds by maximising the proportion of e-scrap into the feed, where possible or establishing complementary dedicated secondary copper processing facilities. There may also be opportunities for new secondary processors to fill gaps in the market where sufficient volume can be captured.

Through a comprehensive understanding of the overall integrated metals production chain, appreciating the need for competitiveness, sustainability and customisation, Outotec is able to work with its customers to deliver the solutions that will be required for the future development of the plants needed to treat ever-increasing amounts of e-scrap whilst minimising the environmental impacts.

References

1. Baldé CP, Forti V, Gray V, Kuehr R, Stegmann P (2017) The global e-waste monitor 2017, quantities, flows & resources. United Nations University (UNU), International Telecommunication Union (ITU) & International Solid Waste Association (ISWA), Bonn/Geneva/Vienna, p 5
2. http://ec.europa.eu/eurostat/web/waste/key-waste-streams/weee. Accessed 23 July 2018
3. Reuter MA, Heiskanen K, Boin U, Van Schaik A, Verhoef E, Yang Y, Georgalli G (2005) The metrics of material and metal ecology, harmonizing the resource, technology and environmental cycles. Elsevier Science, Amsterdam, p 706
4. Glöser S, Soulier M, Espinoz LAT (2013) Dynamic analysis of global copper flows. global stocks, postconsumer material flows, recycling indicators, and uncertainty evaluation. Environ Sci Technol 47(12):6564–6572
5. International Copper Study Group press releases (2013) Data: http://www.icsg.org/index.php/press-releases/finish/170-publications-press-releases/2092-2015-10-03-icsg-factbook-2015, date of release 30/11/2015, 2015 data: http://www.icsg.org/index.php/press-releases/finish/170-publications-press-releases/2462-2017-10-11-icsg-factbook-2017, date of release 26/10/2017 and 2011 data: http://www.icsg.org/index.php/press-releases/finish/170-publications-press-releases/1188-2013-world-copper-factbook, date of release 19/12/2012, all accessed on 24/09/2018
6. Reuter MA, Hudson C, van Schaik A, Heiskanen K, Meskers C, Hagelüken C (2013) UNEP (2013) metal recycling: opportunities, limits, infrastructure, a report of the working group on the global metal flows to the international resource panel
7. Nykänen P (2016) 150 years' evolution toward a greener future, Outotec Oy, Espoo, p 240
8. Kojo IV, Storch H (2006) Copper production with Outokumpu flash smelting: an update. In: Sohn international symposium, international symposium on sulfide smelting 2006, proceedings, vol 8, p 230. TMS
9. https://www.theage.com.au/business/ausmelt-wins-key-japanese-contract-20051025-ge141b.html. Accessed 24 Sept 2018
10. https://www.grm.co.kr/. Accessed 24 Sept 2018
11. Personal communication, Ulf Degerstedt, Boliden Rönnskär Smelter, 24 Sept 2018

Rare Earth Magnet Recovery from Hard Drives by Preferential Degradation

Brandon Ott, D. Erik Spiller and Patrick R. Taylor

Abstract Neodymium recycling by the mineral processing practice of liberation and separation of hard disk drives is envisioned and evaluated. Magnetic material is liberated from the hard drive, constructed mostly of malleable metals, by preferential degradation of the brittle magnet material. The process developed is shown to recover greater than ninety-five percent of the magnet material with a product grade of over 80% magnet material by mass. The process is designed to co-produce stainless steel, aluminum, nickel alloy, carbon steel, and printed circuit board concentrates as contributors to the recycle value of hard drives.

Keywords Recycling · RE magnets · Rare earths · Neodymium

Introduction

Rare earth (RE) magnets are used for their relatively high magnetic field potentials relative to other magnet materials. They are primarily composed of the phase $Nd_2Fe_{14}B$ but often contain amounts of Praseodymium and Dysprosium. The lack of domestic supply contributes to a volatile rare earth market. The demand is expected to increase due mainly to the growth of the electric car market [1]. It is estimated that about 115 million Hard disk drives (HDDs) are decommissioned annually in America and up to 325 million HDDs worldwide per year. About 85% of neodymium produced is consumed in RE magnet applications [2].

RE magnet material occurs along with aluminum, stainless steel, nickel, carbon steel, and printed circuit boards (PCB). The processes developed by this research are evaluated based on estimates of the value of process output streams and estimates of necessary equipment and operating costs to build and operate such a plant.

B. Ott (✉) · D. E. Spiller · P. R. Taylor
Kroll Institute for Extractive Metallurgy, Colorado School of Mines, Golden, CO, USA
e-mail: bott@mines.edu

G. Gaustad et al. (eds.), *REWAS 2019*, The Minerals, Metals & Materials Series,
https://doi.org/10.1007/978-3-030-10386-6_34

Others have conducted research for the dismantling or shredding of HDDs to recover rare earth magnet material. To our knowledge, none of the earlier researchers focused on recovery of RE magnet material along with separated marketable streams of the various metals and PCBs in the HDD [3–7].

Experimental Equipment and Procedures

Experiments were carried out in the Laboratories of the Kroll Institute for Extractive Metallurgy at the Colorado School of Mines. Process equipment, particularly those that will retain RE magnet material are cleaned before and after processing. Crushing was completed with a slow speed Amos mfg. shredder that shears malleable metals as well as fractures brittle materials. The shredder was cleaned before and after use to prevent material hang-up affecting results. There were two grinding mills used in this research for different testing scales. Both of these mills are of a steel construction and so magnet material is demagnetized prior to grinding to enable discharge of RE magnet material. The laboratory scale tests used a 7 × 11 inch rod mill with varied grinding time between 1 and 2 h. Larger tests utilized a one-meter diameter by 1/3 m deep ball mill and 2.5 in. steel balls. Demagnetization is accomplished thermally. This step also burns epoxies and plastics found in the HDD. This demagnetization is carried out in a furnace with airflow and ventilation to a fume hood at about 350 °C. Depending on process specifics, whole hard drives were demagnetized. Various laboratory separations equipment was used. These include high-intensity magnetic drum, low-intensity magnetic drum, Davis tube, rare earth roll magnet, shaking table, eddy current, and vanning plaque. For the Davis tube and shaking table, it was necessary that the material be suspended in water with the use of a common commercial detergent as surfactant. Screening is accomplished with either a Gilson screen or Tyler sieve stack in a ro-tap.

Analytical Procedures and Tools

Analysis of both feed materials and products were completed by several analytical tools at the Colorado School of Mines and at Hazen Research. The results and conclusions were based on ICP-MS and confirmed by ICP-OES results. Fire assay was used to determine the gold and silver composition of process outputs and of printed circuit boards. Circuit boards were prepared by burning at 350 °C for 30 min and then pulverized by a ring and puck pulverizer. Head grades were determined by manual disassembly of hard drives; including removing the magnet from the magnet assembly and the bonded magnet from the spindle motor assembly. The whole of both magnets was digested using aqua regia and diluted for ICP-MS analysis. The RE magnet material is tracked through the various process steps by quantitative analysis of the neodymium composition. The composition of oversize fractions was analyzed

Fig. 1 Sorted metals from shredded hard drive

Table 1 Table of results from several disassembled Hitachi Deskstar HDDs used to compute the head grade of the process (Masses in grams)

Sample	Mass of HDD	Mass of magnet assembly	Mass of magnet	Magnet wt% Nd	Head grade (%)
1	849	76.8	22.1	27.5	2.6
2	850	81.9	22.1	27.7	2.6
3	844	81.3	22.1	27.5	2.6
4	848	81.9	22.1	27.8	2.6
5	844	81.0	21.7	27.6	2.6
AVG	847	80.6	22.0	27.6	2.6
STDEV	2.5	1.90	0.14	0.13	0.00

and used to calculate recovery and grade for aluminum, carbon steel, stainless steel, nickel alloy, and printed circuit boards. This was completed by sorting each fraction by hand and massing the components as part of each fraction. The grade was approximated by the relative composition of each fraction (Fig. 1).

Results and Discussion

The analysis of RE magnet material in the hard drive feedstock was performed by replicate disassembly and analytical chemistry to determine the neodymium content per hard drive (Table 1).

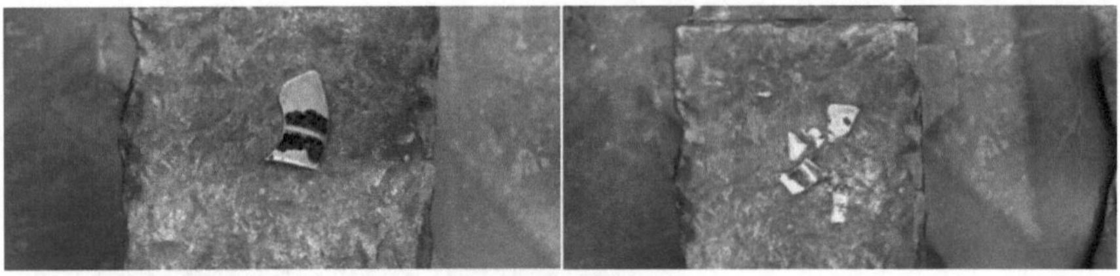

Fig. 2 Results of magnet hammering experiment

Preferential Degradation Theory and Discussion

The goal of this project was the creation of a process that would recover RE magnet material by taking advantage of shredded hard drives as a resource and production of a marketable steel, aluminum, stainless steel, nickel and PCB concentrate liberated in the course of liberation of the magnet material. $Nd_2Fe_{14}B$, the primary phase in the magnet material, is an intermetallic compound that exhibited brittle fracture when loaded mechanically. It is expected that a magnet would fracture readily under impact due to this observed brittle nature (Fig. 2).

A simple experiment was designed to test this expectation by impacting a large piece of magnet material on an anvil using a hammer. The observed result was that the magnet breaks down quite readily upon impact. It is expected from this test that the brittle fracture of magnet material can be achieved by impacting the heterogenous shredder output containing both magnet material and metals. It is expected in this operation that magnet material can be degraded preferentially while the metals remain at large particle sizes exhibiting mostly plastic deformation rather than fracturing. This concept of size reduction of magnet material by impacting the heterogenous shredder output is called preferential degradation for the purposes of this paper. The concept is that the preferentially degraded material can be screened to separate a RE magnet material concentrate. The screen oversize would contain the metals while the undersize would include the crushed magnet material and ground circuit board material. The repeated impacts necessary for preferential degradation were to be achieved using a ball mill (Fig. 3).

Initial Preferential Degradation Flowsheet

Initial experimentation of the preferential degradation flowsheet consisted of testing a 40-hard drive sample. The sample showed that preferential degradation by ball milling could be achieved, degrading the RE magnet material to a finer grain size than the malleable metals.

The undersize output was split and used for testing various additional unit operations for increasing the output grade. The result of the upgrade study shows that RE

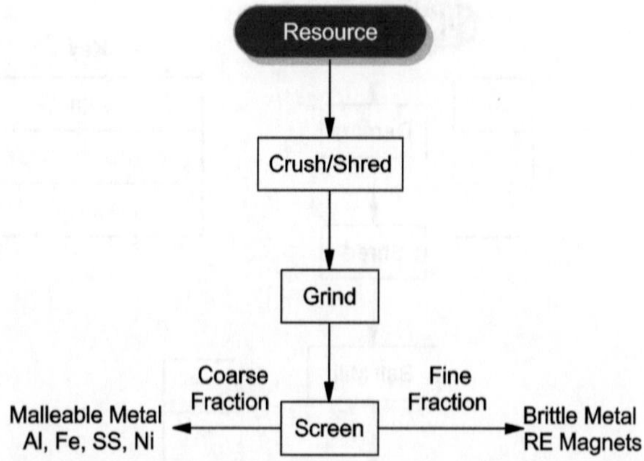

Fig. 3 Concept preferential degradation flowsheet

Table 2 Results from upgrade studies on ball mill fine fraction from Trial 1. Density separations made use of a vanning plaque with different intensities. The magnets (samples 2, 4, 6) were simple separations taking advantage of the several separators available. The mag circuit was a process utilizing the LIMS magnet arranged such that there is a scavenger and cleaner unit operation

Sample	Input magnet grade (%)	Method	Magnet material grade in output (%)	Recovery of magnet material (%)
1	32	Density	47	59
2	32	RE drum	18	46
3	32	Density	59	28
4	32	LIMS	51	81
6	32	Beltmag	25	60
7	32	Mag circuit	59	79

magnet material is liberated in the ball mill grinding operation, additional grinding was not necessary to produce a higher grade output. This research gives evidence that influenced later research direction by identifying the separation forces that may be useful in separating the RE magnet material from expected diluents. The rare earth drum and rare earth roll magnet separator both showed low recoveries and low concentrate grades. It is expected that these separations methods retained, within the process equipment, large amounts of high grade material. The LIMS unit retained material as well but was relatively easy to clean out to recover trapped material. The results of this study indicate that low intensity magnet separation and density separation are potentially useful for treatment of the process outputs to ultimately produce a higher grade product (Table 2).

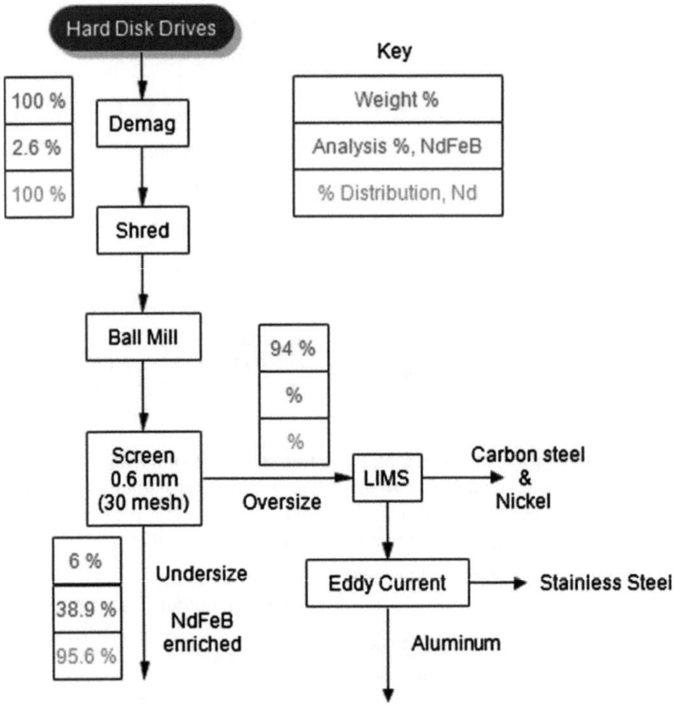

Fig. 4 Preferential degradation flowsheet

Experimental Flowsheet 1

The shredder output is fed directly to the ball mill and ground. The process flowsheet for this is shown in Fig. 4.

The outputs of these two processes were analyzed further to determine the composition and material liberation of the undersize output. The upper part of this flowsheet was tested several times to demonstrate the process viability and confirm grade and recovery results observed. The flowsheet presented above demonstrated recovery of 96% of the RE magnet material in the feed at a product grade of 39 wt% RE magnet material.

Preconcentration by Taking Advantage of Magnetic Behavior

Visual inspection of the shredder product identified clumps of material clustered around RE magnetic material. It was postulated that by separating out these clumps that the magnet material could be concentrated prior to additional processing. As seen in the disassembly of hard drives the RE magnet material is associated with the

Fig. 5 Conceptual preconcentration flowsheet

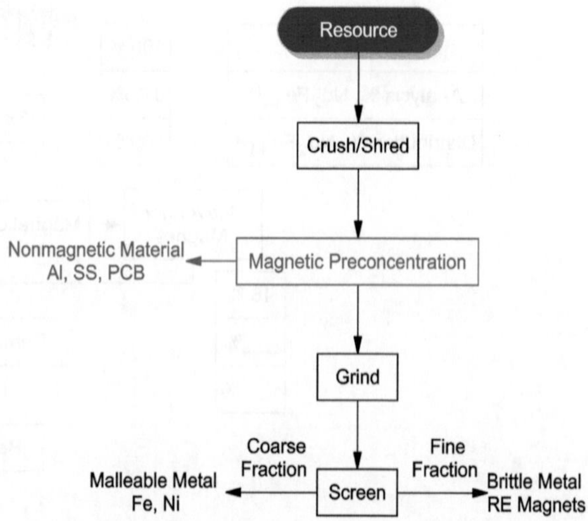

magnet bracket and liberation from that assembly is the largest obstacle to recovery. The magnet bracket is strongly ferromagnetic and therefore segregates to the magnetic fraction when processed by a magnetic drum. It was expected that by magnetic preconcentration (prior to ball milling), barren nonmagnetic material containing little to none of the rare earth in the feed could be removed from the process flow prior to costly demagnetization and grinding unit operations. It was thought that this can be accomplished while simultaneously producing a higher grade RE magnet output. The challenge was accomplishing this without sacrificing recovery (Fig. 5).

This was first demonstrated at laboratory scale with a three-hard drive sample. The laboratory mill results showed an increase in product grade while rejecting 70% of the material prior to grinding. This experiment showed a recovery of 92% of the magnet material with a grade of 75%. The scale-up test of this process design utilized an 80 hard drive feed to approximate a similar ball mill charge to the original preferential degradation flowsheet. The low-intensity magnet drum magnet was used to maximize recovery of RE magnet material to the magnetic fraction. The results of this process can be seen in Fig. 6.

The REE magnet product analyzed at 50 wt% RE magnet material leaving the possibility of further upgrading. One possibility is by further low-intensity magnetic separation which can be simulated by the Davis tube (wet) separation. The Davis tube shows that RE magnet material processed by preferential degradation can be concentrated up to a grade of 80 wt% just by physical separations.

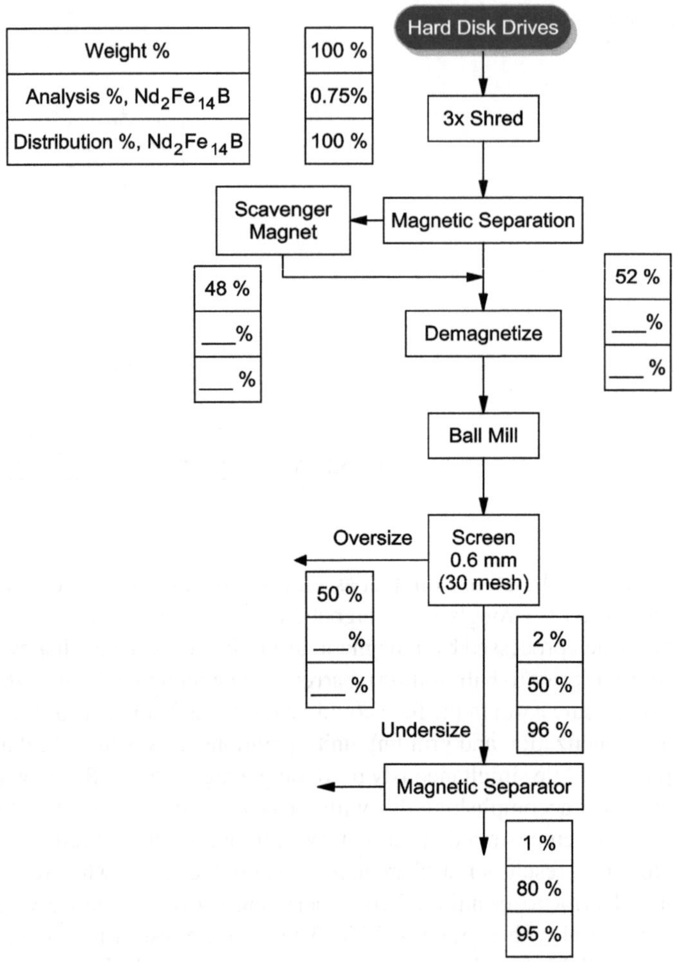

Fig. 6 Preconcentration flowsheet

Discussion of Coproducts Contributing to Viability of Process

Two preferential degradation process variations were evaluated and analyzed. The first is a process where shredder output was fed directly into the ball mill for grinding, illustrated in Fig. 6. The second variation includes a magnetic preconcentration unit operation prior to grinding, illustrated in Fig. 6.

Both process approaches produced a fine RE magnet enriched product and various nonmagnetic products comprised primarily of malleable metals (Aluminum, Nickel, Steel, Stainless Steel as alloys). The coarse fraction was processed differently for the two process variations.

Table 3 Demonstrated material recoveries of coproducts

Material of interest	Fraction in HDD feed (%)	Demonstrated recovery (%)
Al	55	80
Fe	6	95
SS	22	95
Ni alloy	8	90
PCB	4	60
NdFeB	3	95

Recovery of Coproducts from Linear Flowsheet

The linear flowsheet recovered aluminum, nickel, stainless steel and carbon steel to the ball mill product coarse fraction and recovered neodymium, gold and silver from the fine fraction. The gold and silver were liberated from the printed circuit board. Fire assay showed that gold is present at 230 ppm with silver present at 880 ppm in the fine fraction of the ball mill output. The gold and silver may be recovered by any of several processes that are well researched and established. The challenge to gold and silver recovery, as is commonly challenging in recycling, is achieving the throughput necessary to warrant additional unit operations.

The process postulated for the coarse fraction was separation by high-intensity magnet to produce an aluminum concentrate and a mixed ferromagnetic fraction. The nickel alloy was separated from the mixed magnetic fraction using sortation before finally the stainless steel-carbon steel separation was accomplished with low-intensity magnet separation.

Recovery of Coproducts from Preconcentration Flowsheet

The preconcentration flowsheet rejected aluminum, printed circuit boards, and plastics before the demagnetization and ball mill unit operations. This nonmagnetic fraction was concentrated by eddy current separation. The ball mill output coarse fraction included steel, stainless steel, and nickel. The steel and nickel were separated from the stainless steel by low-intensity magnetic drum. The nickel was separated by sortation. The demonstrated recoveries of these materials are shown in Table 3.

Conclusions and Direction for Future Research

The development of a process for the recovery of neodymium-containing magnet material, considered a critical material, from recycling of computer hard drives was evaluated. The process developed takes advantage of the brittle nature of the neodymium magnet material when presented to a grinding mill to produce a fine product composed primarily of powdered magnet material. The rare earth bearing magnet material recovery is shown on mixed hard drive feeds with samples up to 35 kg samples.

The produced $Nd_2Fe_{14}B$ material grade appears sufficient for economic processing to produce pure or mixed rare earth oxides. The total recovery of magnet material is 96% at a grade of 50 wt% magnet material. Maximum demonstrated output grade is 80% magnet material by weight.

Direction for Future Research

Expansion to additional magnet bearing devices should be considered. The magnet's brittle nature is expected regardless of the specific device containing the magnet. It is possible that the preferential degradation type process can be applied to a range of devices for REE recovery. There is particular advantage to the process product being suitable for direct reuse as a bonded or sintered magnet. It is expected that the demonstrated recovery is of material that is significantly oxidized. Consideration should be given to grinding, screening, and other process steps under an inert gas cover gas to maintain the magnet material compound.

Acknowledgements This work was completed with support from the Critical Materials Institute, an Energy Innovation Hub funded by the U.S. Department of Energy, Office of Energy Efficiency and Renewable Energy, Advanced Manufacturing Office.

References

1. Office of Energy Efficiency and Renewable Energy (2010) U.S. department of energy—critical materials strategy. Department of Energy, Washington, DC, United States
2. Sprecher B, Kleijn R, Kramer G (2014) Recycling potential of neodymium: the case of computer hard disk drives. Environ Sci Technol 48(16):9506–9513
3. Abrahami S, Xiao Y, Yang Y (2014) Rare-earth elements recovery from post-consumer hard-disc drives. Miner Process Extr Metall 124(2):106–115
4. Takeda O, Okabe T, Umetsu Y (2006) Recovery of neodymium from a mixture of magnet scrap and other scrap. J Alloy Compd 408–412:387–390
5. Baba K, Hiroshige Y, Nemoto T (2013) Rare-earth magnet recycling. Hitachi Rev 62(8):452–455
6. Bandara H, Darcy J, Apelian D, Emmert M (2014) Value analysis of neodymium content in shredder feed: toward enabling the feasibility of rare earth magnet recycling. Environ Sci Technol 48(12):6553–6560
7. Walton A, Yi H, Rowson N, Speight J, Mann V, Sheridan R, Bradshaw A, Harris I, Williams A (2015) The use of hydrogen to separate and recycle neodymium–iron–boron-type magnets from electronic waste. J Clean Prod 104:236–241

Selective Reduction and Separation of Europium from Mixed Rare-Earth Oxides Recovered from Waste Fluorescent Lamp Phosphors

Mark L. Strauss, Brajendra Mishra and Gerard P. Martins

Abstract Europium is vital to the production of high technology products. In recent years, the worldwide market for europium has been controlled by primary sources in China. However, recycling europium from waste fluorescents is a strategy to supply europium. Waste phosphor powder is recovered from recycled lamps and retorted, sieved, and leached to produce europium and yttrium rich concentrate. Europium is separated from yttrium by reducing Eu(III) to Eu(II) by selective reduction with zinc and precipitated as europium (II) sulfate from solution via sulfuric acid. After building upon previous work and removing unit operations, the optimized conditions for europium sulfate were determined. The effects of varying the entrance pH, quantity of sulfate, choice of inert gas, selective reduction time, and precipitation time were studied upon the final grade and recovery of europium (II) sulfate. The final purity of 93.84% of europium (II) sulfate with a recovery of 84.2% was obtained after using an entrance pH equal to 2.5, 5 times the stoichiometric ratio of sulfate, a 2 h precipitation time, and 30 min selective reduction time.

Keywords Waste fluorescent lamp · Phosphor dust · Rare earths · Recycling · Europium

Introduction

According to USGS [1] 11,000 tons of rare-earth products were consumed in the United States in 2017. Only a minor quantity of rare earth oxides was produced by recycling batteries, magnets and fluorescent lamps. There is zero commercial production of rare earths in the United States. There is the potential that waste phosphor dust could supply some of the US demand for rare earths. According to

M. L. Strauss (✉) · B. Mishra
Worcester Polytechnic Institute, 100 Institute Road, Worcester, MA 01609, USA
e-mail: mlstrauss@wpi.edu

G. P. Martins
Colorado School of Mines (Kroll Institute of Extractive Metallurgy),
1500 Illinois Street, Golden, CO 80401, USA

G. Gaustad et al. (eds.), *REWAS 2019*, The Minerals, Metals & Materials Series,
https://doi.org/10.1007/978-3-030-10386-6_35

the DOE 2011 Critical Materials Strategy Report [2], the demand will increase from 200 tons per year to 220 tons per year in 2020. In 2007, 8000 tons of waste phosphor powder was discarded into landfills after removing mercury, and it is believed that quantity will be higher in 2019. Assuming the concentration of 0.63% europium oxide in the waste powder, there is a resource of 50 tons per year in the United States—assuming no increase in collection. However, the concentration of europium in the dust should have increased since 2007, due to fewer halophosphate based T12 lamps being disposed and a greater ratio of newer T8 lamps being recycled.

Europium was first isolated from samarium using a series of crystallizations experiments by French chemist, Eugène-Anatole Demarçay [3]. The unique chemistry of europium (II) sulfate ($EuSO_4$) could have improved the isolation europium by precipitation. One of the unique chemical properties of europium were discovered in 1906 when Georges Urbain discovered that yttrium oxide doped with europium created a red color. From this discovery, red phosphor was born. Red phosphors are found in all products requiring a screen. The luminescent quality is unmatched and has no replacements. Eduafo and Strauss [4] demonstrated how europium and yttrium follow each other into solution from the waste. The purpose of this research is to develop a method to separate europium from yttrium such that the final product is market ready europium oxide, minimum purity 99.9%. The intermediate product is europium (II) sulfate which can be easily converted to europium oxide.

Molycorp [5] developed a process to recover 99.9% pure europium oxide form Eu–Sm–Gd concentrates separated from monazite. More recently, Preston and Preez [6], Morais and Ciminelli [7], and Rabie et al. [8] demonstrated that zinc metal and sulfuric acid can separate and purify europium from samarium and gadolinium concentrates. The paper proposes using a similar method for europium and yttrium concentrates, as a product of waste lamp phosphor leaching. The equation below demonstrates the equation for converting Eu(III) to Eu(II) via selective reduction.

$$2Eu^{3+} + Zn = 2Eu^{2+} + Zn^{2+} \tag{1}$$

The equilibrium constant, K_c, can be used to describe the thermodynamics of a system. In the equation below the activities of species are replaced with their actual concentrations because there is no simple method to measure the activities of concentrated species in high ionic strength solutions.

$$aA + bB \Leftrightarrow cC + dD$$

$$K_c = \frac{[C]^c[D]^d}{[A]^a[B]^b}$$

In addition, the formation of europium (II) sulfate precipitate, and intermediate in the europium separation experiments is demonstrated in Eq. 2 below.

HSC 5.11 was used to explore the possible reactions for the oxalic precipitation work and selective reduction and precipitation of europium (II) sulfate. Then, using the Gibbs free energy minimization ("Equilibrium Compositions") function, HSC identifies the most plausible chloride complexes after the solvation of europium in hydrochloric acid as $EuCl_3(aq)$, $EuCl_2^+$, $EuCl^{2+}$ as generalized by Eq. (2) below.

$$Eu_2O_3 + 6HCl(aq) = 2EuCl_3(aq) + 3H_2O \quad K_c = 7.36 \times 10^{55} (T = 25\,°C)$$
$$2EuCl_3(aq) + 4HCl(aq) + Zn = 2EuCl_4^{2-}(aq) + ZnCl_2(aq) + 4H^+ \quad K_c = 1.62 \times 10^{16} (T = 25\,°C)$$
$$EuCl_4^{2-}(aq) + H_2SO_4 = EuSO_4 + 2HCl(aq) + 2Cl^- \quad K_c = 1.39 \times 10^{13} (T = 25\,°C)$$

Experimental

Materials

Veolia ES Solutions provided the waste phosphor powder which was treated to create a concentrate of yttrium and europium oxide (~91% pure) based on previous research. Zinc metal (99.8% 20–30 mesh) was provided by Alfa Aesar. Sodium hydroxide (Sigma Aldrich, USA) is dissolved in deionized water. The mixed yttrium and europium product was dissolved in hydrochloric acid and deionized water. The pH was adjusted with NaOH (Sigma Aldrich, USA) dissolved in deionized water. ACS grade 18 M sulfuric acid (Sigma Aldrich, USA) was diluted with deionized water.

Analysis

X-Ray Fluorescence (XRF) (Thermo Fisher Scientific), scanning electron microscopy electron dispersive spectroscopy (SEM-EDS) were used for quantification and identification, and inductively coupled plasma optical emission spectrometry (ICP-OES) (Fischer Scientific). The data was analyzed using Stat-Ease 9.0.5 Design Expert to create contour plots and conduct statistical analysis.

Procedure

A yttrium oxide/europium solution concentrate in solution was separated from lamp waste. The alkalinity of the solution was adjusted with dilute sodium hydroxide between pH 2.5 to 3.25. Next, the solution was added to magnetically stirred vessel. Hydrogen or nitrogen was bubbled through the system until ORP stabilized. Next, 2.5 grams of zinc was added to the mixing vessel. After 0.5–1 h, the reduced europium solution and dilute solution of 3 M sulfuric acid (between 5 times and 15 times the stoichiometric ratio of sulfuric acid), degassed with N_2 or H_2, were mixed together in a cylindrical vessel and bubbled with hydrogen gas. After between 0.5 and 2 h, the solution was filtered by vacuum with Millipore 47 mm filter in a Pall vacuum filtration setup. The precipitate was washed with a 0.001 M sulfuric acid. The precipitate was analyzed by ICP-OES after lithium borate fusion was conducted.

Results and Discussion

Analysis/Results

The starting mixed REO concentrate as well as the precipitate were analyzed by ICP-OES.

The purity of the final product ranged from 84.01 to 93.85% europium (II) sulfate. The calculated recovery was lower than desired in the precipitate. The lower than expected europium (II) sulfate may have been due to competitive reactions or Eu(II) re-oxidizing due to incomplete isolation from oxygen during transferring or precipitation. The purity was determined from ICP-OES by assuming the metals (except for europium (II) sulfate) appear as oxides or sulfates.

The highest precipitation time (2 h) and lowest amount of sulfate (5x) have the largest effect on the recovery of europium (II) sulfate. However, the effect is not statistically significant. According to Le Chatelier's Principle, a larger concentration of reactants (sulfuric acid), should increase the quantity of the products (europium (II) sulfate) in Eq. (4). Preston and Preez [6], found there is a correlation between precipitation time and recovery–which is what is aligned to our results. de Morais and Ciminelli [9] suggests that the effect of sulfate concentration is faintly negative upon the grade and recovery. These results demonstrate a similar effect.

The recovery is maximized with the lowest entrance pH (pH = 2.5). The lower the pH, the greater amount of hydronium ions in the products of Eq. (3). However, as the pH increases, the reaction should shift to the left. This intermediate reaction will lead to more reduced europium and ultimately, more recovery europium (II) sulfate. However, this plot shows the opposite behavior. Rabie et al. [8] showed that the higher the pH the higher the recovery of europium (II) sulfate. The best conditions, determine by Stat-Ease Design Expert 9, are precipitation time = 2 h, sulfate ratio 5X, entrance pH = 2.5, reduction time 0.5 h and an indeterminate answer whether nitrogen or hydrogen gas are more desirable for this process.

References

1. Gambogi J (2018) USGS minerals year book. Rare Earths 703:2011–2012
2. Bauer D, David D, Mckittrick M (2011) U.S. department of energy critical materials strategy
3. Emsley J (2001) Nature's building blocks: an A–Z guide to the elements
4. Eduafo PM, Strauss M (2015) Experimental investigation of recycling rare Earth metals from waste fluorescent lamp phosphors. Rare Met Technol, pp 253–259
5. Gupta CK (2005) Extractive metallurgy of rare earths
6. Preston JS, Preez AC (1996) The separation of europium from a middle rare earth concentrate by combined chemical reduction, precipitation ind solvent-extraction methods, pp 93–101
7. Morais C, Ciminelli VST (2001) Recovery of europium by chemical reduction of a commercial solution of europium and gadolinium chlorides. Hydrometallurgy 60(3):247–253
8. Rabie KA, Sayed SA, Lasheen TA, Salama IE (2007) Europium separation from a middle rare earths concentrate derived from Egyptian black sand monazite. Hydrometallurgy 86(3–4):121–130
9. de Morais CA, Ciminelli VST (2002) Europium recovery by photochemical reduction from Eu and Eu–Gd chloride solutions. Sep Sci Technol 37(14):3305–3321

Part VIII
REWAS 2019: Secondary and Byproduct Sources of Materials, Minerals, and Metals: Circularity and Materials Availability

Circular Cities, E-Mobility and the Metals Industry—A World in Transition

Christina Meskers, Mark Caffarey and Maurits Van Camp

Abstract The transition to a circular, sustainable society is well on its way. Cities are the place where this is happening bottom-up, supported top-down by the UN Sustainable Development Goals. E-mobility, circular economy and IoT are happening today. Circular cities integrate all aspects of life, connecting across people, economic actors, institutions and geographies. Circular cities are powered by renewable energy and responsible, sustainable materials; have closed resource cycles and are smart. The technical, industrial, economic, cultural and social systems meet, interact and challenge each other. The raw materials industry is an enabler of this transition in society. It provides materials, technologies and solutions. Concurrently, the industry itself needs to adapt to be an essential part of the transition to the future society. Using the e-mobility (batteries) value chain as example, recent and future developments in mobility and responsible material supply will be illustrated, including the impacts and challenges.

Keywords Sustainable development goals · Clean (e-)mobility · Materials value chain · Metals industry

Introduction

The current take-make-dispose economy is a linear model of resource consumption: raw materials are employed to make a product, and when this product reached the end of its useful life, it is disposed, and thus lost. However, with the growing population, especially in the cities, and the scarcity of natural resources it is not possible to maintain the current lifestyle. Hence, there is a need for a shift towards a circular economy that is based on a more sustainable management and efficient use of natural

C. Meskers (✉) · M. Van Camp
Umicore, Watertorenstraat 33, 2250 Olen, Belgium
e-mail: Christina.Meskers@eu.umicore.com

M. Caffarey
Umicore USA Inc, 3600 Glenwood Ave # 250, Raleigh, NC 27612, USA

G. Gaustad et al. (eds.), *REWAS 2019*, The Minerals, Metals & Materials Series,
https://doi.org/10.1007/978-3-030-10386-6_36

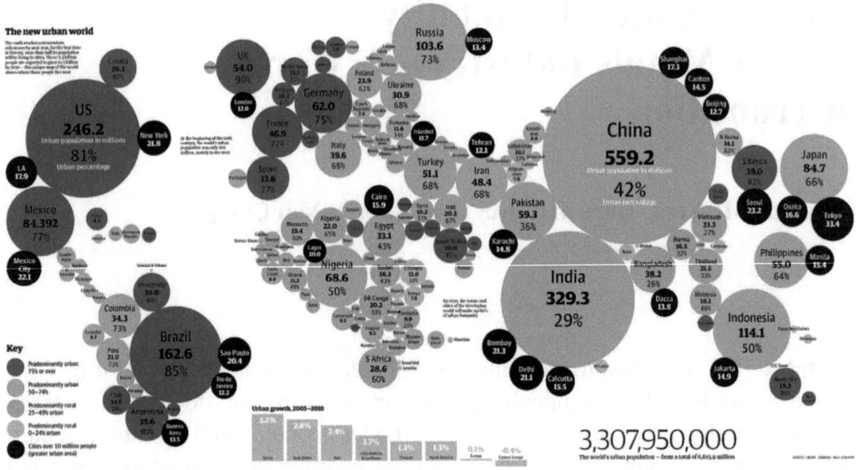

Fig. 1 Percentage of people living in cities [3]

resources, in combination with reuse, restoration and regeneration, and elimination of waste through smart design [2].

For centuries, cities have been the place where people moved in search for a better future. They are hubs of opportunity for education, innovation and entrepreneurship and the citizens have direct influence on how the city is governed and organised. Therefore change can happen fast, compared to the speed of change on a national level. Today, many people already live in cities (see Fig. 1). In Europe, the majority of people lives in cities, and in emerging economies like Brazil (85%), Indonesia (50%) and China (42%), urbanisation is taking place in a rapid pace. Of the 33 megacities (>10 million inhabitants), two-thirds of them are outside Europe and North America [7]. Cities also have challenges such as overcrowding, traffic, air pollution, waste management and income equality. Currently, cities account for 60–80% of the energy consumption and 75% of the carbon emissions, whereas they only occupy 3% of the Earth's Land [6]. Therefore, cities will be the living labs where new concepts and systems will be developed and adopted first. As cities keep on growing, the need for a 'better life' drives a transition that is taking place today, all around the world. A transition towards cities that is circular, sustainable, connected and smart.

Guided by the Global Goals for Sustainable Development (see Fig. 2), cities will use, among others, smart, digital, connected systems for housing, mobility, energy and resources. This means, i.e. that cities and citizens take action to provide adequate and affordable housing and public transport, besides reducing the environmental impact of cities. The latter is done by efficient use of natural resources, reducing the waste generation, as well as by appropriate waste management and air quality control. In addition, sustainable cities are about connecting people and people to their community, by providing access to safe green and public spaces, and with social and environmental links between urban, per-urban and rural areas.

Fig. 2 Sustainable development goals [5]

Materials, and the metal materials industry, have a large responsibility in this transition, as they are part of the solution to the environmental and societal challenges. Indeed, a sustainable raw materials and recycling industry are essential for the realisation of circular cities as it provides the materials and recycling solutions necessary to make the transition happen, in a sustainable and responsible manner.

Clean Mobility

Since all the Sustainable Development Goals are connected to each other and represent a grand challenge that cannot be solved by one actor or sector alone, they require the collaboration between citizens, government, industry and academia, and other stakeholders to be achieved. Taking the example of mobility, we see that citizens all around the world ask for action to create clean air, as the tolerance for pollution is decreasing rapidly. For example, cities like Oslo, Antwerp and others across Europe create zones in which only low-emission vehicles are allowed. At the same time, policymakers make emission norms more stringent. In Europe, new directives are introduced for even lower CO_2 emissions, and low-emission vehicles are awarded 'super credits'. In Asia, countries like India and China introduced, respectively, Bharat Stage 6 and China 6, new norms that are revolutionary in terms of emission standards for vehicles, and which result in an even faster global transition.

Pushed by the legislative changes and societal developments, car original equipment manufacturers (OEMs) adapt their vehicle portfolio—including both e-mobility and cleaner internal combustion engine (ICE) vehicles–embracing clean mobility solutions. The electrification trend has evolved rapidly. Merely, 10 electrified vehicle (EV) models were on the market in 2012, and only 6 years later, this number has grown to over 50 models.

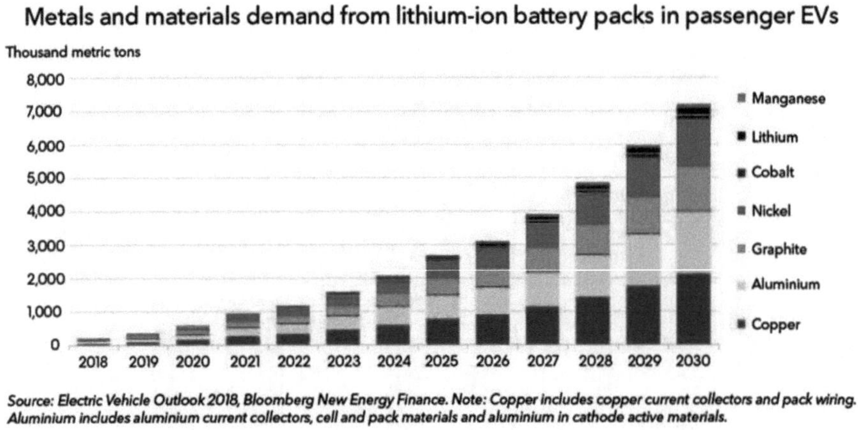

Fig. 3 Metals and materials demand from lithium-ion battery packs in passenger EVs [1]

Last year, the sales of EVs reached a record of 1.1 million, and this number is estimated to further increase to 11 million in 2025 and even to 30 million in 2030 [1]. Bloomberg's forecast further predicts that by 2040 33% of the global feet will be electric.

From Battery EVs (BEV) for long or mid-range distances to hybrid or plug-in hybrid EVs ((p)HEV) to fuel cell EVs (FCEV); the EVs are available in a wide range of size and shapes to accommodate specific (individual) demands, for light-duty and heavy-duty vehicles (LDV and HDV), mining vehicles and off-road applications. Nevertheless, the ICEs will remain an important part of the mobility mix. ICEs with catalytic converters will need to meet stringent emission norms. The vehicle technologies use in the future will thus be a mixture of technologies because of the mixed challenges the society is facing and the clear trend of product customization. Hence, technological flexibility is key during the clean mobility transition.

Impact of the Mobility Transition on the Metals Industry

The transition towards clean mobility has undoubtedly a huge impact on the mining and metals industry.

First, the strong growth of EVs entails a dramatic increase in the demand for battery packs and the materials used to make these see Fig. 3. This includes the battery material elements such as Li, Co and Ni, while also basic materials such as Al and Cu are impacted.

To account for this continuous increase of the material demand, both primary production and recycling of manufacturing scraps and end-of-life products are important sources of material. Indeed, looking at the supply and demand outlook for 2030, it is expected that there will be some imbalances—especially for cobalt, but also for nickel on the longer term—when considering only mine production [4]. Thus, recycling of production scraps and end-of-life products is essential. Today there are many small electronic devices (cell phones, tablets) containing batteries that are not collected and recycled. If the batteries from these devices would be collected and recycled, they would make a valuable contribution to the future materials supply. The challenge is here not so much technological, recycling processes are in place, as it is about influencing the behaviour of individual people.

When talking about clean mobility, the entire energy, materials and product value chain needs to be considered.

For ICE vehicles, the well-to-wheel efficiency is an important metric. The well-to-wheel energy efficiency can be divided into two parts: (i) the well-to-tank efficiency, i.e. the energy needed to convert fossil fuel into usable fuel (or electricity) and (ii) the tank-to-wheel efficiency, i.e. the energy needed to transform the fuel (or electricity) into motive energy. Regarding the tank to wheel efficiency, a BEV is far superior compared to an ICE: a BEV has a 90% efficiency when converting chemical energy into motion, whereas an ICE only has an efficiency of 25–30%. Moreover, for a BEV, there are only CO_2 emissions associated with the conversion process of fossil fuel into electricity. If the BEV is powered by electricity from renewable sources, the CO2 emissions associated with the well to tank efficiency are further lowered.

For sustainable and clean mobility, we also need to consider the entire materials' value chain, from (urban) mine to wheel. To mine, refine, produce and recycle the materials in a responsible, sustainable way is the key task for the metals and materials industry. It will transform the industry. The sustainable development goals can be used as a long-term compass and way to communicate and connect to other stakeholders and citizens. Innovative and entrepreneurial people are enablers of the transformation.

To make this more tangible, consider Umicore's activities. Umicore is a global material technology and recycling company. It provides the key materials for rechargeable batteries used in among others consumer electronics and (B)EVs, and provides the active materials for catalytic converters used in ICE vehicles. Furthermore, it provides recycling solutions for rechargeable batteries and catalytic converters, closing the materials loop. In this way, Umicore has a unique value chain integration.

Umicore's approach is to start by ensuring supply via responsible sourcing. This has two parts.

(1) Responsible sourcing of primary materials using the Sustainable Procurement Framework for Cobalt. This Framework, the first one of its kind, is third-party validated and aligned with the OECD framework, and assures all upstream practices are in line with Umicore's standards. This includes among others no child or forced labour; focus on environmental protection, health and safety; complying to the law. The same Sustainable Procurement principles apply for sourcing of lithium, nickel and manganese.

(2) A closed-loop model, using the recycled metals from our own industrial operations. Our innovative state-of-the art (recycling) plants with the highest environmental standards may recover high percentage yields of non-ferrous metals which are then supplied to our production processes. For spent rechargeable batteries the

non-ferrous metals are Li, Ni, Co and Cu and for spent catalytic converters these are platinum group metals. Through our know-how we produce best in class materials technology and product quality (e.g. battery cathode materials) which are then supplied to our customers covering different technology applications for mobility. Once the materials reach end-of-life and are supplied to us, the loop starts again and again. This also applies to intermediates and production scrap created along the value chain. For spent EV batteries, Umicore also has a battery dismantling step in Hanau, where the battery modules and cells can come back to us to recycle.

Another key aspect is to have sustainable processes and operations. These will provide the company with the Societal License to Operate. Many dimensions come together in this key aspect. For example, resource-efficient use of materials (reagents), energy and water and minimization of emissions. Despite its growth, Umicore managed to considerably reduce energy consumption (−21% vs. 2015 benchmark), and emissions to air and water (−41% and −69%, respectively vs. 2015 benchmark). Also, the possibility to put the products from the processes on the market, in other words, License to Market or License to Manufacture, is included. Legislation such as REACh, product standards etc., plays a role. A dialogue with various stakeholders is part of the License to Manufacture. Lastly, the creation of a positive connection to the wider society and being seen as a part of the community is essential for obtaining the Societal License to operate. The Sustainable Development Goals and societal transitions such as circular cities and clean mobility provide an opportunity to connect with the community and individual citizens and stakeholders and show how the metals and materials industry is, on the one hand, facilitating the transitions by providing its products and services and on the other hand is participating in the societal transitions by changing its way of working.

References

1. Bloomberg New Energy Finance (2018) Electric vehicle outlook 2018. Bloomberg
2. Ellen MacArthur Foundation (2013) Towards the circular economy
3. Employment Conditions Abroad (2016) Global mobility and the world context in 2016. https://www.eca-international.com/insights/articles/may-2016/global-mobility-and-the-world-context-in-2016
4. McKinsey & Company (2017) The future of nickel: a class act. Basic Materials
5. United Nations (2018a) Global goals. Retrieved from sustainable development goals. http://www.unfoundation.org/features/globalgoals/the-global-goals.html
6. United Nations (2018b) Sustainable cities and communities. Retrieved from sustainable development goals. https://www.un.org/sustainabledevelopment
7. World Economic Forum (2018) These are the megacities of the future. Retrieved from world economic forum. https://www.weforum.org/agenda/2018/10/these-are-the-megacities-of-the-future, 10 Oct

The Role of Scrap Recycling in the USA for the Circular Economy: A Case Study of Copper Scrap Recycling

Phillip J. Mackey, V. Nubia Cardona and L. Reemeyer

Abstract Recycled metals have become a significant source of supply, with a lower environmental impact than primary metals. Factors such as geography, trade regulations and economics on domestic recycling and international trade in scrap all influence the extent of recycling. The role of scrap recycling in the USA and its contribution to the Circular Economy with respect to metals is discussed. At one time, almost all the metal scrap arising in the USA was recycled and treated within the country. Since about the early 1980s, there has been a decline in the proportion of scrap metal recycled and treated in-country in the USA in favour, due to apparent cost benefits, of exporting to other countries such as China for treatment. It is considered that subsequently skills and technology in this area will require re-building to enable metal scrap to be efficiently treated at home using the best available technology. With the recent ban by China on certain metal imports, each country will now need to handle more of its own scrap metal. The opportunities in the USA as a consequence of this are discussed in the paper. The paper also describes the results of a case study for copper scrap recycling. As part of this, the technology, energy consumption and GHG emissions are examined in terms of the grade of scrap copper for selected smelting process routes. It was found that the energy consumption in copper scrap recycling is far lower than that required to produce copper from as-mined ore, while at the same time, it is quite dependent on the grade of actual scrap treated. It is concluded that enhanced copper recycling from secondary materials would be beneficial to the USA, driven by modern technologies as discussed in the paper, as well as encouraging policies from the regulators.

P. J. Mackey (✉)
PJ Mackey Technology Inc., Kirkland, QC H9J 1P7, Canada
e-mail: pjmackey@hotmail.com

V. N. Cardona
Deltamet Consultants, Pointe-Claire, QC H9R4G8, Canada
URL: https://www.deltametconsultants.com

L. Reemeyer
Resourceful Paths, Vancouver, BC, Canada
e-mail: laurie@resourcefulpaths.com

G. Gaustad et al. (eds.), *REWAS 2019*, The Minerals, Metals & Materials Series,
https://doi.org/10.1007/978-3-030-10386-6_37

Keywords Copper · Scrap copper · Circular economy · Energy consumption · Scrap recycling · Smelting copper scrap and grade of copper scrap

Advancing the State of Prospective Materials Criticality Screening: Integrating Structural Commodity Market and Incentive Price Formation Insights

Michele Bustamante, Tanguy Marion and Rich Roth

Abstract Broad, screening-style assessments of criticality highlight economically important materials facing significant threats to secure, sustainable supply. The studies identify many valuable metrics influencing the likelihood and impact of constraint; however, approaches to aggregating these metrics vary and lack unifying, causally grounded strategies. The present work seeks to advance the state of criticality screening by targeting this gap with structural commodity market insights. Grounded in the principle of price as indicative of economic scarcity, the model uses a small number of high-level metrics to drive changes in future incentive price (i.e., cost of opening new mines) by influencing structural markets elements in different ways (i.e., shape of long-run supply, demand, and incentive curves). Metrics like resource cover, ore grade loss, barriers to entry, secondary supply, end-use substitution and dematerialization rates, and all help to approximate changing supply–demand balance and inform criticality. The approach balances strengths of rapid, broad screening with insights of more detailed market modeling.

Keywords Criticality · Screening · Supply · Demand

Introduction

Studies of material criticality serve an important role of highlighting economically important materials facing significant threats to secure, sustainable supply. Criticality in supply chains can threaten essential sectors, such as national defense, as well as budding technological developments, such as renewable energy devices and advanced vehicles. The literature on criticality has grown substantially in the past 10 years since the National Research Council (NRC) report first introduced the use of a two dimensional, multi-indicator assessment framework [5]. Since then, frameworks have expanded, both in terms of indicators used within each dimension and even the number of dimensions considered. This style of assessment has been quite valuable

---

M. Bustamante · T. Marion (✉) · R. Roth
Materials Systems Laboratory, Massachusetts Institute of Technology, Cambridge, USA
e-mail: tanguym@mit.edu

G. Gaustad et al. (eds.), *REWAS 2019*, The Minerals, Metals & Materials Series,
https://doi.org/10.1007/978-3-030-10386-6_38

because it enables a broad scope of materials to be considered on the basis of common factors. This facilitates triaging of more detailed analytical work so that research as a whole can be conducted efficiently, focusing primarily on those at greatest risk. This is also particularly important for use in future-looking assessments, where conditions change rapidly.

Many relevant metrics influencing the likelihood and impact of constraint have been used. However, since the goal is to deliver a single criticality score upon which to triage the group being evaluated, the metrics used must be aggregated in some way. Some studies utilize averaging of all metrics within categories [1, 3], some utilize a weighting scheme informed by stakeholders [2], and still, others utilize different approaches, such as weighting by number of disruption events caused [4]. Overall, they vary throughout the literature and lack unifying, causally grounded strategies. Regardless of the exact method used, these approaches all share a common problem: reliance on subjective aggregation schemes that are not informed by the way these features influence the structure of material markets in practice.

The authors of this paper propose a shift toward a new approach that leverages existing strengths but also targets the aforementioned challenge with broad screening-style assessments of criticality. The framework is grounded in the principle of price as indicative of economic scarcity. Although not a perfect indicator, the price is certainly an important driver of both supplier and consumer behavior and are more likely to reflect criticality moving forward.

Method

Markets which are typically considered as critical are those that have outsized demand growth, rapidly diminishing existing supply, or limited opportunities for new supply. Typical criticality metrics seek to address these through a variety of weighting functions or other approaches, yet are quite subjective in the choices of weights and seldom pay sufficient attention to interactive effects. Economic price formation theory provides an alternative approach where these same criticality concepts are used to estimate future price movements. Price changes are reflective of the changing market balances which are often a considerable part of the underlying criticality concerns.

While there are multiple approaches based on economic theory, a common method uses an incentive model. This takes the form of estimating a need for new or incentive supply in response to demand growth combined with the loss of existing supply capacity. Prices are then forecast as the value needed to incentivize enough new mines to meet the anticipated supply gap.

On the demand side, criticality metrics have often attempted to address unexpected or newly increased rates of demand growth that might arise from new technological applications or the need for specific function properties of materials that lead to increased substitution in existing applications. On the supply side, criticality concerns have focused on two general concepts, rapid decline of the existing supply base and insufficient future realizable sources of supply. All of these concepts are explicitly considered in incentive price formation models.

For criticality assessments to be useful in a prospective fashion, they must consider the current state of key drivers and how those drivers are changing. Future supply of a material is determined by the current state and changes to physical availability, rate of use, mining and processing technology, and other social and political barriers to mine openings. Future demand of a material is influenced by end-use volumes and intensities and amount and qualities of available substitutes.

The proposed model uses a small number of high-level metrics to drive changes in future incentive price by influencing structural markets elements in different ways. The metrics used are similar to those used in existing studies, but will be combined in different ways. The metrics are used to create a mock incentive supply curve and show how it is expected to change over time. Simultaneously, other metrics will determine how demand evolves in the same time horizon. Assessing the gap that results from comparing evolving demand and incentive supply curve shape suggests changing incentive price and can be an important indicator of criticality.

Foundational elements of the model are:

- Incentive supply curve—the supply curve is a foundational element of commodity supply analysis. It is a monotonic, cumulative visualization of the amount of material available to produce currently and at what cost. Similarly, an *incentive supply curve* is a representation of the material available in the next time period, which could be made available and at what *incentive price*.
- Demand—quantity currently demanded.

Proposed supply side driving metrics include:

- Ore grade loss—ore grade loss refers to the change in the amount of desired mineral present in ore being mined and is the main driver of loss of existing supply.
- Resource cover—resource cover is a measure of how many years of unmined supply exists and indicates a limit on available new supply.
- Barriers to entry—this includes technological complexity, geographic constraints and a variety of other factors which limit the ability to bring in new supply as needed.
- Secondary supply—the need for new supply can be offset by increased availability of recycled material.

Proposed demand side metrics include:

- New applications are driven by technological change.
- End-use substitution driven by functional properties and economics.
- Dematerialization enabled by technology and design improvements.

Each of these factors is used to forecast the elements of an incentive price formation model. While it is infeasible to actually predict commodity prices based on these limited, high-level supply and demand metrics, the movements of price are quite indicative of the ways in which market balances will change and thus provide a consistent representation of key elements of criticality.

Remarks

The framework is still being refined, but together the metrics used help to approximate changing supply–demand balance and inform criticality. However, quantification of these metrics and the way they are combined require careful consideration to avoid the same pitfalls as prior criticality assessment methods. Once finalized, the framework is going to be demonstrated using various metals such as copper, lithium, cobalt, tin, and silver. These results will be presented at the REWAS 2019 conference for discussion with other experts.

References

1. Erdmann L, Graedel TE (2011) Criticality of non-fuel minerals: a review of major approaches and analyses. Environ Sci Technol 45(18):7620–7630
2. Gleich B et al (2013) An empirical approach to determine specific weights of driving factors for the price of commodities—a contribution to the measurement of the economic scarcity of minerals and metals. Resour Policy 38(3):350–362
3. Graedel TE, Nuss P (2014) Employing considerations of criticality in product design. JOM 66(11):2360–2366 (Web)
4. Hatayama H, Tahara K (2018) Adopting an objective approach to criticality assessment: learning from the past. Resour Policy 55:96
5. National Research Council (2008) Minerals, critical minerals, and the U.S. economy (Web)

Mining Value from Waste Initiative: Towards a Low Carbon and Circular Economy

Janice Zinck, Bryan Tisch, Terry Cheng and Rory Cameron

Abstract The concept of tailings reprocessing is not new. However, due to the complexities ranging from S&T to material handling to policy and regulations, reprocessing successes have been limited. Several factors have been identified by stakeholders as requirements for success, such as economic considerations, engineering challenges, and policy and regulatory implications. Innovation underlies each of these factors. One of the biggest challenges is to reduce the environmental liabilities associated with mining waste, for which the financial securities may not be sufficient to ensure adequate protection. The recovery of metal values from tailings is potentially attractive and economically viable, particularly when combined with a concomitant reduction in environmental liabilities. Recently, Natural Resources Canada embarked on a pan-Canadian effort entitled 'Mining Value from Waste' to develop tools, technologies, and policies to de-risk and accelerate demonstration and full-scale waste reprocessing/repurposing projects with the goal to reduce mine waste liability and environmental impact, while providing local and national value.

Keywords Reprocessing · Repurposing · Tailings · Recovery · By-products · Liability · Bioleaching

Introduction

It is estimated that the liability for managing mine wastes in Canada and the US exceeds $50 billion [16]. These wastes represent a huge liability for both mining companies and governments. The liability associated with the management of tailings impoundment areas (TIA) alone is significant. Specifically, the cost of remediation for TIA failures worldwide is estimated to be US$ 6 billion in this decade alone [2]. In 2016, Mining Watch Canada analysed Auditor General for Canada reports from three mining jurisdictions, British Columbia, Quebec and Ontario, and stated

J. Zinck (✉) · B. Tisch · T. Cheng · R. Cameron
Green Mining Innovation, CanmetMINING, Natural Resources Canada, Ottawa, ON, Canada
e-mail: janice.zinck@canada.ca

G. Gaustad et al. (eds.), *REWAS 2019*, The Minerals, Metals & Materials Series,
https://doi.org/10.1007/978-3-030-10386-6_39

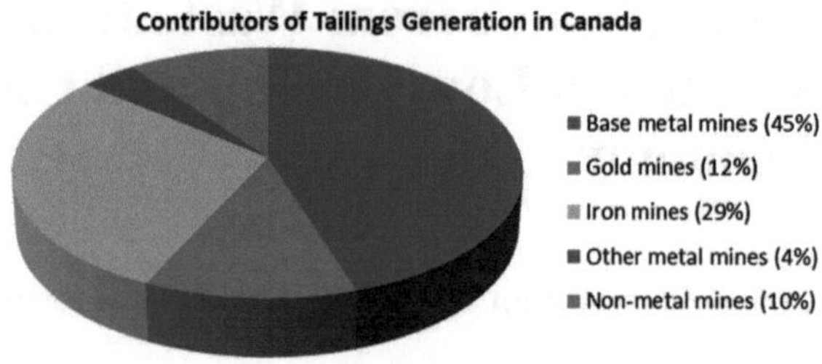

Fig. 1 Tailings types in Canada by weight (6 billion total tonnes)

that the total environmental liability for mine site clean-up was estimated to be over C$7 billion; of which C$4 billion was unsecured by financial securities held by the provinces [9].

The Kam Kotia mine site near Timmins, Ontario, Canada was considered to be one of the worst contaminated sites in Ontario. Beginning around 2000, the Ontario government spent in excess of C$135 million in care, maintenance and remediation to address this site [17]. Most of this cost was associated with stabilization of the tailings impoundment and treatment of acidic tailings seepage water.

The Canadian Federal Contaminated Sites Action Plan (FCSAP) is a 15-year/$4.54 billion programme that was established in 2005 by the Government of Canada to address the risks that these sites pose to human health and the environment, and to reduce the associated financial liability. Many of the sites included in this programme are from mining and contain tailings. Under FCSAP, the estimated cost to remediate just one site in the Yukon, the Faro Mine, exceeds $1B.

In addition to government liabilities, tailings and other mine wastes pose a long-term liability to mine operators. Besides holding long-term insurance bonds, mine operators are actively engaged in developing best practices in tailings management [8, 11–14] to mitigate the risks associated with these wastes. Furthermore, mine waste management efforts and objectives are typically formulated around reclamation and mine closure, not targeted at reaping potential economic benefits through waste reprocessing and repurposing.

In general, tailings represent over 95% of the rock mass mined and because of the diversity of ore bodies and technologies employed to process them, the resulting tailings are very much site-specific. The rate of tailings generated at Canadian mines is on the order of 200–250 million tonnes per annum, with the cumulative sum of the tailings from the last three decades amounting to over 6 billion tonnes [1]. Records indicated that over 6 billion tonnes of tailings were generated in Canada over the last 30 years; of which 90% originated from metal mines with base metals, iron and gold mines as the major contributors (Fig. 1) [1]. The typical tonnage of tailings generated in each commodity industry varies largely depending on the primary metal mined. For example, about 400 tonnes of tailings are generated per kg of gold produced, 125 tonnes per tonne of copper produced, and 2 tonnes per tonne of iron produced. These ratios will continue to climb as high-grade ores become scarce, while the continuous advancement of technologies will allow the industry to process lower grade ores [6].

The cost of waste reprocessing is often considered to be prohibitive and the process problematic. As a result, technologies for metal recovery from wastes are rarely adopted. However, with increasing environmental pressures and mining costs, the option for metal recovery from mine waste becomes more attractive, especially when coupled with the revenue from the recovered metals. With this in mind, there is a need for technologies that can cost-effectively recover metals from mine wastes. Some technologies exist to reprocess tailings particularly for metal recovery (e.g. gold and silver) but a holistic approach to look at tailings as a source of secondary metals and other valuable industrial minerals has not been undertaken.

Program Rationale

By coupling the high liabilities associated with mine waste management areas and the decline in easy to access, high-grade ore deposits, the option to reprocess 'older' tailings becomes an attractive and potentially economically viable alternative to a new mine development. With a growing global middle class, it is likely that there may be a supply risk for some minerals and metals that will increase in the coming years. In addition, finding high-grade ores in politically stable regions is more difficult and permitting time for new projects is lengthening [10], while many historic tailings contain metal grades higher than currently mined today.

Cheng et al. [6] conservatively estimated that the total metal value in Canadian gold tailings is on the order of $US10 billion. Eighty per cent of this could be easily recoverable. Despite the significant metal value contained in mine waste, reprocessing of these wastes is unconventional. In general, the remaining metal value present in tailings is considered a process loss that does not warrant any further effort to recover. This is mostly due to economics, environmental liabilities and regulation or technological deficiencies.

The development of a cash-flow model, which will become part of the prefeasibility study tools to provide techno-economic analysis of the selected process options for a particular tailings site, is underway at CanmetMINING (Natural Resources Canada). The preliminary model depicts a hypothetical tailings reprocessing project with 50 million tonnes of gold tailings and an average gold grade of 0.5 g/t. Assuming the mine life of this tailings reprocessing project to be 8 years, with 80% gold recovery at a gold price of US$700 per ounce, the discounted cash flow rate of return (DC-ROR) is estimated to be 17%. It is important to note that the cash flow analysis shows DC-ROR is highly sensitive to gold price, gold grade of the tailings and gold recovery, and is moderately sensitive to reprocessing (operating) cost and least sensitive to capital investment.

Mining Value from Waste Program

Launched in August 2018 at the Canadian Energy and Mines Ministers Conference (EMMC), the Mining Value from Waste programme (Fig. 2) is a pan-Canadian initiative focused on reducing the environmental, social and economic footprint of mine wastes, such as tailings, and examining options to obtain value from these wastes, both by recovering valuable metals and by using the wastes as resources in other applications. The objective of the project is to develop tools, technologies and policies to de-risk and accelerate demonstration and full-scale waste reprocessing and repurposing projects, with the goal of reducing mine waste liability and providing value (cleaner environment, jobs, resources, etc.).

While the concept of tailings reprocessing is not new, successes to date have been limited in Canada and elsewhere due to complexities related to science and technology, materials handling, as well as policy and regulations. Mining Value from Waste stakeholders recognize that to achieve the goal of developing onsite process demonstrations (and eventually full tailings reprocessing) will require addressing several factors including: economic considerations such as reducing capital and operating costs; engineering challenges related to processing a complex and variable low-grade feed at a high throughput; and policy and regulatory factors. To address these factors in an integrated way, and to move from concept to reality, and to a circular and low carbon economy, requires a multi-disciplinary and multi-stakeholder approach.

In advance of the program launch, a stakeholder workshop was held [17] to discuss current projects being undertaken in the area of waste reprocessing/repurposing and define areas of collaboration and future work. It was recommended that a holistic approach be taken, which would include the following:

- Waste reprocessing
- Repurposing options
- Novel materials from waste components
- Processing to eliminate tails and other waste

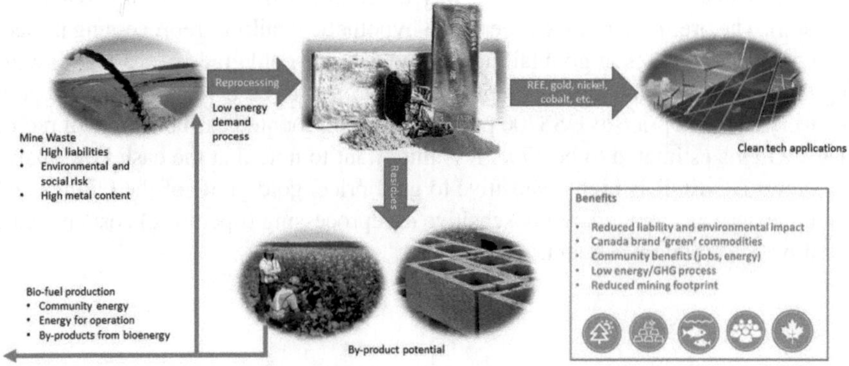

Fig. 2 Mining value from waste programme concept

- Tailings and mine waste lifecycle
- Mine waste management options to facilitate reprocessing at a later date
- Policy and regulatory reviews

The program has identified three types of tailings, which pose high risks to the environment and/or contain high economic values that make them commercially attractive. These are: (i) reactive sulphide tailings prone to acid mine drainage, (ii) gold tailings with low sulphide content and (iii) tailings that pose high liability to the environment (Table 1).

CanmetMINING is already undertaking several projects under the Mining Value from Waste Programme. These are detailed below.

Reprocessing of pyrrhotite rich tailings [3, 4]: Large quantities of pyrrhotite are produced during physical separation of nickel ore. The pyrrhotite contains nickel and cobalt and is highly reactive, producing significant acidic drainage. Reprocessing of this material will result in lower treatment cost and significant revenue generation.

Achievements include:

- Completion of stirred-tank bioleaching experiments to examine the effect of chemical additives and flotation on sample beneficiation.
- Preliminary techno-economic evaluation (TEA) and genomic characterization

Rare earth elements and scandium from coal ash [5]: Coal and coal ash contain significant quantities of rare earths and other critical elements. The rare earths and scandium are relatively readily extractable and could provide a secure source of critical metals from a waste source.

Achievements:

- Several coal ash sources examined
- Scandium is 70–90% of the contained value and able to leach 30–50% with dilute sulphuric acid
- Residue can be used in supplementary cementing materials for civil engineering.

Iron removal using biomass-based technologies to produce a saleable product: This project emphasizes the removal of iron from base metal mine wastes. Considering that the base metal tailings can contain high levels of iron oxides that could be economically exploited upon the development of an efficient iron recovery technology. This project focusses on the coupling of biomass-based technologies to facilitate iron removal by magnetization. The magnetic iron produced may be suitable as a saleable product, but also as an important sink for trace metal contaminants in tailings.

Achievements include:

- Successful magnetization of relevant ferric oxide (hematite), hydroxide (goethite), and sulphate (jarosite) as well as oxidized pyrrhotite tailings in the presence of biomass at low temperature.
- Detailed characterization of the reactive process during phase transformation.

Table 1 Tailings types for consideration (From Cheng et al. [6])

Tailings type	Economic driver	Environmental benefit	Technical merit
Reactive sulphide tailings (e.g. pyrrhotite-rich tailings)	High Reactive sulphide tailings, especially those generated several decades ago, often contain base metals/ precious metals/strategic metals with metal grades comparable to new exploration projects	High Reactive sulphide tailings require perpetual containment in costly engineered cover systems to avoid AMD generation. Tailings reprocessing can eliminate/reduce the volume of the reactive tailings by converting part of them to benign residues	High
Gold tailings with low sulphide content	High Tailings with gold grade as low as 0.2–0.3 g/t can be commercially attractive provided that the operating cost of the novel reprocessing technology is significantly reduced. (less than $10 per tonne)	High Through reprocessing of the tailings and recovery of the gold values, the resulting residues have great potential to become benign tailings, which can be repurposed for backfill or construction use	High
High liability tailings (e.g. high arsenic tailings)	High The liability cost and the future remediation cost are very high for contaminated tailings sites	High Bio-mineralization can be used to stabilize deleterious elements such as arsenic. Microbial transformation of arsenic in the subsurface can provide a long-term arsenic stabilization solution, thereby protecting the environment	High

Demonstration and optimization of the alternative binder technology: This technology could be used to place enormous quantities of tailings and slag back underground, thereby decreasing surface liabilities. Further, the alternative binder 'recipe' results in significant greenhouse gas reduction and cost savings for the mining industry, due to the replacement of Portland cement.

Achievements include:

- Recipe development for various waste types
- Slag characterization
- Binder strength on par if not greater than Portland cement

Recovery of metal values from gold tailings [7]: The focus of the project is to develop novel processes to recover metal values, particularly gold from historic gold mine tailings and to reduce the environmental liabilities associated with these wastes. The project focusses on the development and deployment of innovative tailings reprocessing solutions, specifically in the areas of reactivation of sulphide minerals, recovery of metal values, and removal of contaminants.

Achievements:

- Process flow sheet developed
- Successful gold recovery (>80%) without the use of cyanide

Bioenergy production from mine waste areas [15]: This project examines the opportunity to use municipal waste and other residues on mines sites to reclaim mining lands and grow energy crops. It employs the strategy of utilizing one waste to reclaim another with the aim of generating bioenergy from these sites.

Achievements:

- Successful field trials employing hybrid willows, canola and corn
- Data suggest oilseed production on mine tailings could generate approximately 3600 litres per hectare, and a gross profit of approximately $900 per hectare per year
- A preliminary bioenergy feasibility study showed a combustion option to be the best, and that <400 ha of hybrid willow could provide enough electricity to power a typical water treatment plant.

Conclusion

The Mining Value from Waste program has the potential to be a game changer in the way we look at mining. The approach strives for green extraction of metals with the goal being value generation, along with associated liability and environment impact reduction. As we look to transition to a low carbon and circular economy, we need to examine novel approaches to secure mineral and metal resources. Considering the volume and variety of mine waste available, the opportunity for reprocessing and repurposing is real and potentially immense. Through collaboration, existing technical, engineering and regulatory challenges can be overcome to achieve 'mining value from waste'.

References

1. Barter S (2011) Tailings inventory project 2011. CanmetMINING internal project
2. Bowker LN, Chambers DM (2015) The risk, public liability, and economics of tailings storage facility failures. Center for Science in Public Participation. http://www.csp2.org/technical-reports
3. Cameron R, Lastra R, Thibault Y, Morin L, Gould D (2017) Stirred-tank bioleaching of pyrrhotite-rich tailings sample. Paper presented at the annual conference of metallurgists of CIM, Vancouver
4. Cameron R, Yu B, Baxter C, Plugatyr A, Lastra L, Dal-Cin M, Mercier P, Perreault N (2018a) Extraction of cobalt and nickel from a pyrrhotite rich tailings sample via bioleaching. Paper presented at extraction 2018, Ottawa, Canada, August 2018
5. Cameron R, Thibault Y, Lien Y, Hayward B, Lastra R, Chaulk J (2018b) Developing a process to recover rare earth elements and scandium from coal fly ash. Paper presented at resources for future generations (RFG), 16–20 June 2018, Vancouver, British Columbia, Canada
6. Cheng T, Kassimi F, Zinck J (2016) A holistic approach of green mining innovation in tailings reprocessing and repurposing. Paper presented at Tailings & Mine Waste' 16, Keystone, Co. USA
7. Cheng TC, Tomlinson M, Di Feo A, Lastra R (2019) reprocessing of historic tailings from the Aunor mine in Timmins. Paper presented at the 51st Canadian mineral processor conference, Ottawa (in press at time of submission)
8. Environment Canada (2009) Environmental code of practice for metal mines. https://www.ec.gc.ca/lcpe-cepa/default.asp?lang=En&n=CBE3CD59-1
9. Mining Watch (2016) Analysis—environmental liability of mine sites in British Columbia. https://miningwatch.ca/sites/default/files/2016-05-30-bcminingliability-analysis_0.pdf
10. Ramji K (2016) Strategic risk assessment and management of Indigenous issues in the extractive sector. CIM J 7(1):52–60
11. The Mining Association of Canada (2011a) Developing an operation, maintenance and surveillance manual for tailings and water management facilities. http://www.mining.ca/
12. The Mining Association of Canada (2011b) A guide to the management of tailings facilities. http://www.mining.ca/
13. The Mining Association of Canada (2011c) A guide to audit and assessment of tailings facility management. http://www.mining.ca/
14. The Mining Association of Canada (2015) MAC tailings management initiatives. http://www.mining.ca/
15. Tisch B, Beauchemin S, Langley S, Metsaranta J (2018) Fast growing willows and organic covers. Paper presented at resources for future generations (RFG), 16–20 June 2018, Vancouver, British Columbia, Canada
16. Zinck J (2004) Emerging technologies in mine waste management. Paper presented at the annual conference of metallurgists of CIM, 43, Waste processing and recycling in mineral and metallurgical industry, international symposium, vol 5, pp 49–66
17. Zinck J, Tisch B (2018) Mining value from waste workshop report. Toronto, CanmetMINING internal report, Natural Resources Canada, 19 Dec 2017, 152 pp

Exploring Drivers of Copper Supply and Demand Using a Dynamic Market Simulation

Jingshu Zhang, Omar Swei, Richard Roth and Randolph Kirchain

Abstract Several authors have suggested that supplies for key non-renewable resources may soon cease to expand at the same rate as demand. Inevitably this would lead to some peak and then decline in production. This work describes the development and application of a fully dynamic model of the copper market. This model includes both a probabilistic simulator of future supply expansion based on published data on copper deposits and a novel model of copper demand that includes both short-term and long-term elasticity of demand—both due to self-price and aluminum price. Using this model in long-term simulations suggests significant copper reserve depletion would not be expected to occur in this century. The model predicts a time frame to significant depletion that is more than 50% further out in the future than previously reported results that did not include these effects.

Keywords Price elasticity · Economics · Criticality

As world population grows and, more importantly, gains wealth demand for materials of all kinds is growing rapidly. As shown in Fig. 1, the use of some materials has grown nearly exponentially. In fact, for the material of focus of this paper, copper, we now produce same amount every 3 years as was produced in the first 50 years of the last century [1].

The intensity of use is a sign of economic progress and those materials do bring important benefit through the globe. Nevertheless, that same intensity of use has raised some concerns of both the environmental and economic sustainability of the metals and minerals supply [2–4]. Broadly, these concerns are referred to as scarcity, which in the authors' opinion often leads to some misunderstanding. Scarcity is popularly understood to refer to running out of something. In economically driven markets scarcity will have an effect long before the last ton is extracted from the earth. As a material becomes more scarce, generally, its extraction will become more and more challenging. This increase in extraction cost, as well as the normal allocative

J. Zhang · R. Roth · R. Kirchain (✉)
Materials Systems Laboratory, Massachusetts Institute of Technology, Cambridge, MA, USA
e-mail: kirchain@mit.edu

O. Swei
Department of Civil Engineering, University of British Columbia, Vancouver, BC, Canada

G. Gaustad et al. (eds.), *REWAS 2019*, The Minerals, Metals & Materials Series,
https://doi.org/10.1007/978-3-030-10386-6_40

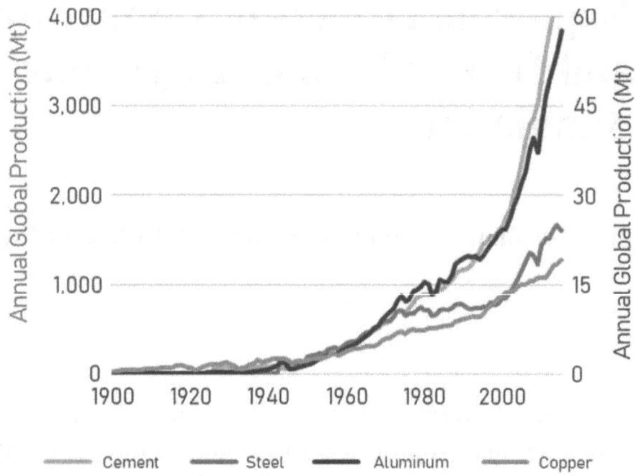

Fig. 1 Annual global production in million metric tons (Mt) of four major materials: cement, steel, aluminum, and copper. Note that, cement and steel are plotted on left axis. Aluminum and copper are plotted on right axis [1]

function of markets, will mean that scarce resources will become ever more costly. The first effect of scarcity is felt when some application is no longer economical because the material from which it is made is too expensive.

In the end, this more nuanced appreciation of scarcity leads to the same outcome—some function can no longer be met with the technologically superior option because that option is now too expensive. For copper, we could imagine real consequences of this. Would a device become less electrically efficient because it no longer used copper as a conductor? Would a machine become less thermally efficient because it no longer used copper to transfer heat?

Given the economic importance of copper, it is not surprising that several authors have explored its market function. For instance, Aguirregabiria and Luengo [5] developed a dynamic structural model that incorporates information regarding production delay, unit cost heterogeneity, and market concentration. Because of the preliminary nature of the available report, no estimates of long-term copper use are provided.

Northey et al. [6] created an agent-level supply model based on a detailed database of copper mines and deposits coupled with a univariate model of future demand. Using these models, they have conducted a century-long analysis to investigate scarcity risk at various levels of reserve size. As we will see later, however, they work has not incorporated price mechanism, and has not allowed for explicit materials substitution as a result of changing commodity prices.

Glöser et al. [7, 8] have developed a sophisticated copper market model that takes into account the delay in adjustment of supply and price mechanism and a detailed physical material flow model. Although this model represents a significant advance, there is no reported application of it to long-term use or scarcity questions.

We build on these previous works to better understand the scarcity trends in the copper market. Specifically, we develop a novel set of econometric models of demand that provide insight into both short-run and long-run elasticities both for self-price and substitute-price elasticity. (We limit substitute-price affects to aluminum price in the analysis presented here.) Figure 2 shows the results of these analyses for the

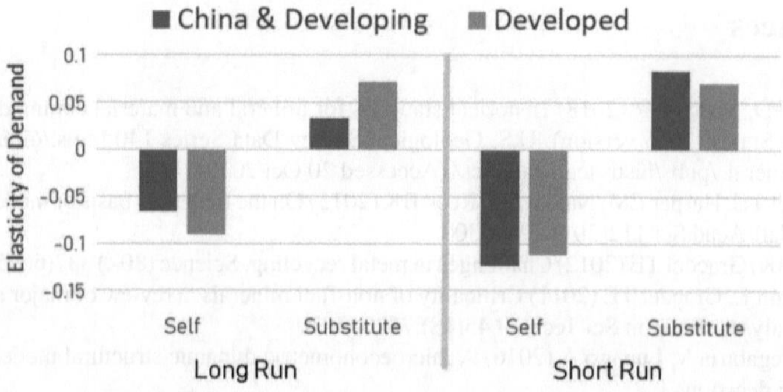

Fig. 2 Elasticities of demand for copper in electrical applications, inferred from auto-regressive distributed lag modeling of past cost use and prices. Short-run elasticities apply to demand shifts within 1 year. Long-run elasticities apply to demand shifts longer than 1 year out. Self-elasticities are a response to copper price. Substitute-elasticities are a response to aluminum price

electrical products sector. In this work, short-run elasticities apply to demand shifts within one year. Long-run elasticities apply to demand shifts longer than one year out. Self-elasticities are demand responses to copper price. Substitute-elasticities are a response to aluminum price.

From these elasticities, we see that China and developing world demand is more responsive in the short term to changes in price (short-run elasticities are larger in magnitude), but less responsive in the long run (long-run elasticities are smaller in magnitude).

Additionally, in this model, we incorporate algorithms that allow supply to expand in response to price. This structure is analogous to that proposed by Aguirregabiria and Luengo, but incorporates not only known projects, but a statistical analysis of known deposits as reported by the USGS [9–11]. The model considers decisions at each prospective mine site and evaluates like return on investment. Mines only open when they would be economically attractive.

By incorporating information on how price would likely alter both future demand and future supply, we see that the likely future of copper will extend well beyond some other previous predictions. Specifically, our model predicts that if demand is assumed to be inelastic and the supply does expand with price, then demand would roughly equal current reserves by the end of the century. However, we see that price effects would be expected to decrease future demand by more than 20% compared to levels one would find from extrapolations based on today's use. That drop results from substitution away from copper to other materials like aluminum if copper prices were to rise. Additionally, we see that economically driven expansion of supply could easily double available copper compared to today's economically viable reserves.

In the end, economically sensitive models will increasingly be important tools for understanding long-term trends in materials use.

References

1. Kelly TD, Matos GR (2018) Historical statistics for mineral and material commodities in the United States (2016 version): U.S. Geological Survey Data Series 140 https://minerals.usgs.gov/minerals/pubs/historical-statistics/ (Accessed 20 Oct 2018)
2. Graedel TE, Harper EM, Nassar NT, Reck BK (2015) On the materials basis of modern society. Proc Natl Acad Sci 112(20):6295–6300
3. Reck BK, Graedel TE (2012) Challenges in metal recycling. Science (80-.) 337(6095):690–695
4. Erdmann L, Graedel TE (2011) Criticality of non-fuel minerals: a review of major approaches and analyses. Environ Sci Technol 45(18):7620–7630
5. Aguirregabiria V, Luengo A (2016) A microeconometric dynamic structural model of copper mining decisions
6. Northey S, Mohr S, Mudd GM, Weng Z, Giurco D (2014) Modelling future copper ore grade decline based on a detailed assessment of copper resources and mining. Resour Conserv Recycl 83:190–201
7. Glöser S, Soulier M, Tercero Espinoza LA (2013) Dynamic analysis of global copper flows. global stocks, postconsumer material flows, recycling indicators, and uncertainty evaluation. Environ Sci Technol 47(12):6564–6572
8. Glöser-Chahoud S, Hartwig J, Wheat ID, Faulstich M (2016) The cobweb theorem and delays in adjusting supply in metals' markets. Syst Dyn Rev 32(3–4):279–308
9. Mosier D, Singer V, Donald A (2009) Volcanogenic massive sulfide deposits of the world: database and grade and tonnage models
10. Cox D, Lindsey D, Singer D, Diggles M (2003) Sediment-hosted copper deposits of the world: deposit models and database
11. Singer D, Berger V, Moring B (2005) Porphyry copper deposits of the world: database, map, and grade and tonnage models

Toward a Solid Waste Economy in Colombia: An Analysis with Respect to Other Leading Economies and Latin America

José J. Rúa-Restrepo, Gloria I. Echeverri and Henry A. Colorado

Abstract This investigation aims to analyze critically the historical situation, current, and potential trends of the main solid wastes in Colombia, not only from a detailed and internal point of view, but also in the Latin American countries. To give a better context and understanding of the issue, some data is also studied in comparison with some leading economies worldwide. Most countries worldwide including Latin America still work with the linear economy model, where the wastes are not intended to be minimized, re-used, or considered in the initial design as is in the circular economy. To implement the circular economy, one of the major challenges in many countries is the quantification and thus reliable and verifiable data for waste, therefore being one of the main goals of the current investigation in Colombia, particularly focused in the main solid wastes. In addition, important clues have been found in relation with the waste, economy, population, gross internal product, regulation, and society practices. Results from the current investigation can be used for similar economies and for countries with comparative waste numbers to Colombia.

Keywords Sustainable economy · Circular economy · Solid waste · Colombia · Latin America

J. J. Rúa-Restrepo · H. A. Colorado
CCComposites Laboratory, Universidad de Antioquia UdeA,
Calle 70 No. 52-21, Medellín, Colombia

J. J. Rúa-Restrepo · G. I. Echeverri
Grupo Pluriverso, Universidad Autónoma Latinoamericana
Unaula, Medellín, Colombia

H. A. Colorado (✉)
Universidad de Antioquia, Facultad de Ingeniería, Bloque 20,
Calle 67 No. 53-108, Medellín, Colombia
e-mail: henry.colorado@udea.edu.co

G. Gaustad et al. (eds.), *REWAS 2019*, The Minerals, Metals & Materials Series,
https://doi.org/10.1007/978-3-030-10386-6_41

Introduction

Today, the estimation is that the world generates about 1.3 billion tons yearly of solid wastes, and it is expected to double this number for 2025 with the population increase [1], which is considered now a worldwide problem [2]. In Europe, the uncontrolled growing of wastes derived from the world consuming its natural resources faster has forced that politics are all adapted to promote the use and exploitation of these wastes with the optimal strategy according to the socioeconomic projections [3]. It is known the relation among the solid waste generation, the population increase, and the industrial production [4] and Bruvoll and Ibenholt [5]: the increase in the population has generated an increase in the raw materials and correlation with the economic level, production, and generated wastes overall. This has been validated by Berglund and Söderholm [6], using econometrics for a flow analysis of the used raw materials in the paper industry, showing a relation among the generated waste, the management, and the economy, for industry and communities [7]. In order to minimize the waste, because targeting this one to zero is impossible itself by the thermodynamics laws, the strategy of minimizing the waste has been compared to recycling [8], finding the first as the more effective solution. The waste minimization concepts [9] have been collected by the circular economy (CE) model [10], as an alternative to the take-make-dispose alternative [11] and to take actions to preserve and make sustainable the available the planet Earth resources [10, 12–16].

Important research of the circular economy has been conducted in Asian nations, led by China [17–19], which shows a positive interest on the CE as their population and industrial production numbers have been increased significantly in the last few years, having a lot of public health issues [15]. Circular economy directly contributes to the sustainable development [20], but deserve a complete change not only in the procedures and technical aspects including the exploitation of natural resources, design, logistics, and manufacturing, but also in the politics and the economic model [21], and in the society role as well [22].

Organizations such as the Organization for Economic Co-operation and Development (OECD) [23] have been informing about the potential world-scale catastrophic scenario that can overcome a weak take in action and poor regulation in the correct implementation of the needed environmental and economical politics. The manufacturing sector plays a significant position in the accomplishing of these goals [24], with a main role of the plastic industry as a derived from a very contaminant oil industry [25], in the last years always being the focus of public attention, since the derived materials are typically not properly exploited or recycled [26], pero con alto potencial de aprovechamiento [27].

In particular in Colombia, and in other developing countries in Latin America and the Caribbean and also in Africa, the problems are even worst because the regulation is weak, the politicians are not really interested, and also because the waste statistics are not well-known. The numbers regarding the correct disposition and exploitation of waste with respect to the total amounts generated constitute the starting point for the circular economy of plastic materials [28–30].

In this paper, several data is presented and also other is generated mainly about the solid waste in Colombia and the main regulations. In addition, comparison with other nations are presented and most importantly presented in the Latin American context. Some details of the plastic waste generation and recycling are also discussed in relation with the Gross Domestic Product (GDP) and the population, all this conducted in order to set the bases for the calculation of a realistic and reliable circular economy model.

Materials and Methods

The materials analyzed in this investigation correspond mainly to plastic, organic, paper, and ordinary solid wastes. In order to establish feasible statistics about these wastes, not only local but also international databases have been studied, which include provided and official data. The main sources are detailed below.

American Society of Testing Materials—ASTM

The ASTM-D5231-92 [31] established a characterization methodology for the generated wastes from the residential and urban areas. Commercial and industrial wastes are also included. The mean composition determination for the municipal solid waste (MSW) starts with the collection and manual classification of wastes for a selected period of time of minimum a weak. Three main procedures are involved:

- Random sample collection in site.
- Collection time of data.
- Sectoral analysis for the characterization of different parts of the simple.

United Nations for Environmental Protection (UNEP)

UNEP has a detailed procedure for the solid waste management that includes its characterization and quantification [32]. Among other information, the solid wastes must be divided into the following types:

- Municipal solid wastes (residential and commercial)
- Construction and demolition wastes
- Industrial solid wastes (nonhazardous)
- Hazardous wastes (industrial, health, laboratory, and construction and demolition wastes).

The collection of information for the must follow the following steps:

- Step 1: Classification of the characteristics to measure
- Step 2: Procedure for sample collection
- Step 3: Waste quantification
- Step 4: Analysis methods.

World Bank (WB)

By the What a Waste [1] source, the World Bank (WB) studies the socioeconomic model, the demographic growing, the level of life, and the income level, all against the solid waste generation. The results are clear and reveal that more plastic waste is generated in the nations with a highest per capita income, where more materials are consumed and therefore more waste is expected, see Fig. 2. For Fig. 3, it can be seen how the waste is sectorized, it is also possible to observe the quantity of them and the classification of them in different types according to the countries analyzed, and it should be noted that this documentation is the result of the consolidation of data obtained during several years of research until 2012, including only member countries of the OECD. Analysis of the flow and econometrics using the Hodrick and Prescott (HP) can be used to determine the relationship between the flow and the reduction in the quantity of waste in a measure of time [33].

Hodrick and Prescott analysis use time series for analyze the data, using as a main parameter the period measured, such as year, each 3 months, and each 6 months, in order to give an approximation or an estimate of the behavior of the variables [34]. For the trend estimation, HP filter takes each point of data for every year, after it builds a trend according to the algorithm [35] and afterward describes the behavior of the variables and according to the inclination of the curve it is possible to predict data for ten years in the future. In the research, the data from different sources (see Fig. 5) is analyzed with HP filter, the variation of the algorithm uses an alpha of 100, since the data is given annually and the results can be detailed in Figs. 6 and 7, in which a cycle of behavior is presented, and the behavior of the softened curve explains the model that allows to describe the future data for the selected variable. Plastics are used as a variable of study.

Results and Analysis

Figure 1 shows a general overview of the OCDE countries for the solid waste data between 2010 and 2012 from D-Waste Atlas 2013 [36]. Figure 1a shows the solid waste per capita, which is very high in North America and Australia. Figure 1b shows the plastic waste generated relative to the total production, high in more distributed countries and also high in South America. Figure 1c shows the organic

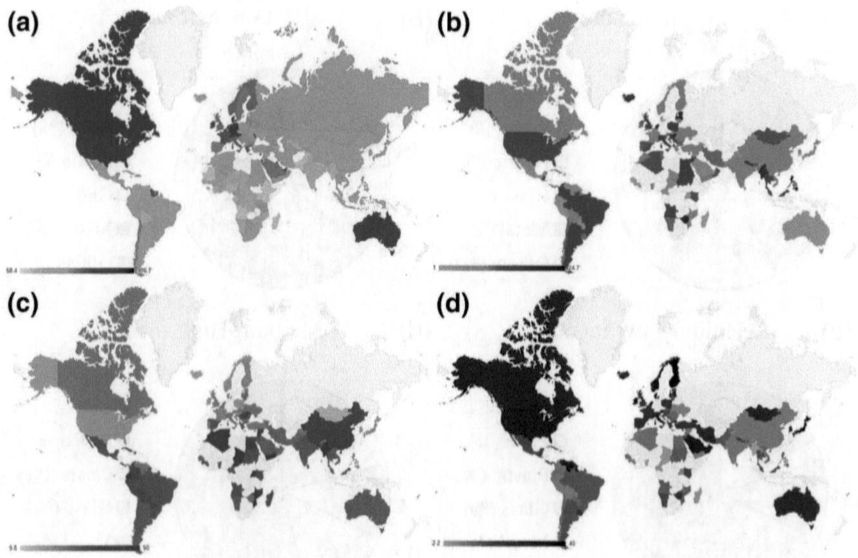

Fig. 1 Sectorized generation of Global Solid Waste (GSW): **a** Per capita solid waste generation (Kg/Año). **b** Plastic waste (%) regarding to total generation. **c** Organic waste (%) regarding to total generation. **d** Paper and cardboard (%) regarding to total generation. *Source* D-Waste Atlas (2013)

waste generated, also with a very important contribution from Latin and Central America. Finally, Fig. 1d shows the paper and cardboard waste generated, very high in the more developed countries, including USA, Canada, Australia, and Europe. These maps clearly show two faces: one is revealing an environmental problem, the other one is an opportunity for innovation and creative ways to use and recycle wastes. Particularly in Latin America and the Caribbean (LAC), an economics that includes solutions for the organic and plastic wastes can be a route toward the development.

The Situation with Respect the Leading Economies

The solid waste generation and the global income economy of the country have been summarized in Fig. 2 with data from World Bank. In this representation, solid waste was characterized by the type (organics, paper, plastic, glass, metals, and other solid wastes) and summarized by the type of economy of the country in general: high, medium-high, medium-low, and low-income country. In general, it is clear and accepted the inverse relation between the country income and the amount of waste generated. However, specifically low-income countries produce more organic waste than high-income countries; and high-income countries produce more plastic waste. In general, people with better economy have a higher product consumption of

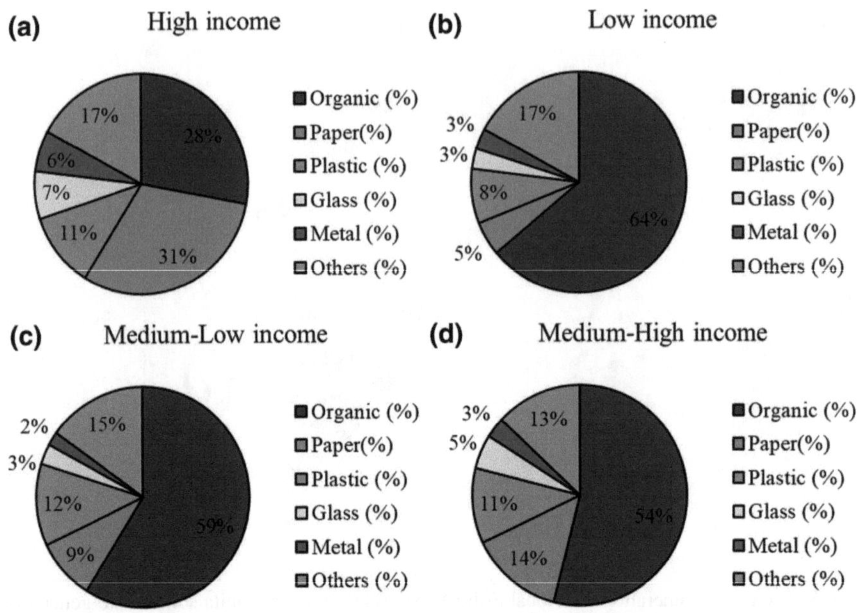

Fig. 2 Global waste generated by stream of waste and by economic income. **a** High income. **b** Medium-high income. **c** Medium-low income. **d** Low income. *Source* What a waste (2012)

materials and parts. These graphs also reveal that low-income countries need to work more in the use of organic wastes, which is better treated in the leading economies.

Figure 3a also supports Fig. 2, since countries as USA and Australia, show the lowest organic waste generated, but the highest numbers in terms of the plastic and paper wastes. On the other hand countries with weak economies show the organics as its main solid waste. Medium-high income countries show a high organic waste volume, and similar amount of plastic waste among them, with Denmark out of the trend with just 1 wt% of plastic waste, which is of course a consequence of good practices not only for recycling and utilization of the generated wastes but also a decrease in the solid waste generation.

Figure 3b suggests how each country manages its waste given by the numbers obtained directly from the landfills, with information including the classification and use of waste. Some countries such as Japan and Denmark have significant numbers of energy recovery in incineration plants of these wastes. The other part of the pie is revealed by countries with 100 wt% of wastes disposed of only in landfills, which suggests no programs and strategies for classification, recuperation and use of the solid wastes. Countries including Colombia, Costa Rica, and Uruguay, still have an important open to air landfills, which in most cases is the result of lack of interest of the government in a better solution. In Colombia, the regulation and politics for the solid waste management are proved to be not very efficient as there are several serious and active conflicts with the communities in the areas of interest. Moreover,

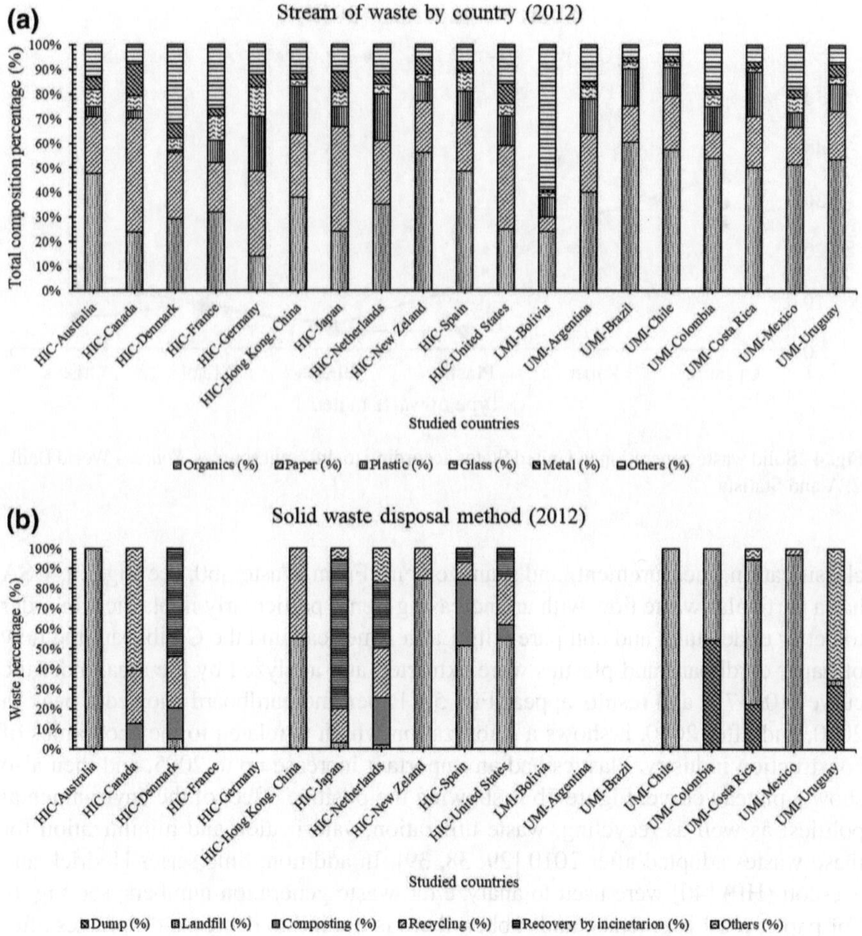

Fig. 3 Waste typology by country regard to income level: high income (HIC), low medium income (LMI) and upper medium income (UMI). **a** Stream of waste by country. **b** Solid waste disposal method. *Source* What a waste (2012)

the solid waste statistics and the official information provided the governing agencies is still in many cases in consolidation, as regulation just started to be applied and therefore in sectors like plastics, not much information is available, and the USA Environmental Protection Agency (APA) is being taken as a reference but still with a lot of limitations.

Figure 4 shows waste data for three different agencies in the same year: EPA, World Bank (WB), Statista. EPA and WB statistics show very close trends in the flow of waste materials, particularly in plastics and organics. STATISTA data present significant differences in the same waste types, which can be associated with differences in the information consolidation by each organization, with differences in the

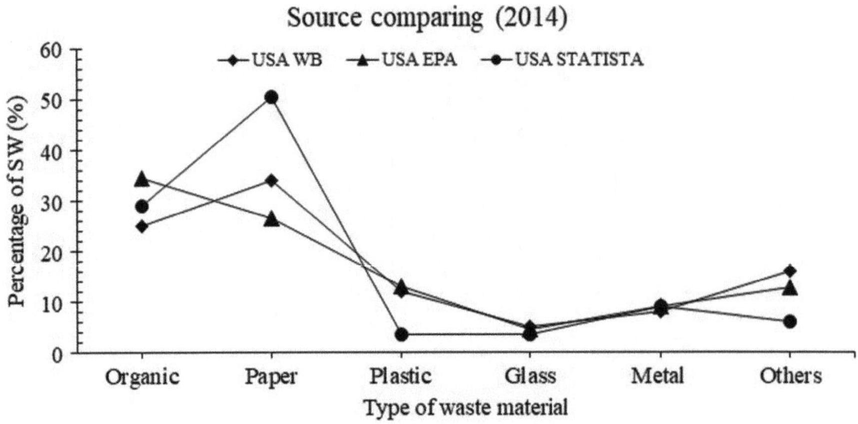

Fig. 4 Solid waste generation in United States according to different sources. *Sources* World Bank, EPA and Statista

classification, measurement, and management. From Waste 360, see Fig. 5a, USA has a particular waste flow with an increasing trend, particularly in plastics. In order to better understand and compare with Latin American and the Caribbean, the flow of paper cardboard and plastics were extracted and analyzed by the area under the curve [10, 37]), and results appear Fig. 5b. Paper and cardboard showed a peak in 2000, and after 2010, it shows a stabilization which is related to the economics of construction industry. Plastics had an important increase up to 2005, and then also show a plateau curve. Figure 5b is showing the positive effect of the environmental politics; as well as recycling, waste utilization, valorization and minimization for these wastes adopted after 2010 [29, 38, 39]. In addition, time series Hodrick and Prescott (HP) [40] were used to analyze the waste generation numbers, see Fig. 6. For paper, plastics, metals, and rubber, there is a trend to decrease the wastes after 2010, perhaps, as a result of improvement in waste management, and also due to a lack in the economy growth. The HP cycle also allows to predict future numbers, as a softened curve, see Fig. 7. This curve reveals a constant growing from 1960 with a slight decrease in the growth after 2010 (Table 1).

The Situation in Latin America

In South America, the data from the Inter-American Development Bank (IDB) for solid wastes was compared with the information from the Gross Domestic Product (GDP).

Figure 8a shows a strong relation between the GDP and the generated solid wastes for different countries in Latin America ordered by the amount of solid wastes. The two curves have the same trend. Figure 8b shows the wastes per capita generation,

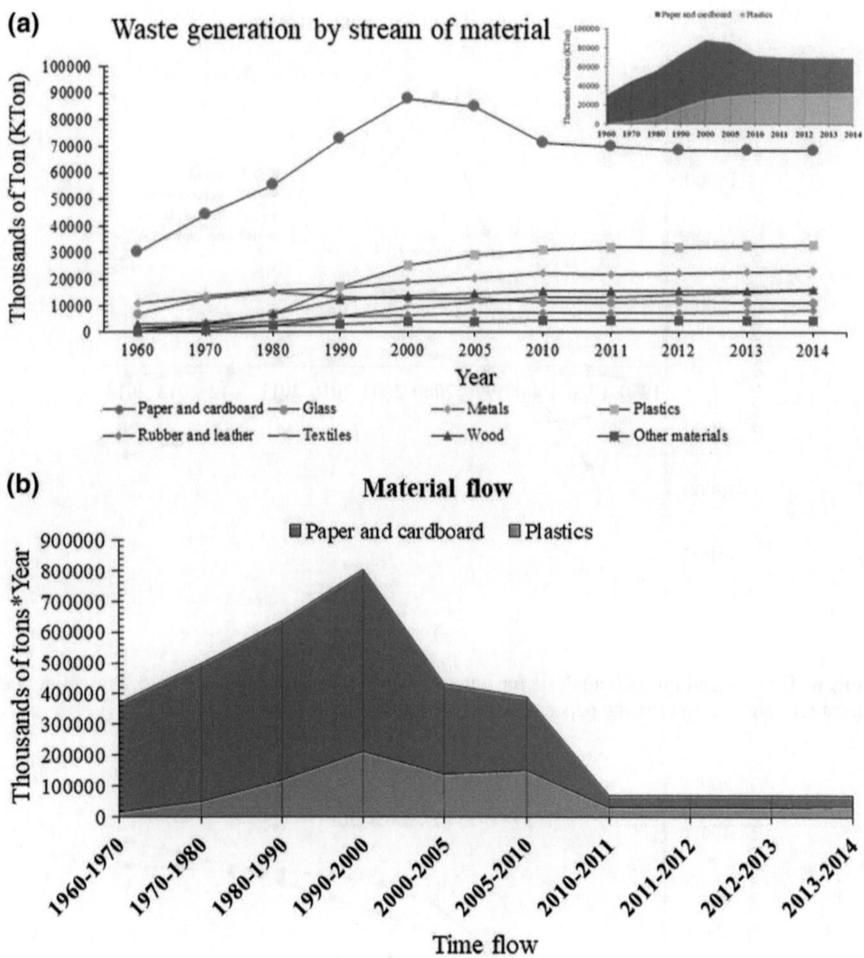

Fig. 5 Waste generation trend according to EPA and Waste 360.org análisis: **a** Historical generation by material flow. **b** Area under the curve

with also some relation with the GDP. Brazil is leading in the amount of wastes generated which is consistent with its largest population of the continent. However, when the waste amount is divided over the GDP, Bolivia pass to lead the statistics with 105 tons/year per million dollars, against lowest one, Venezuela, with 27 tons/year per million dollars. Figure 8c is a summary of the wastes by type, where organics, as shown before, are very significant in Latin America. The graph also reveals the progress in Brazil besides the large industry compared with other Latin American countries, have a better waste management plans.

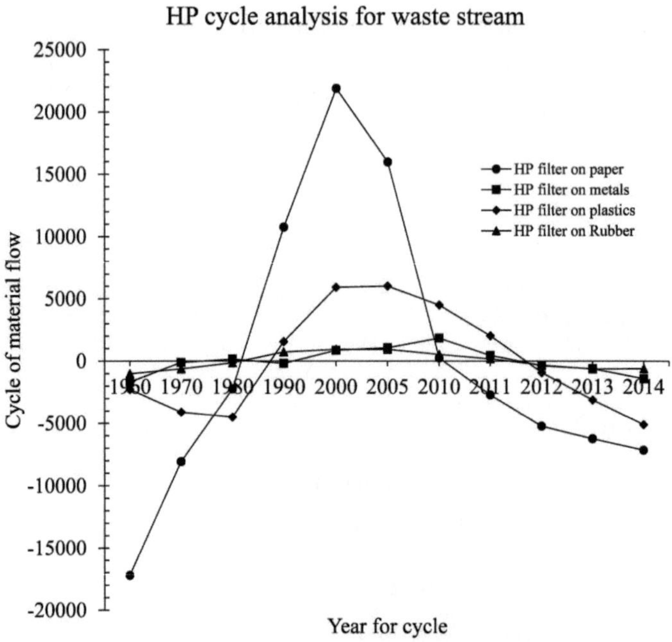

Fig. 6 Hodrick and Prescott analysis for time series of waste generation in Fig. 5. Presenting the most relevant stream of waste paper, plastics, metals, and rubber

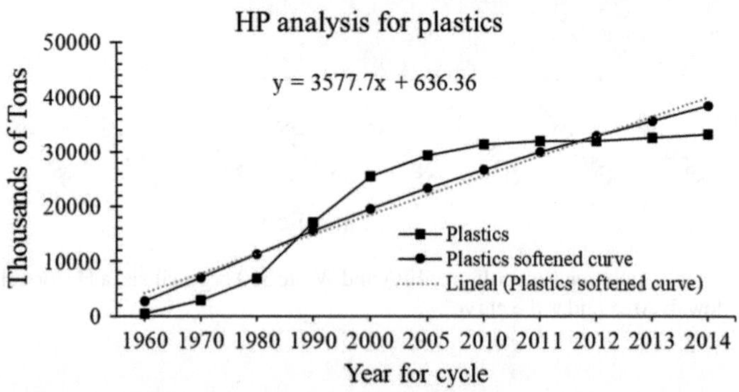

Fig. 7 Examples of softened curve for plastic waste stream

Figure 9 shows waste statistics for Colombia data from WB and BID, which is very consistent. Both sources share information from The Pan American Health Organization (PAHO) and the World Health Organization (WHO).

Figure 10 is a summary of the waste utilization rates in Colombia by the type of waste. A lot of work needs to be done for improving the use of organics, cement and concrete, and plastics.

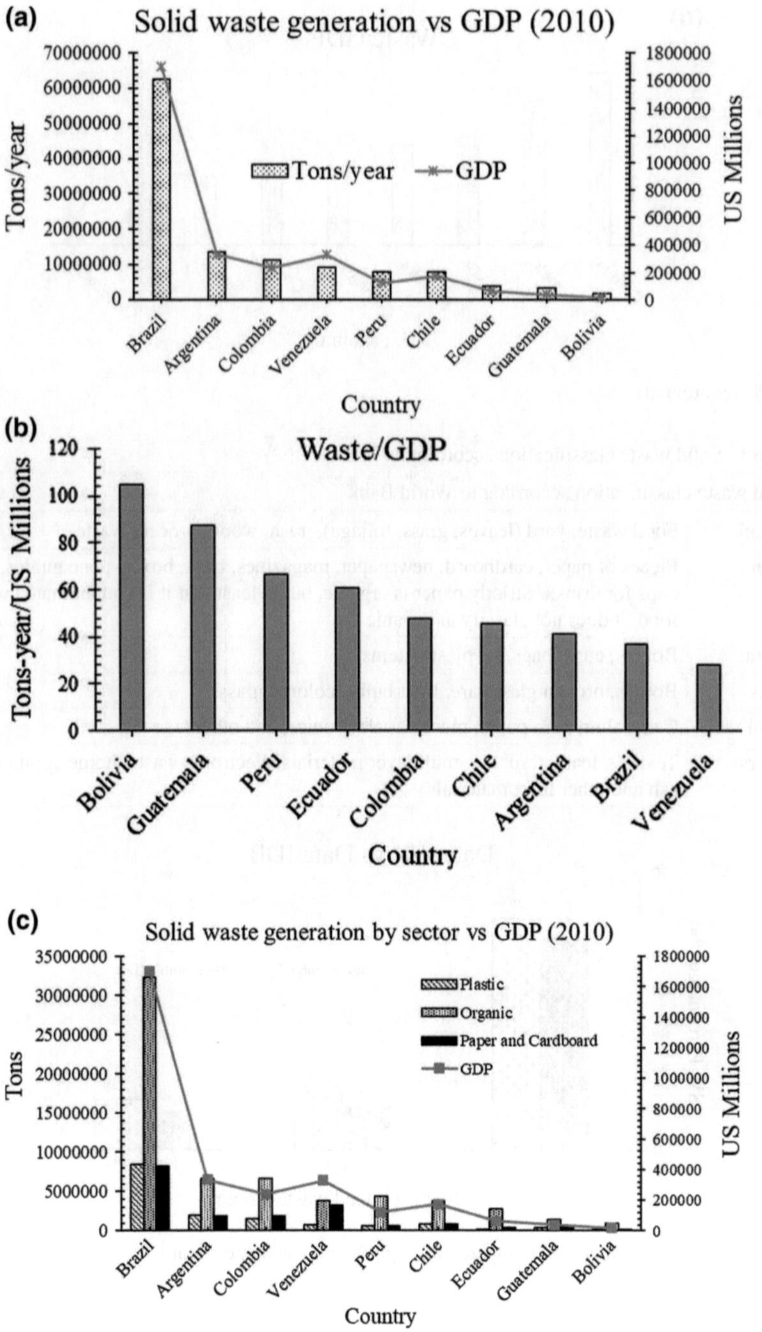

Fig. 8 Solid waste generation by country according to IDB and Waste Atlas: **a** solid waste versus GDP. **b** Solid waste versus per capita GDP. **c** Solid waste by stream of waste. **d** Generation radius versus GDP

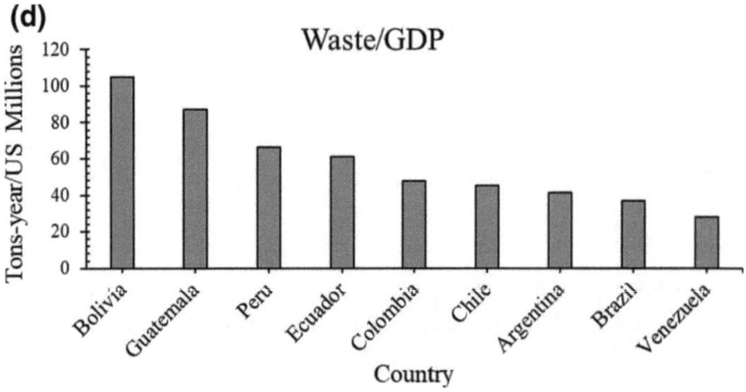

Fig. 8 (continued)

Table 1 Solid waste classification according to WB

Solid waste classification according to World Bank	
Organic	Food waste, yard (leaves, grass, foliage), trash, wood, process waste
Paper	Pieces of paper, cardboard, newspaper, magazines, bags, boxes, pone guides, paper cups for drinks. Strictly paper is organic, but at least that it is contaminated with food, it does not classify as organic
Plastic	Bottles, cups, bags, all plastic items
Glass	Bottles, broken glassware, light bulbs, colored glass
Metal	Cans, aluminum paper, pieces of electronics, and others
Others	Textiles, leather, rubber, multilayer materials, electronic waste, home appliances, ash and other inert materials

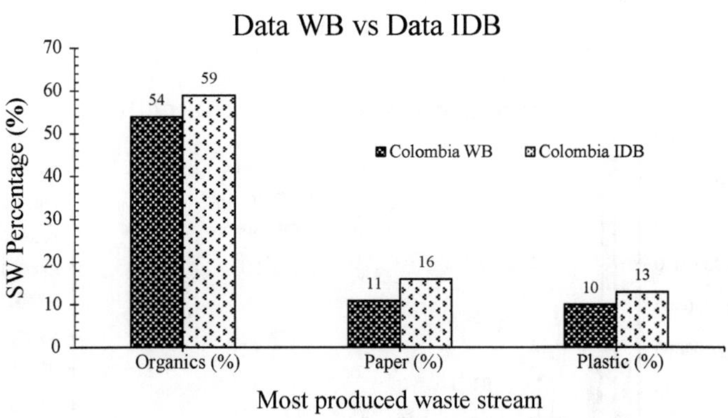

Fig. 9 Sources consulted and compared through the mayor waste stream

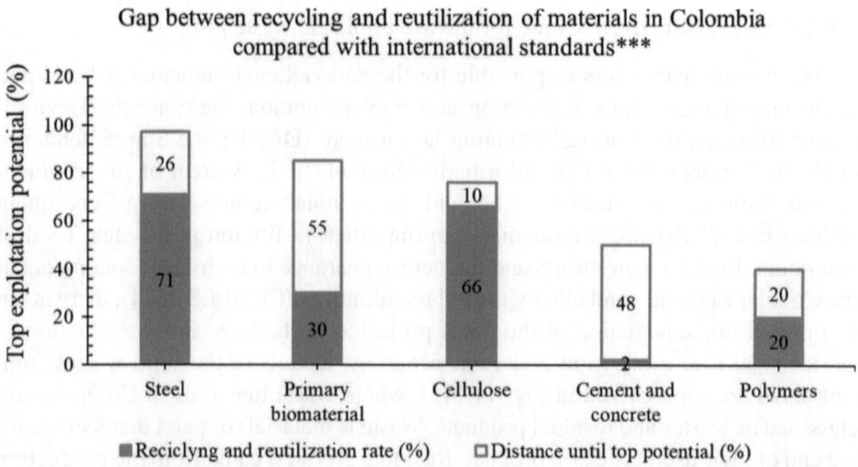

Fig. 10 Comparative exploitation rates of waste in Colombia with other countries. *Source* Tecnalia

The Situation Inside Colombia

In Colombia, there enough legislation and policies about the solid waste management and waste utilization, however, a poor law enforcement and real implementation mainly at the high-level government make this initiative obsolete. Some of the landmark laws and other regulations are listed below.

- Law 99 December 22 of 1993 (Ley 99 de diciembre 22 de 1993). For the creation of the Environmental Minister [41].
- Law 142 of 1994 (Ley 142 de 1994). For establishing the regimens for the home public services, which includes the public waste management service [42].
- Decree 605 of 1996 (Decreto 605 de 1996). For an adequate public service of the waste services [43].
- National policy for the integral waste management, 1997.
- Resolution 201 of 2001 (Resolución 201 de 2001). Establishes the conditions for the elaboration, update, and evaluation of environmental management's plans.
- Decree 1713 of 2002 (Decreto 1713 de 2002). Over the final disposition of solid wastes.
- Decree 005 January 7 of 2003 (Decreto 005 de enero 7 de 2003). To prevent the sanitary emergency in the city of Medellin.
- Resolution 1045 September 26 of 2003 (Resolución 1045 del 26 de septiembre de 2003). Method for the elaboration of Comprehensive Solid Waste Management Plans (PGIRS).
- Resolution 008 of 2004 (Resolución 008 de 2004). Master plan for the Metropolitan Area of Medellin for the Comprehensive Solid Waste Management.
- GTC 86: 2003-10-22. Guide for the implementation of Comprehensive Solid Waste Management GIR–[44].

- GTC 53-2: 2004-07-28. Guide for the use of plastic waste [44].

Among the main actors responsible for the national environmental policies and waste management plans elaboration and implementation, there are the Environmental Minister, the National Planning Department (DANE), the Superintendence of Public Services, the Single Information System (SUI), System of Environmental and Economic Accounts (SCAE), and the National Administrative Department of Statistics (DANE). Corantioquia is an important institution in the state level of Antioquia. Besides their efforts and the new appearance in high-level documents of the Circular Economy and other world-class initiatives (CONPES 3874), there is not an optimal implementation of the goals projected in the laws and the situation of wastes is still far away from a circular economy. Details of the solid wastes from Colombia are summarized in Fig. 11 [45], where waste materials in Colombia are classified as wastes and residual products. Waste is materials or parts that went, until the end of their useful life as a product. Residual products came from the production and immediately can be used in the same process, or even commercialized as raw semi-virgin materials for other products or processes. Figure 12a shows details about this flow, in which the residual products are produced in highest amount than wastes. The increase in the waste generated can be associated with an increase in the GDP for a larger industrial production and consumption, see Fig. 12b. Figure 12c shows the solid waste disposal growth rate in the main cities in Colombia from 2012, showing stability in the trend, associated with the beginning of the implementation of diverse national level strategies in the Compes developing plan (plan de desarrollo Compes).

Conclusion

The presented results showed some differences for the solid waste numbers not only for the case of Colombia but also for other nations from local and international databases. Particularly in the case of Colombia, the environmental topics was not even a topic of discussion in the government 10 years ago, however, even the lack of technology of the country, there is now an important social and international pressure for the improvement in the environmental regulations, policies and the main problem, their implementation. The main problem is that the case of Colombia is not a particularity, is the situation of if not half, more than half of the countries worldwide, and without a real interest of the government in investment in topics as the circular economy rather than just mentioning in campaigns, the situation will worsen with days and the population, pollution, and other derived environmental effects will increase. With the potential adhesion of Colombia to the OECD, an effort that the past and current government is pursuing, perhaps the country can progress in the current situation and give better number for 2020.

This work has provided and also summarizing some of the most representative data from Colombia in terms of solid wastes, but most importantly with respect to

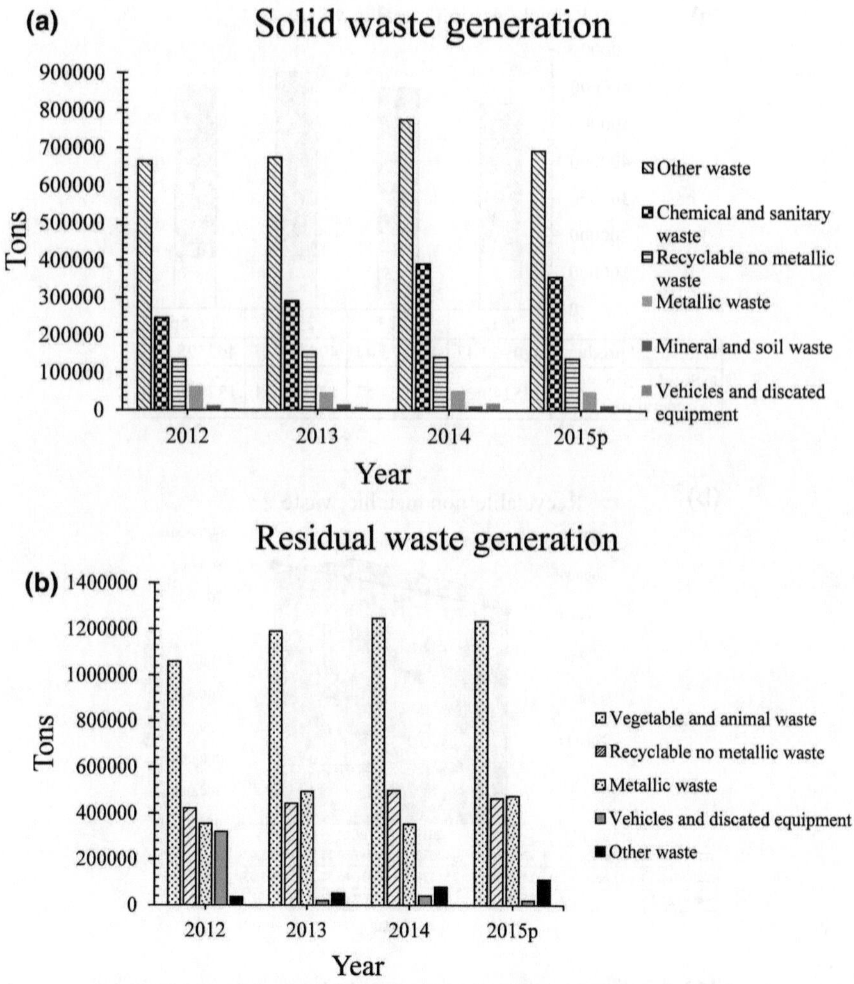

Fig. 11 Generation and classification of Waste in Colombia **a** solid waste generation. **b** Residual products generation. Environmental accounts (Environmental system of information-Colombia, 2017) [45]

leading economies and Latin America. It is clear than today more than before wastes had a profound impact in the economy of a country, and appropriate resources must be involved in order to adapt all the country statements and align the industrial and residential sectors in order to implement the circular economy model. In the case of Colombia, organic and plastic wastes represent an opportunity also generating new industries, which certainly is convenient for an economy mostly based in agriculture and raw materials sells. Finally, the study presented here has revealed a lack of communication in the data managed by the official agencies in Colombia, which certainly is affecting the implementation of correct strategies and plans to decrease

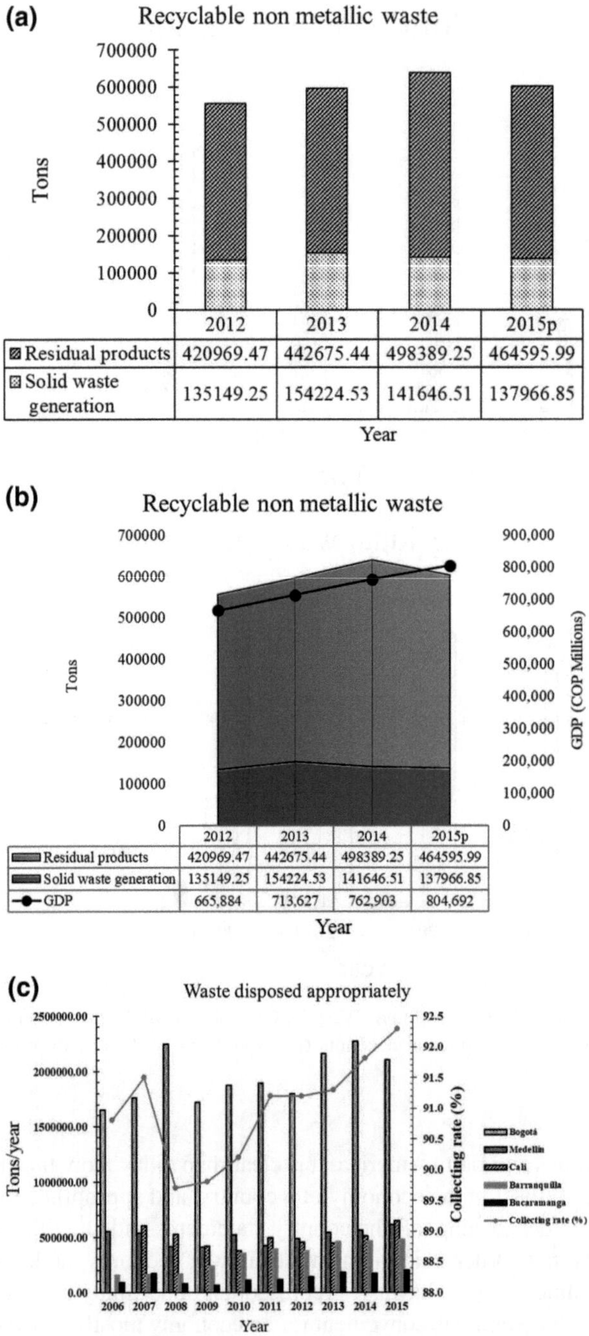

Fig. 12 Waste and collection rate. **a** Comparative between residual products and solid waste 2012–2015. **b** Flow rate residuals waste and GDP **c** waste disposed versus collection rate

the negative consequences of the non-properly managed wastes, and its adverse influence in the public health. This also diminishes the economic competitiveness of the country with respect to other economies and certainly bad preparation for environmental contingencies. A better plan is needed, and the circular economy is the way, with all society sectors working together.

References

1. Hoornweg D, Bhada-Tata P (2012) What a waste: a global review of solid waste management
2. Koroneos CJ, Nanaki EA (2012) Integrated solid waste management and energy production-a life cycle assessment approach: the case study of the city of Thessaloniki. J Clean Prod 27:141–150
3. Andersen FM, Larsen H, Skovgaard M et al (2007) A European model for waste and material flows. Resour Conserv Recycl 49:421–435
4. Alfsen K, Bye T, Holmoy E (1996) An applied general equilibrium model for energy and enviromental analyses
5. Bruvoll A, Ibenholt K (1997) Future waste generation: forecasts on the basis of a macroeconomic model. Resour Conserv Recycl 19:137–149. https://doi.org/10.1016/S0921-3449(96)01189-5
6. Berglund C, Söderholm P (2003) An econometric analysis of global waste paper recovery and utilization. Environ Resour Econ 26:429–456. https://doi.org/10.1023/B:EARE.0000003595.60196.a9
7. Hamer G (2003) Solid waste treatment and disposal: effects on public health and environmental safety. J Biotechnol Adv 22:71–79
8. Tonglet M, Phillips PS, Bates MP (2004) Determining the drivers for householder pro-environmental behaviour: waste minimisation compared to recycling. Resour Conserv Recycl 42:27–48
9. Schandl H, Schaffartzik A (2015) Material flow analysis. In: International encyclopedia of the social & behavioral sciences, 2nd edn. pp 760–764
10. Adams R, Jeanrenaud S, Bessant J et al (2016) A review on circular economy: the expected transition to a balanced interplay of environmental and economic systems. J Clean Prod 8:1–17. https://doi.org/10.1016/j.jclepro.2015.12.042
11. Ness D (2008) Sustainable urban infrastruture in China: towards a factor 10 improvement in resource productivity through integrated infrastructure systems. Int J Sustain Dev World Ecol 15: 288–301. https://doi.org/10.3843/SusDev.15.4:2a
12. Mazzantini U (2014) Rivoluzione a Davos, il big business mondiale vuole l'economia circolare
13. Lett LA (2014) Las amenazas globales, el reciclaje de residuos y el concepto de economía circular. Rev Argent Microbiol 46:1–2
14. Park JY, Chertow MR (2014) Establishing and testing the reuse potential indicator for managing wastes as resources. J Environ Manage 137:45–53
15. Geng Y, Fu J, Sarkis J, Xue B (2012) Towards a national circular economy indicator system in China: an evaluation and critical analysis. J Clean Prod 23:216–224. https://doi.org/10.1016/j.jclepro.2011.07.005
16. Su B, Heshmati A, Geng Y, Yu X (2013) A review of the circular economy in China: moving from rhetoric to implementation. J Clean Prod 42:215–227
17. Zhijun F, Nailing Y (2007) Putting a circular economy into practice in China. Sustain Sci 2:95–101
18. The Ellen MacArthur Foundation (2012) Towards a circular economy—economic and business rationale for an accelerated transition. Greener Manag Int 97. 2012-04-03

19. Yuan Z, Bi J, Moriguichi Y (2006) The circular economy: a new development strategy in China. J Ind Ecol 10:4–8
20. Group OW, Goals SD, Group OW, et al (2015) Sustainable development goals and targets. United Nations
21. Porter RC (2005) The economics of waste. In: resources policy. pp 141–142
22. Lu Y, Nakicenovic N, Visbeck M et al (2015) Five priorities for the UN sustainable development goals. Nature 520:432–433
23. Publishing O (2008) OECD environmental outlook to 2030. Organisation for Economic Co-operation and Development
24. Kalpakjian S, Schmid SR (2014) Manufacturing engineering and technology. Manuf Eng Technol
25. Harper CA (2006) Handbook of plastics technologies. IEEE Electr Insul Mag 22:53
26. PlasticsEurope (2007) The compelling facts about plastics: an analysis of plastics production, demand and recovery for 2005 in Europe
27. Warren LM, Burns R (1988) Processors make a go of mixed-waste recycling. Plast Technol 34:41–42
28. Ellen MacArthur Foundation (2016) The New Plastics Economy: Rethinking the future of plastics. Ellen MacArthur Found 120. https://doi.org/10.1103/Physrevb.74.035409
29. Vélez SLP, Vélez AR (2017) Recycling alternatives to treating plastic waste, environmental, social and economic effects: a literature review. J Solid Waste Technol Manag 43:122–136
30. Tokai A, Furuichi T (2000) Evaluation of recycling policies for PET bottles based on multiattribute utility indices. J Mater Cycles Waste Manag 2:70–79. https://doi.org/10.1007/s10163-999-0021-6
31. ASTM D (2003) Standard test method for determination of composition of unprocessed municipal solid waste. ASTM Int 5231–5292
32. Manual T (2009) Developing integrated solid Waste Management Plan. In: Prepared by United Nations Environ Program
33. Kim H (2004) Hodrick-Prescott Filter. Business 2004
34. Ahumada H, Garegnani ML (1999) Hodrick-Prescott filter in practice. n IV Jornadas Econ Monet e Int (La Plata, 1999). 10.1.1.121.3365
35. Harvey A, Trimbur T (2008) Trend estimation and the Hodrick-Prescott filter. J Japan Stat Soc 38: 41–49. https://doi.org/10.14490/jjss.38.41
36. ISWA; University of Leeds; Sweepnet; Wtert; Swapi (2013) Atlas D-waste. In: Waste atlas. http://www.atlas.d-waste.com/
37. Nielsen RW (2016) Mathematical analysis of historical income per capita distributions. arXiv: 1603.01685
38. Anon (2004) Wood-plastic compounds: when plastic touches wood. Kunststoffe-Plast Eur 94: 38–39
39. Neale CW, Hilyard NC, Barber P (1983) Observations on the economics of recycling industrial scrap plastic in new products. Conserv Recycl 6:91–105
40. Phillips PCB, Jin S (2015) Business cycles, trend elimination, and the HP filter
41. Congreso de Colombia (1993) Ley 99. Congr Colomb 44. https://doi.org/10.1017/CBO9781107415324.004
42. COLOMBIA.CONGRESO DE LA REPÚBLICA. Ley 142 de 1994. Por la cual se establece el régimen de los servicios públicos domiciliarios y se dictan otras disposiciones. (1994) Ley 142 de 1994
43. Presidencia de la República de Colombia (1996) Decreto 605 de 1996
44. ICONTEC: Instituto Colombiano de Normas Tecnicas y Certificacion
45. Sistema de Información Ambiente de Colombia (2017) Postconsumo - IDEAM. In: Minist. Ambient. y Desarro. Sosten. Colomb. http://www.siac.gov.co/residuospostconsumo

Cobalt Criticality and Availability in the Wake of Increased Electric Vehicle Demand: A Short-Term Scenario Analysis

Danielle Beatty, Xinkai Fu, Michele Bustamante, Gabrielle Gaustad, Callie Babbitt, Randolph Kirchain, Richard Roth and Elsa Olivetti

Abstract Understanding cobalt criticality and availability is necessary as lithium-ion battery demand, particularly for electric vehicles, is projected to increase exponentially throughout the next decade. With these increases in demand, supply concentration and mining limitations could have a significant impact on cobalt-dependent firms, sectors, and emerging technologies. Increased recycling, cobalt refining capability, and possible substitution away from cobalt in high demand sectors form the basis of modeling scenarios; these scenarios are created to identify the implications of changes in the cobalt supply-demand balance to 2030. Based on these, cobalt faces a tight but not impossible short-term market, where recycling and additional cobalt sources should be pursued and will become increasingly important out to and past 2030.

Keywords Cobalt metal · Availability · Criticality · Scenario analysis · Supply · Demand · Electric vehicle

Understanding cobalt criticality and availability is necessary as lithium-ion battery demand is projected to increase exponentially throughout the next decade. Lithium-ion batteries, concentrated in consumer electronics and electric vehicles (EVs), are currently the largest cobalt demand sector (accounting for 53% of the global cobalt demand). The market for electric vehicles is expected to see significant increases between 2020 and 2026, as costs for electric vehicles equalize with those for traditional ICE vehicles. Along with cost, country-specific policy decisions are limiting the production and sale of traditional vehicles in favor of a more sustainable alternative, such as electric vehicles. High complex annual growth rates for electric vehicle

---

D. Beatty
University of Utah, Salt Lake City, USA

X. Fu · M. Bustamante · R. Kirchain · R. Roth · E. Olivetti (✉)
Massachusetts Institute of Technology, Cambridge, USA
e-mail: elsao@mit.edu

G. Gaustad · C. Babbitt
Rochester Institute of Technology, Rochester, USA

G. Gaustad et al. (eds.), *REWAS 2019*, The Minerals, Metals & Materials Series,
https://doi.org/10.1007/978-3-030-10386-6_42

sales are projected globally and electric vehicle demand (sales) could reach over twenty-nine million vehicles in 2026.

With these increases in demand, supply concentration and mining limitations could have significant impact on cobalt-dependent firms, sectors, and emerging technologies. Increased recycling, cobalt refining capability, and possible substitution away from cobalt in high demand sectors form the basis of modeling scenarios; these scenarios are created to identify the implications of changes in the cobalt supply-demand balance to 2030. Four distinct scenarios are considered, including substitution away from cobalt in the superalloy demand sector, decreased mining efficiency due to political instability in the highly concentrated cobalt supply chain, increased mining efficiency in all cobalt mines worldwide, and secondary supply from electric vehicle batteries reaching their end of life.

Due to the projected increase in demand over the coming decade, lithium-ion batteries will continue to be the largest end-use sector of cobalt. With expected lifetimes between eight and twelve years and high cobalt recovery rates at end of life, EV lithium-ion batteries will provide a significant source of secondary supply for cobalt. Due to relatively high recycling rates (up to 50%), recycling programs of small consumer electronics may also play a future role in secondary cobalt supply. Short-term projections indicate that battery recycling is not currently economically or systematically developed to make an impact on cobalt supply up to 2030. Despite this, recycled cobalt from used batteries will be a significant cobalt source in the long term, potentially matching seventy percent of primary supply by 2030.

Superalloys are the largest sector of non-battery demand, accounting for sixteen percent of total demand (when batteries are included) and thirty-four percent of total non-battery demand; smaller demand sectors include hard materials, catalysts, and magnets. To explore demand sector dependence on cobalt, a scenario involving stagnated superalloy demand is developed. Although superalloys constitute the second largest sector of cobalt demand, this scenario results in only a small decrease in non-battery demand for cobalt. Despite this, if the market is tight enough even small changes may play a large role in the cobalt supply-demand balance in the short term.

Primary supply of cobalt is heavily geographically concentrated, both for mining/production and processing of the mined materials. Current estimates put approximately sixty percent of all mined cobalt production in the Democratic Republic of Congo (DRC); this value is expected to reach upwards of sixty-five percent before 2030. No more than five percent of primary cobalt is supplied by any other country. Fifty-eight percent of the produced cobalt is processed in China; ninety-one percent of this originates in the DRC. Along with these heavy supply concentrations, primary cobalt supply is also restricted by its by-product status: cobalt is mined mainly as a by-product of nickel, copper, or platinum group metals and very few mines produce cobalt as their primary source of revenue. Because of this, cobalt supply may not respond to the expected increases in cobalt demand; heavy supply concentration may also effect the supply-demand balance in the short term. To explore this, two scenarios are developed to assess the effects of both increases and decreases in primary supply.

Mine location, ownership, and metal by-product status may all effect cobalt mining efficiency globally. To assess additional cobalt primary supply, cobalt mining efficiency is increased for all mines globally; results indicate that increased mining efficiency as a result of increased cobalt demand, could supply up to twenty percent more cobalt by 2028. To explore significant cobalt supply decreases, supply is limited for three mines in the DRC; a scenario such as this could arise with political instability in this copper-belt region. While over seventy mines are currently producing cobalt, limiting the production in only three could show a thirteen percent decrease in cobalt supply within the next decade. Supply levels from currently operating mines are unlikely to meet the exponential increases projected for cobalt demand from the electric vehicle market. To meet this supply-demand gap, many additional mines are expected to open or increase cobalt production within the next decade.

There are a number of additional areas of interest that are likely to affect the cobalt supply-demand balance, both in the short term and out past 2030. These include the degree of integration of firms along the cobalt supply chain and the potential for hedging and speculation to influence supply. Other sources of cobalt should also be taken into account, including potential undersea cobalt resources, and secondary supply from consumer electronics; evolution of electric vehicle battery chemistry should also be considered when assessing cobalt demand in the long term.

Based on the scenarios explored, cobalt faces a tight but not impossible short-term market. Political instability in the DRC, low values of secondary supply entering the market, and demand increases in the battery and non-battery sectors increase the gap between supply and demand. However, high values of cobalt recovery and secondary supply from EVs bring supply within reach of demand to 2030. With extreme demand increases and continued geographical supply concentration over the next decade, secondary supply from cobalt-containing products reaching end of life and additional new cobalt resources may be necessary for supply to meet demand. Recycling, cobalt refining capability, possible substitution away from cobalt in high demand sectors, and additional cobalt resources should be top priorities both politically and technologically in the years to come.

Part IX
REWAS 2019: Secondary and Byproduct Sources of Materials, Minerals, and Metals: Poster Session

Distribution and Chemical Species of Chromium in the EAF Dust from Stainless Steel Plant

Zhi Li, Guojun Ma and Xiang Zhang

Abstract Electric Arc Furnace (EAF) dust generated in the production process of stainless steel is a hazard to the environment and human health due to its large amounts of toxic substances, such as Cr(VI). Cr in the EAF dust is mainly in the form of a residue according to sequential extraction procedure test, and only small amounts of Cr are mobility. Although the formation mechanism of the EAF dust has been gradually understood, the distribution and chemical species of Cr in EAF dust are rarely studied. In this paper, the distribution and chemical species of Cr in EAF dust were studied. The results show that Cr(III) and Cr(VI) are mainly present in the small particles. After sputtering about 140 s with Ar^+ ion beam, the Cr(VI) on the surface of dust particles disappears, and the content of Cr(III) reaches 100%, indicating the Cr(VI) in the EAF dust comes from the oxidation of lower valent Cr-containing substances and enrich on the surface of the dust particles.

Keywords EAF dust · Cr(VI) formation · Cr distribution · Stainless steel

Background

Chromium comes from chromium ore, which is widely used in metallurgy, refractories, and chemistry. The component of the chromium ore used in metallurgy industry contains Cr_2O_3 of more than 46% and Cr/Fe of 2, respectively [1]. Due to its corrosion resistance, chromium is widely used in the product of stainless steel in the form of

Z. Li · G. Ma (✉) · X. Zhang
Hubei Provincial Engineering Technology Research Center of Metallurgical Secondary Resources, Wuhan University of Science and Technology, Wuhan 430081, China
e-mail: gma@wust.edu.cn

Z. Li · G. Ma · X. Zhang
Key Laboratory for Ferrous Metallurgy and Resources Utilization of Ministry of Education, Wuhan University of Science and Technology, Wuhan 430081, China

Z. Li · G. Ma · X. Zhang
State Key Laboratory of Refractories and Metallurgy, Wuhan University of Science and Technology, Wuhan 430081, China

G. Gaustad et al. (eds.), *REWAS 2019*, The Minerals, Metals & Materials Series,
https://doi.org/10.1007/978-3-030-10386-6_43

Fe–Cr alloy. Stainless steel is classified into chromium stainless steel, chrome nickel stainless steel, chrome nickel molybdenum stainless steel and chromium manganese nitrogen stainless steel according to its chemical composition. In consideration of reducing production cost and the consumption of nickel, the chromium stainless steel becomes one of the most developed varieties. According to the corresponding standards [2–4], the content of chromium is higher than 10.5 wt% in chromium stainless steel.

In China, the production of stainless steel has increased from 7.3 million tons in 2001 to 24.93 million tons in 2016 [5]. With the increase of product of stainless steel, the environmental problems in the vicinity of the stainless steel plants have been concentrated. One of the concerns is the disposal of EAF dust, which is considered as one kind of carcinogenic materials as it contains 0.14–0.6 wt% Cr(VI) . It exceeds the limit values of many countries [6]. Typically, 18–33 kg of dust generate per ton stainless steel produced [7]. The dust generated through the homogeneous or heterogeneous nucleation because of the splash of liquid steel and slags directly fly-off from the charge as well as volatilization of metals [8]. However, the distribution of Cr in the dust, especially Cr(VI) , in different particle size has been rarely studied. As the hazards of waste depend on the concentrations of hazardous elements as well as its particle size since the small dust particles are easier to spread in air and dissolve in water [9], more attention should be paid to the control and collection of small particles of dust.

In consideration of the mobility of metals in the EAF dust, some regulations should be satisfied during the treatment process of EAF dust. Some researchers [10, 11] utilized TCLP to determine the toxicity of EAF dust and found that EAF dust is the hazardous waste due to exceeding the regulation limits of Cr. However, the mobility of the Cr-containing species and chemical fractionation are still required to study. To understand this, the sequential extraction method was adapted according to the procedure of Ma et al. [10], namely F1-soluble, F2-exchangeable species, F3-bound to carbonates, F4-bound to amorphous Fe–Mn oxides, F5-bound to crystallized Fe oxides, F6-associated to organic matter and sulfides, and F7-residual which bound to spinel group minerals and silicates.

In this paper, various techniques, such as Inductively Coupled Plasma (ICP), X-ray Photoelectron Spectroscopy (XPS), Scanning Electron Microscopy (SEM), and thermodynamic calculation, have been used to study the distribution and chemical species of Cr in the EAF dust from stainless steel plant, in order to understand the formation of Cr(VI) in the EAF dust.

The Existence Form of Cr in EAF Dust

By studying the process of smelting Fe–Cr alloys, Ma [7] and Sedumedi et al. [12] reported that Cr(III) is hard to be oxidized since the smelting process is carried out in a reducing atmosphere and the oxygen partial pressure in the smelting furnace is less than 10^{-8} atm. However, while the oxygen partial pressure increases in the off-gas duct, the oxidation of Cr(III) may occur. Compared with the total amount of Cr(VI) generated in acidic slag and alkaline slag [13], it indicates that alkaline media would promote the formation of Cr(VI).

According to Cheng et al. [14], the Cr oxidation dynamics analysis indicates that Cr_2O_3 is the main oxidation product at low temperature, which will be decomposed to CrO and CrO_3 with the increase of temperature. At high temperature, CrO_3 is the main product. In the EAF dust, Cr mainly exists in the forms of Cr(III) and Cr(VI), and Cr(VI) is in the forms of CrO_3, $Cr_2O_7^{2-}$ and CrO_4^{2-} [15]. However, Peng and Peng [8] reported Cr exists in the form of CrO and $FeCr_2O_4$ by XRD analysis. Ma et al. [16] made a thermodynamic equilibrium calculation of Cr–Fe–Zn–Mn–Al–Ca–Ni–O–Cl system, which includes main components in the EAF stainless steel dust. The results indicated that Cr is in the form of $FeCr_2O_4$ with temperature higher than 1327 °C, and would transform into Cr_2O_3 when temperature is lower than 1327 °C. The evolutionary behavior of Cr in Cr–O_2 system with alkaline oxides was also simulated, demonstrating that CrO_4^{2-} is easily formed with the existence of alkaline oxides.

In order to further study the variation of Cr–Fe–C–O system with temperature, oxygen partial pressure and carbon content, the predominance area of Cr in the Cr–Fe–C–O system was calculated in Fig. 1. From the Fig. 1, the increase of temperature will make the predominance area more complex. In 1700 °C, CrO(l) is generated when $P(O_2) > 10^{-6.2}$ Pa, and Cr_3O_4(s) would be generated with the increase of the oxygen partial pressure. When the oxygen partial pressure continues to increase to $10^{-5.8}$ Pa, Cr_2O_3(s) is generated. The Cr is stable in the form of $FeCr_2O_4$(s) with the oxygen partial pressure more than $10^{-4.5}$ Pa. Cr(VI) would be produced when the oxygen partial pressure reaches 9.77 Pa. In the process of stainless steel smelting, the equilibrium oxygen potential of carbon is related to carbon content and temperature (Eq. 1) [17]:

$$\lg\left(PO_2/P^{\theta}\right) = -2\lg\omega[C] - 0.28\omega[C] - 14558/T - 3.6752 \quad (1)$$

The temperature of stainless steel smelting is about 1600 °C, and carbon content at the endpoint $\omega[C] = 0.04$–0.08% [18]. When the endpoint carbon content is controlled at 0.04%, the oxygen partial pressure is 2.23 Pa calculated by Eq. (1). Combined with Fig. 1, the Cr is mainly in the form of $FeCr_2O_4$(s) in the EAF dust.

Experimental

The EAF dust from a domestic stainless steel plant is rufous with small particles and easy to be accumulated. The chemical composition of the EAF dust was analyzed by Inductively Coupled Plasma (IRIS Advantage ER/S, Thermo Elemental, USA) shown in Table 1. It shows that the EAF dust has high levels of Cr, and more than 4 wt% alkaline oxides.

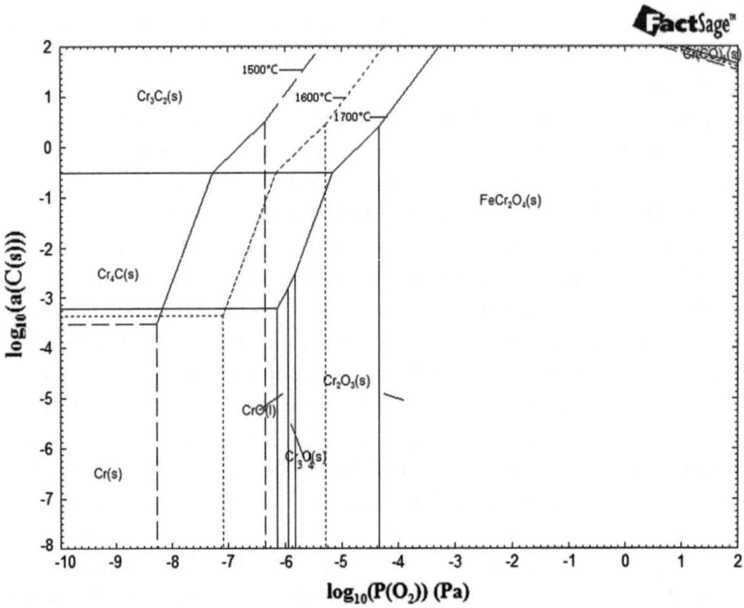

Fig. 1 Predominance area of Cr in the Cr–Fe–C–O system

Table 1 The chemical composition of EAF dust (wt%)

SiO_2	Al_2O_3	Fe_2O_3	CaO	MgO	K_2O	Na_2O	MnO	Cr	NiO	Zn	Cd	Pb
12.56	3.52	49.93	1.96	1.21	2.29	1.20	1.98	8.20	2.35	1.28	0.004	0.30

Phase and Microstructure Analysis of EAF Dust

X-ray diffractometer (X'Pert PRO MPD) was used to analyze the crystalline phases of the EAF dust with Cu Kα radiation, tube voltage 40 kV, tube current 40 mA, 2θ scanning range 10°–70° and continuous scanning mode. The microstructures of EAF dust were observed by Scanning Electron Microscopy (PHILIPS XL30 TMP) with an energy dispersive spectrometer. Before sample preparation, the EAF dust particles was placed in the oven at 110 °C to dry for 2 h, and then it was placed on the conductive tape and observed after spraying gold.

Distribution of Cr in Different Particle Size

The EAF dust was sieved by the dry method with ultrasonic vibrating screen (ZD-250A). The sample was sieved into five different particle sizes of >40, 30–40, 10–30, 5–10, <5 μm. The content of Cr was analyzed by ICP, and the content of Cr(VI)

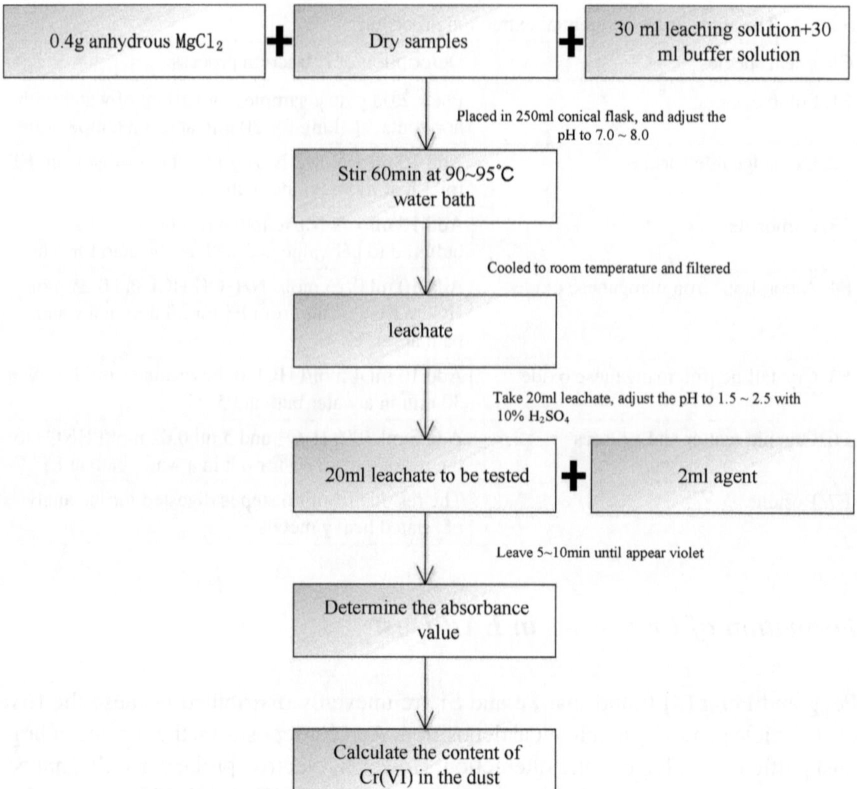

Fig. 2 The steps of determining the content of Cr(VI) in the EAF dust

in the dust of different particle size was analyzed by the EPA method 3060A. The specific steps of measurement of the content of Cr(VI) in the EAF dust are shown in Fig. 2. In this experiment, the standard curve equation is y = 0.6974x + 0.001, R^2 = 0.9994 (y is the concentration value of Cr(VI), x is the absorbance value).

Chemical Species of Cr in EAF Dust

Table 2 is the typical seven extraction procedure for residual elements. After each extraction step, the mixture was centrifuged at 2500 rpm for 10 min. The residual was then washed with 10 ml ultrapure water and centrifuged repeatedly. The supernate of each extraction step was used to determine the concentrations of heavy metals with ICP.

Table 2 The steps of the sequential extraction procedure

Chemical species	Description of extraction process
F1. Soluble	Place 2.00 g dust samples into 10 ml of water with horizontal shaking for 30 min at room temperature
F2. Exchangeable species	Add 10 ml 1 N $Mg(NO_3)_2$ into the residue from F1 for 1 h at room temperature
F3. Carbonate	Add 10 ml 1 N NaAc to the residue from F2 buffered to pH value is 5 with acetic acid for 5 h
F4. Amorphous iron manganese oxide	Add 10 ml 0.25 mol/l $NH_2OH \cdot HCl$ and 0.25 mol/l HCl to the residue from F3 for 30 min in a water bath at 50 °C
F5. Crystalline iron manganese oxide	Add 10 ml 4 mol/l HCl to the residue from F4 for 30 min in a water bath at 95 °C
F6. Organic matter and sulfides	Add 5 ml 30% H_2O_2 and 5 ml 0.02 mol/l HNO_3 to the residue from F5 for 6 h in a water bath at 85 °C
F7. Residue	The residue from F6 step is digested for the analysis of related heavy metals

Formation of Cr Species in EAF Dust

Peng and Peng [8] found that Fe and Si are unevenly distributed because the EAF dust particle grows by mechanical deposition. And Cr deposits on the surface of large dust particles by electrostatic interaction. However, electron probe can only analysis the bulk composition of the dust particles. A XPS (ESCALAB 250) was used to investigate the elements distribution on the surface of dust. The high kinetic energy Ar^+ ion beam was employed to sputter the surface of the dust particles continually, and then to analyze the surface elements of dust at interval.

Results and Discussion

Phase and Microstructure of EAF Dust

Figure 3 is the crystalline phases of EAF dust. It can be seen that Cr(III) is mainly in the form of $FeCr_2O_4$. This assists with the thermodynamic calculation on the predominance area of Cr species in the Cr–Fe–C–O system. In addition, Fe_3O_4 and NiO are also found in the EAF dust.

Figure 4a shows a typical spherical particle in EAF dust with a particle size of about 8 μm. The energy spectrum analysis shows that it is $CaO \cdot SiO_2$, which may be derived from the splash of the slag. In addition, a large number of well-crystallized particles (as shown in Fig. 4b) can be found in EAF dust. These particles are mainly composed of Fe, Cr, and O, which can be inferred to be $Fe(Fe, Cr)_2O_4$ by energy

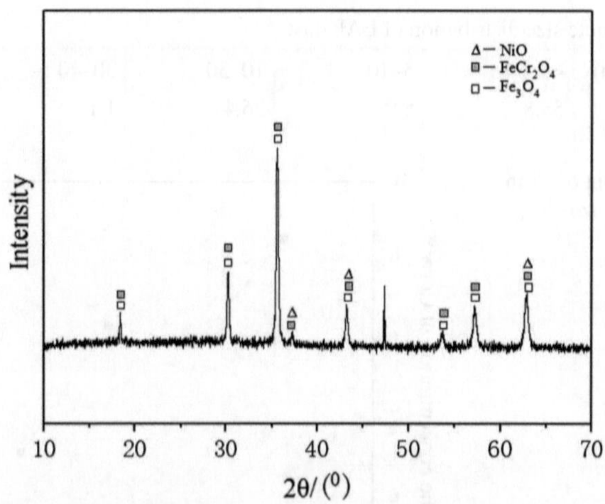

Fig. 3 Crystalline phases of EAF dust

Fig. 4 Microstructure of EAF dust (a spherical dust; b octahedron crystal; c cube crystal)

spectrum analysis. According to Peng and Peng [8], if there are not enough solid particles to agglomerate in a heterogeneous nucleation manner, the metal volatilized from the molten bath can form EAF dust by homogeneous nucleation, which means that volatized substances collide and bond each other. When the particles grow into about 0.02–1.0 μm, they are heterogeneously nucleated and mechanically aggregated to form dust [8]. The formation of regular octahedral spinel particles in Fig. 4b may be formed by the vaporization of Fe and Cr elements at high-temperature zone of the smelting furnace and reaction in the off-gas pipe. Cube-shaped particles can also be observed (Fig. 4c). It is a Ni–Fe–Cr–K–Al–O based material (4.67% Ni, 39.12% Fe, 9.01% Cr, 6.85% K, 2.25% Al, and 38.10% O).

Table 3 The particle size distribution of EAF dust

Particle size (μm)	<5	5–10	10–30	30–40	>40
Percent (wt%)	58.8	5.9	26.4	1.1	7.8

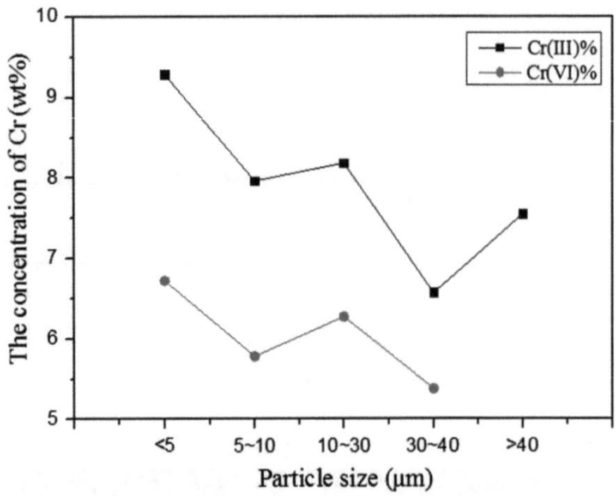

Fig. 5 The amount of Cr in different particle size

Distribution of Cr in Different Particle Size

The distribution of particle size is shown in Table 3. The particle size of EAF dust is mainly less than 5 μm (up to 58.8 wt%), there are only 8.8 wt% EAF dust of which the size is larger than 30 μm.

The result about the amounts of Cr(III) and Cr(VI) in different particle size is depicted in Fig. 5, interpreting the Cr(III) and Cr(VI) trend to accumulate in the small particles and they both have same tendency with the change of particle size. From the thermodynamics and dynamics analysis in the Cr–Fe–C–O system [13, 14], Cr(III) is easier to be oxidized while the oxygen partial pressure is higher than 9.77 Pa. When the EAF dust reaches into the off-gas pipe, the oxygen partial pressure would be higher than 9.77 Pa, indicating that the oxidation of Cr(III) would happen. From Table 1, there are alkaline media in the duct which can promote the oxidization of Cr(III) [16]. In addition, the smaller dust particle has larger specific surface area, which is easier to absorb Cr(III) and make the reaction between Cr(III) and oxygen to form Cr(VI) . Therefore, Cr(VI) has the same tendency to accumulate in the small particles.

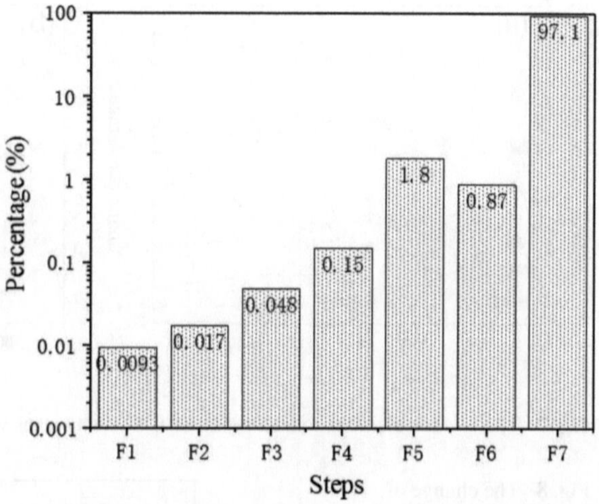

Fig. 6 Fractionation of Cr in EAF dust

Chemical Species of Cr in EAF Dust

Figure 6 shows the fractionation of Cr in EAF dust. It indicated that Cr tightly bound to stable phases extracted mainly in crystalline Fe–Mn oxide, sulfides and spinel group. More than 97% Cr are bound to spinel and there are only 0.0743% chromium extracted from F1 to F3 steps, which is mobile when EAF dust contacts with water or the pH value changes. This is further demonstrated that Cr is mainly presented in the spinel phase in the form of Cr(III). However, it is still paid more attention to that the Cr(III) species can transform into Cr(VI) species with the present of CaO and oxygen. It seems that the best way to treat these Cr-containing dusts is to recycle them back to the stainless steelmaking process through the existing technologies.

Formation of Cr Species in EAF Dust

The change of content of surface element is depicted in Fig. 7. Except O(32 wt%), F(10 wt%) and Si(12 wt%), the contents of elements are lower than 4 wt% on the surface of EAF dust. With the increase of Ar^+ sputtering time, the contents of Fe, Cr and O on the surface have an increasing trend, and other elements, such as Zn and Cl, do not change significantly. During the smelting process, the rupture of CO bubbles results in that the molten steel and slag are splashed to form the core of the dust particles. With the particles flowing up, the temperature decreases and the oxides with high smelting points would be condensed first. With the principle of nucleation, these oxides would become the crystallization nuclei. And then nuclei continue to grow through homogeneous or heterogeneous nucleation. Fe and Cr are mainly in the form of oxides with high smelting points, and they would become the nuclei of dust

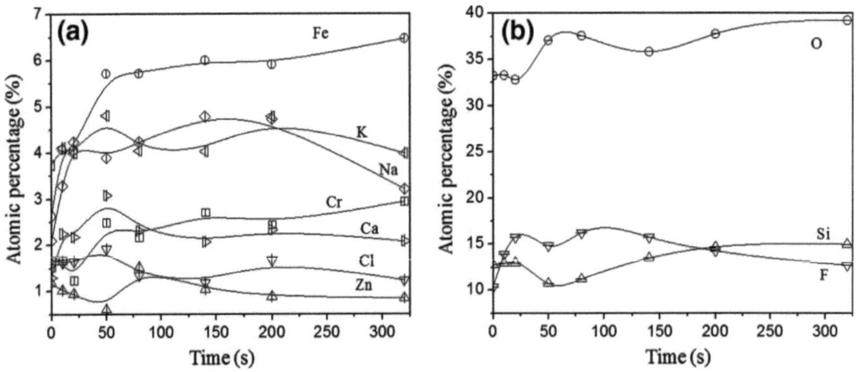

Fig. 7 The surface element concentration of EAF dust varies with Ar^+ sputtering time

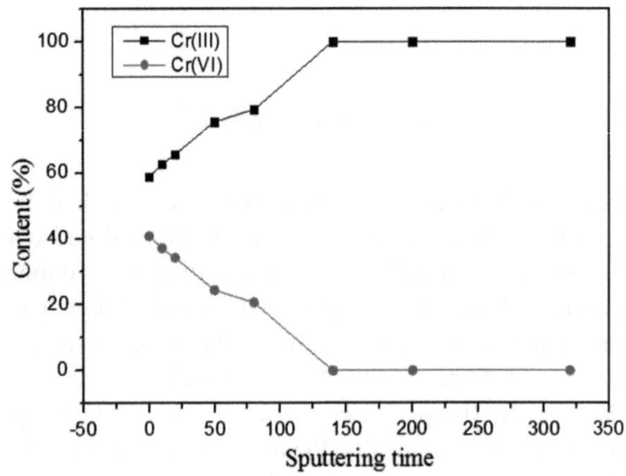

Fig. 8 The change of content of Cr at different sputtering times

causing the surface content of Fe and Cr is lower than the ones inside. The contents of Zn and Cl did not change much, indicating that they are evenly distributed during the whole process of grow of dust.

At different Ar^+ sputtering time, the contents of Cr and Cr-containing compositions have been analyzed by XPS. Figure 8 is the change of content of Cr at different sputtering time, indicating that the content of Cr(VI) decreases gradually with the increase of the sputtering time, on the other hand, the content of Cr(III) increases. When the sputtering time is 140 s, the content of Cr(VI) on the surface is nearly zero.

As the smelting process of stainless steel requires decarburization reaction, the smelting furnace is mainly in high temperature and reduction atmosphere, Cr(III) is difficult to be oxidized. According to the thermodynamics calculation and dynamics analysis [13, 17], if the oxygen partial pressure is higher than 9.77 Pa at high temperature, the Cr(VI) would be generated. It proves that Cr(VI) would derive from the oxidization of Cr(III). Part of Cr form Cr_2O_3 or $FeCr_2O_4$ as the nuclei and grow through homogeneous or heterogeneous nucleation. And while the particle size of Cr_2O_3 is large, the dust will be formed by mechanical deposition. With the oxygen partial pressure become high gradually in the off-gas duct, the Cr(III) on the surface is easier to be oxidized to Cr(VI) . Thus, with the increase of sputtering time, the Cr(VI) on the surface disappears firstly, and the content of Cr(III), which is not oxidized inside, is getting higher and higher until content of Cr(III) is 100%.

Conclusions

(1) The Cr contained in EAF dust is mainly in the form of $FeCr_2O_4$. The amount of Cr(VI) would decrease with the increase of size of EAF dust particle because the smaller dust particle has larger specific surface area, which is easier to make the oxidation reaction of Cr(III) to form Cr(VI) .
(2) Cr in EAF dust is mainly in the form of Cr(III) existing in spinel residual.
(3) The contents of Cr(VI) and Cr(III) have opposite trend, with the increase of Ar^+ sputtering time. The former would decrease from 40 to 0%, and the latter would increase from 60 to 100%. Since the oxygen partial pressure becomes higher and higher from smelting bath to off-gas duct, the Cr(III) on the surface would be oxidized to Cr(VI) .

References

1. Yan YF, Chen JX, Hu L (2007) Chromium metallurgy. Metallurgical Industry Press, Beijing
2. GB/T 12230, General purpose industrial valves-Specification of stainless steel castings
3. ASTM A351, Standard specification for castings, austenitic, for pressure-containing parts
4. JIS G5121, Corrosion-resistant cast steels for general application
5. International Stainless Steel Forum (ISSF). Meltshop production statistics 2001 to 2006. http://www.worldstainless.org/crude_steel_production/meltshop_production_2001_2016
6. Vargas ASD, Ângela B, Masuero, Vilela ACF (2006) Investigations on the use of electric-arc furnace dust (EAFD) in Pozzolan-modified Portland cement I (MP) pastes. Cem Concr Res 36(10):1833–1841
7. Ma GJ (2005) Cr (VI)-containing electric furnace dust and filter cake: characteristics, formation, leachability and stabilization. University of Pretoria, Pretoria, South Africa, pp 25–98
8. Peng B, Peng J (2003) Stainless steel electric arc furnace dust physical and chemical properties and formation mechanism. North China Univ Technol 15(1):34–40
9. Davis A, Ruby MV, Bergstrom PD (1992) Bioavailability of arsenic and lead in soils from the Butte, Montana, mining district. Environ Sci Technol 26(3):461–468
10. Ma GJ, Fan W, Xue ZL, Wang W, Tang H (2010) Leachability and fractionation of heavy metals in stainless steelmaking plant dusts. Acta Metallurgica Sinica (English Letters) 3(4):267–276
11. Laforest G, Duchesne J (2006) Characterization and leachability of electric arc furnace dust made from remelting of stainless steel. J Hazard Mater 135(1):156–164
12. Sedumedi HN, Mandiwana KL, Ngobeni P, Panichev N (2009) Speciation of Cr(VI) in environmental samples in the vicinity of the ferrochrome smelter. J Hazard Mater 172(2):1686–1689
13. Beukes JP, Dawson NF, Zyl PGV (2010) Theoretical and practical aspects of Cr(VI) in the South African ferrochrome industry. J S Afr Inst Min Metall 110(12):743–750

14. Fan W, Ma GJ, Xue ZL, Yang F, Fang Y (2012) Formation of Cr(VI)-bearing phases in the EAF dust from stainless steel plant. J Mater Metall
15. Cheng JF, Xu MH, Zeng HC, Feng R, Qiao Y (2002) The study of oxidation dynamic of Cr at high temperature. Chin J Electr Eng 22(8):135–138
16. Ma GJ, Su WH, Xue ZL, Xu ZH (2009) Microstructure and occurrence of chromium of the stainless steel plant dusts. China Iron and Steel Association, Beijing, China, 12 Nov 2009
17. Wu YJ, Jiang ZH, Liang LK, Jiang MF, Huang ZZ, Chen ZP (2003) Analysis of correlative thermodynamic problems in stainless steel smelting process(II)-dephosphorization of chromium-containing hot metal. J Iron Steel Res 15(4):12–17
18. Zhang JY, Xing XR, Song B (2004) Metallurgical physical chemistry. Metallurgical Industry Press, Beijing, pp 317–318

Effect of Coal Ratio on Preparation of Si–Ti–Fe Alloy by Carbothermic Reduction with Coal Fly Ash

Kun Wang, Yan Liu, Song Qi, Jun Hao, Zhi-he Dou, Li-ping Niu and Ting-an Zhang

Abstract With the development of electric power industry, the discharge of coal fly ash increases year by year. At present, one problem of coal fly ash is low-value utilization. The improper treatment of coal fly ash will also cause harm to air, water, and soil as well as human health. To test a process including high-value utilization and reduction of environmental pollution, a Si–Ti–Fe alloy was prepared in an intermediate frequency induction furnace by carbothermic reduction using coal fly ash as raw material. Coke was used as reducing agent. In this paper, the effect of coal ratio on metal recovery rate, composition, and microstructure was studied. The suitable coal ratio was proposed for future process development which could expand the utilization of coal fly ash.

Keywords Coal ratio · Si–Ti–Fe alloy · Coal fly ash · Carbothermic reduction

Introduction

Coal fly ash is mainly derived from thermal power plants and is a major industrial solid waste [1]. With the development of electric power industry, the discharge of coal fly ash increases year by year. Such a large amount of coal fly ash has not been treated in time, so it can only be piled and stored [2]. However, it will cause water and soil pollution, lead to occupancy of vast lands [3], disrupt ecological cycles, and cause serious pollution of environment [4]. At present, the method of coal fly ash treatment is often low-value application and the valuable metals in coal fly ash are not fully extracted, recovered, and utilized [5, 6].

K. Wang · Y. Liu (✉) · S. Qi · J. Hao · Z. Dou · L. Niu · T. Zhang
Key Laboratory of Ecological Metallurgy of Multi-metal Intergrown Ores of Ministry of Education, Special Metallurgy and Process Engineering Institute, Northeastern University, Shenyang 110819, Liaoning, China
e-mail: liuyan@smm.neu.edu.cn

G. Gaustad et al. (eds.), *REWAS 2019*, The Minerals, Metals & Materials Series,
https://doi.org/10.1007/978-3-030-10386-6_44

Comprehensive utilization of coal fly ash has been studied when environmental and resource problems have been paid more attention around the world [7]. Therefore, in order to improve the valuable utilization value of coal fly ash, a great many efforts have been undertaken recently to study coal fly ash. For example, the coal fly ash could be used in mine backfill materials and soil amelioration [8, 9], foam glass recovery [10], filler in polymers [11], glass ceramics [12], CO_2 capture and wastewater treatment [13], mullite ceramics [14], etc.

Besides the above applications, in recent years, researchers and entrepreneurs pay more and more attention to extract valuable metals from coal fly ash. The recovery of valuable metals from coal fly ash is mainly divided into two methods: physical and chemical methods [15]. The physical method generally uses magnetic separation to recycle iron oxide in coal fly ash. Chemical methods include leaching, sintering, gas–solid reaction, and direct reduction. Lin [16] uses a chemical recovery method to extract Al_2O_3 and SiO_2 from coal fly ash. Result shows that the chemical enrichment method can reduce the silica content in raw materials and the obtained materials can be used for alumina production. Pickles [17] uses the plasma method for metal recovery in coal fly ash. The result shows that 90% iron can be recovered by this method.

Apart from wet extraction, many domestic and international researches have been done on the extraction of valuable metals from coal fly ash by smelting at high temperature. In the 1990s, Zhang and Qiu [18] proposed the use of coal gangue to manufacture aluminum–silicon–iron alloy by an electrothermal process and verified its feasibility, which has significant social, environmental, and economic benefits. Yang et al. [19] studied carbothermic reduction for manufacturing aluminum–silicon alloy and proposed that the suitable reaction temperature of carbothermic reduction with Al_2O_3, SiO_2 for making aluminum–silicon alloy was 2000–2100 °C.

In order to extract the valuable metals silicon, titanium, and iron in coal fly ash, a Si–Ti–Fe alloy for steelmaking deoxidizing agent was prepared in an intermediate frequency induction furnace by carbothermic reduction. Coal fly ash was used as raw material, and coke powder was used as reducing agent. The effect of different coal ratios on the metal recovery rate was studied. The composition and microstructure of the alloy and slag were analyzed, which could provide the basis for further optimizing the experimental in future.

Experiment

Experimental Materials

The coal fly ash used in the experiment is from Henan Province, and the main components of it are shown in Table 1.

The phase compositions of coal fly ash are analyzed by X-ray diffractometer, and the result is shown in Fig. 1.

Table 1 The main components of coal fly ash (%)

Composition	SiO_2	Al_2O_3	Fe_2O_3	CaO	K_2O
Content	59.74	25.71	4.01	2.88	2.10

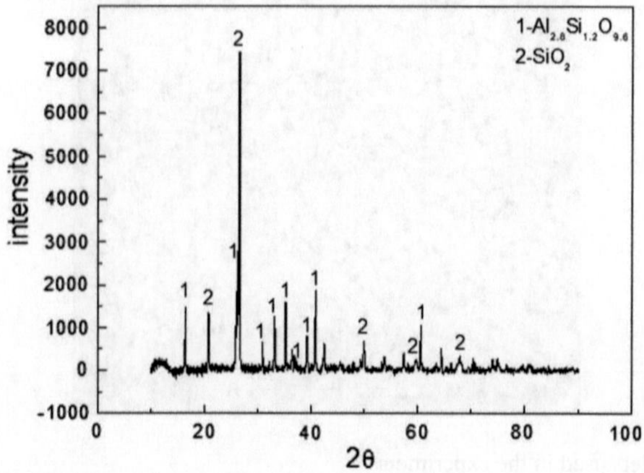

Fig. 1 The X-ray diffraction pattern of coal fly ash

Table 2 The main components of coke (%)

Compositions	Fixed carbon	Volatiles	Ash	Moisture
Content	74.78	5.12	19.1	1.0

The result shows the existence of two forms of SiO_2 in coal fly ash. Part of SiO_2 exists in the form of quartz, and the other is combined with Al_2O_3 to form a mullite phase. No free Al_2O_3 exists. There are no obvious diffraction peaks of other metal oxides, so their content must below.

Coke from Hebei Province was used as reducing agent. The components are shown in Table 2.

Experimental Method

The coal fly ash and coke powder were mixed evenly according to stoichiometric ratio and pelletized in pelletizer. The pellets were manufactured with a diameter of 5–10 mm shown in Fig. 2. The intermediate frequency induction furnace was used to smelt the dried pellets. After feeding was done, keep warm and smelt for a period of time. The metal and slag were separated after cooling completely, and the Si–Ti–Fe alloy product was obtained.

Fig. 2 Dried pellets used in the experiment

The slag was analyzed by X-ray diffractometer (XRD) . The chemical composition analysis was used for the alloy components. The microstructure of the alloy was analyzed by scanning electron microscopy (SEM).

Experimental Equipment

The equipment used in the experiment are as follows: mixer, pelletizer, drying oven, high purity graphite crucible, and intermediate frequency induction furnace. The type of intermediate frequency induction furnace used in the experiment is SPZ-160, which is shown in Fig. 3.

Results and Discussion

Description of Experimental Phenomena

After the induction furnace was turned on, and the temperature had reached about 2073 K, pellets were added slowly. They pulverized and then melted to form a hot bath. The pellets floated on the surface of the molten pool and then rapidly pulverized as more pellets were added. A lack of powder produced by the pulverization was

Fig. 3 SPZ-160 intermediate frequency induction furnace

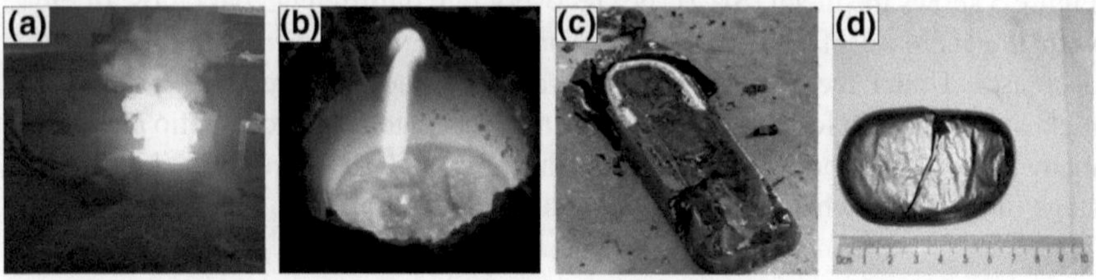

Fig. 4 The experimental phenomena, metal, and slag after cooling solidification

entrained out of the crucible along with the gas, which may be unreacted carbon powder.

Throughout the whole feeding process, the surface of the molten pool was in a state of rolling with a lot of smoke. Many large bubbles constantly floated to the molten pool surface and combusted with bright flames, as shown in Fig. 4a. It can be judged that CO gas was generated during the reaction process. In Fig. 4b, the end of the reaction, the smoke, and CO bubbles gradually decreased. When the experiment was over, the crucible was taken out with crucible forceps to cast the molten bath into the preheated crucible mold. After cooling, the metal and slag were separated as shown in Fig. 4c, d. Figure 4d shows that the separation of metal and slag is very well.

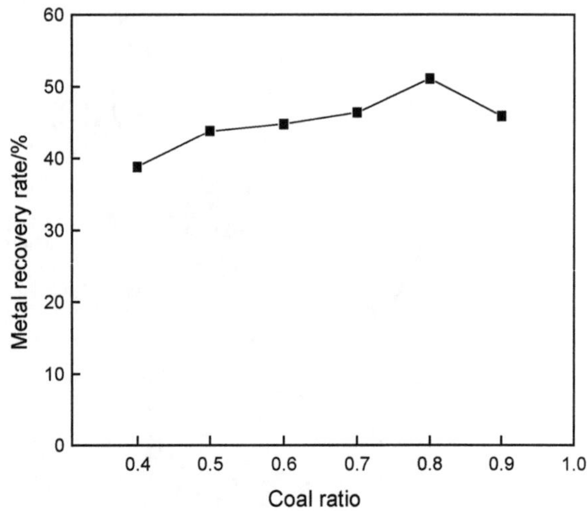

Fig. 5 The metal recovery rate of different coal ratios

Effect of Coal Ratio on Metal Recovery Rate

Figure 5 shows the metal (Si+Ti+Fe) recovery rate of different coal ratios. In Fig. 5, with the increase of coal ratio, the metal recovery rate increases first and then decreases. Under the condition of coal ratio 0.8, the metal recovery rate was the highest. The reason for decrease of metal recovery rate at high coal ratio may be that many high melting point compounds such as SiC and TiC were formed in the early stage of smelting, which hindered the reduction reaction.

Effect of Coal Ratio on Si–Ti–Fe Alloy Composition

Figure 6 shows Si–Ti–Fe alloy composition with different coal ratios. In Fig. 6, the aluminum content in the alloy is less than 1%. Aluminum can be generated only by reaction between alumina and carbon when the temperature is higher than 2223 K. It is difficult to reach such a high temperature in an intermediate frequency induction furnace; therefore, alumina is hardly reduced to generate aluminum, so the aluminum content in the alloy is very low.

Iron has the highest content in alloy which fluctuates at around 50%. From thermodynamic calculation, it is known that when the temperature is larger than 900 K, the standard Gibbs free energy of all possible reactions in the Fe_2O_3–C system is less than 0, which means that iron oxides begin to react with carbon. Therefore, iron is the most easily reduced metal in coal fly ash.

The silicon dioxide content in coal fly ash is nearly 60%, so the reaction of silicon dioxide with carbon has a great influence on the silicon content in alloy. As shown in Fig. 6, the silicon content fluctuates between 30 and 43%. As the coal ratio

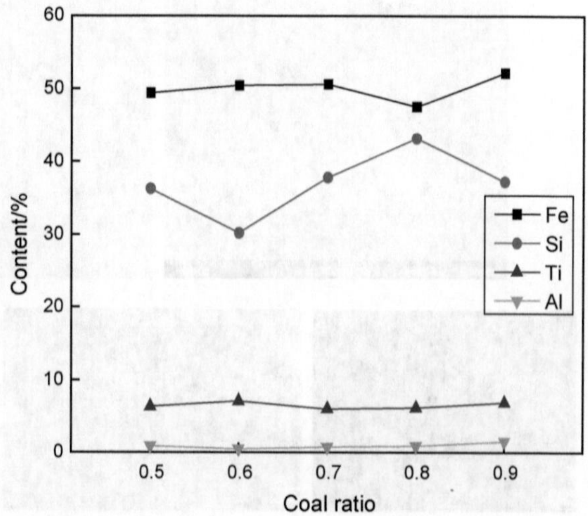

Fig. 6 The Si–Ti–Fe alloy composition of different coal ratios

increases, the silicon content in the alloy increases first and then decreases. This is because silicon dioxide reacts with carbon first to generate silicon carbide. The generated silicon carbide reacts with silicon dioxide to generate silicon. When coal ratio is high, more silicon carbide is generated and the amount of silicon dioxide involved in reaction decreases; therefore, less silicon is generated. In addition, the volatilization of silicon at high temperature also led to a reduction in recovery.

The thermodynamic calculation result shows that titanium dioxide reacts with carbon to produce titanium carbide at a low temperature, the initial reaction temperature of which is 1580 K. However, titanium dioxide reacts with carbon to form titanium at an initial reaction temperature of 2050 K. Therefore, in order to reduce the generation of titanium carbide and increase the recovery rate of titanium, the feeding temperature should be much higher than 1580 K, which is determined at about 2073 K based on previous studies. Titanium content in alloy is between 6 and 7.2%. A small amount of titanium dioxide reacts with carbon to form titanium carbide that has a high melting point of 3413 K and exists in the slag.

The Microstructure of Si–Ti–Fe Alloy

Figure 7 shows the microstructure analysis of Si–Ti–Fe alloy. The SEM surface scanning results clearly show the phase interface of the alloy. The black and dark gray areas are silicon-rich phase. The gray-white area is iron-rich phase. Both above elements have a large distribution area, which also indicates that the contents of silicon and iron elements in alloy are very high. The narrow needle-like and reticular area is the titanium-rich phase. It indicates that titanium may combine with silicon and iron to produce the silicon–titanium–iron phase in alloy. It can be seen from the

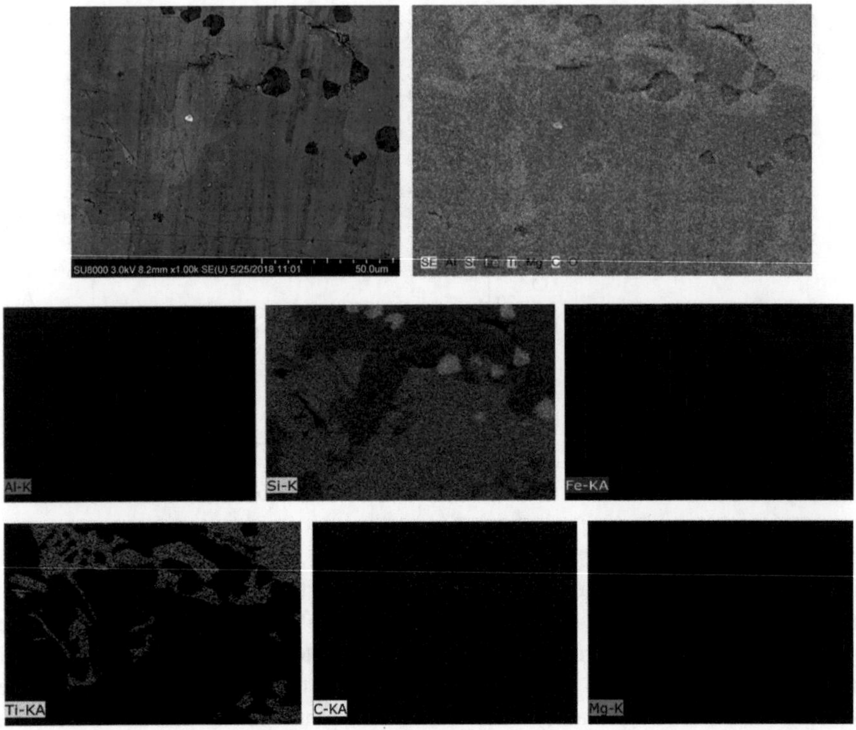

Fig. 7 The microstructure analysis of Si–Ti–Fe alloy

surface scanning results that only a small amount of aluminum exists in the alloy indicating that alumina is very hard to reduce for aluminum preparation. There is also a small amount of carbon in the alloy, possibly in the form of carbide.

The XRD Analysis of Slag

Figure 8 shows the XRD analysis result of slag. The result shows that Al_2O_3 and MgO are contained in slag, and the peaks of them are very strong, which could reflect that the reduction effects of Al_2O_3 and MgO are not good. Fe–Si phase also exists in slag; it is because of incomplete separation of metal and slag. Moreover, the carbide produced in carbothermal reaction is eventually contained in slag. In later research, the characteristics of the slag will be studied to achieve the purpose of comprehensive utilization.

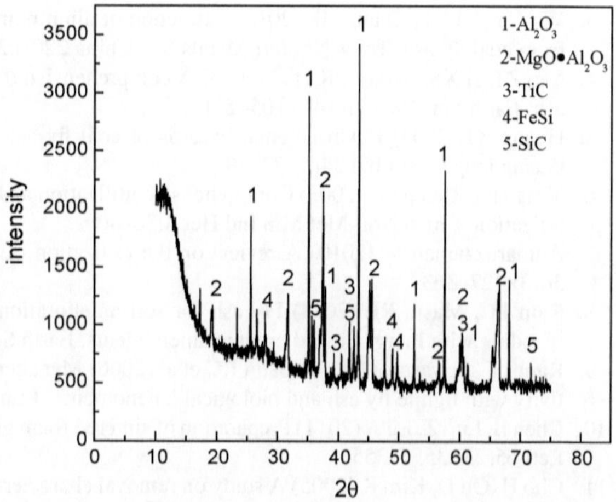

Fig. 8 XRD analysis of slag

Conclusions

In this paper, the effect of coal ratio on the metal recovery rate and compositions was studied by carbothermic reduction with fly ash. The microstructure of alloy and composition of slag were analyzed. The conclusions are as follows:

(1) Appropriate increase of coal ratio is beneficial to increase the metal recovery rate, but too much coal ratio produces much carbide and inhibits metal recovery rate.
(2) Components and microstructure analyses show that the aluminum content in alloy is very low, indicating that it is difficult for alumina reacting with carbon to produce aluminum.
(3) Comprehensive consideration of metal recovery rate and experimental cost, the suitable coal ratio recommended is 0.8, the reaction temperature is 2073 K, and the reaction time is 60 min. The suitable experimental condition can be used for utilization of coal fly ash at larger scale.

Acknowledgements This work was financially supported by the National Natural Science Foundation of China (Nos. U1402271, U1710257), the National key research and development plan (2017YFC0210403-04).

References

1. Lu GZ, Zhang TA, Wang L et al (2014) Direct spray pyrolysis of aluminum chloride solution for alumina preparation. J Cent South Univ 21(12):4450–4455
2. Ma BY, Ren XM, Yin Y et al (2017) Effects of processing parameters and rare earths additions on preparation of Al_2O_3–SiC composite powders from coal ash. Ceram Int 43:11830–11837

3. Wu CY, Yu HF, Zhang HF (2012) Extraction of aluminum by pressure acid-leaching method from coal fly ash. Trans Nonferr Metals Soc China 22(9):2281–2288
4. Yao ZT, Ji XS, Sarker PK et al (2015) A comprehensive review on the applications of coal fly ash. Earth Sci Rev 141(141):105–121
5. Huang XJ (2007) Environmental hazards of coal fly ash and new technological utilization. Guangdong Chem Ind 34(5):77–79
6. Yang HC, Zheng SL (2003) Comprehensive utilization and prospects of fly ash and its characterization. China Non-Met Min Ind Her 4:38–40
7. Ahmaruzzaman M (2010) A review on the utilization of fly ash. Prog Energy Combust Sci 36(3):327–363
8. Ram LC, Masto RE (2014) Fly ash for soil amelioration: a review on the influence of ash blending with inorganic and organic amendments. Earth Sci Rev 128(1):52–74
9. Ram LC, Srivastava NK, Tripathi RC et al (2006) Management of mine spoil for crop productivity with lignite fly ash and biological amendments. J Environ Manag 79(2):173–187
10. Chen B, Luo Z, Lu A (2011) Preparation of sintered foam glass with high fly ash content. Mater Lett 65(23):3555–3558
11. Cho H, Oh D, Kim K (2005) A study on removal characteristics of heavy metals from aqueous solution by fly ash. J Hazard Mater 127(1):187–195
12. Rawlings RD, Wu JP, Boccaccini AR (2006) Glass-ceramics: their production from wastes-a review. J Mater Sci 41(3):733–761
13. Zheng Y, Jensen AD, Windelin C et al (2012) Review of technologies for mercury removal from flue gas from cement production processes. Prog Energy Combust Sci 38(5):599–629
14. Blissett RS, Rowson NA (2012) A review of the multi-component utilisation of coal fly ash. Fuel 97(7):1–23
15. Nagano T, Chen FY (1992) Technical trend of recovering valuable materials from fly ash. Comprehensive utilization of fly ash 02:47–50
16. Lin IJ, Malts N, Shindler Y (1998) The complex chemical treatment of alumina–silica-containing materials. J Mater Synth Process 6(1):27–35
17. Pickles CA, Mclean A, Alcock CB et al (1990) Plasma recovery of metal values from fly ash. Can Metall Q 29(3):193–200
18. Zhang MJ, Qiu ZX (1992) Manufacturing Si-Al-Fe alloy directly from coal waste and fly ash by electrothermal process. Ferroalloy 04:19–23
19. Yang D, Feng NX, Wang YW et al (2010) Thermodynamic analysis and experimental test of coarse Al–Si alloy prepared by carbothermal reduction of bauxite tail. Non-ferr Min Metall 26(01):38–43

Effect of Contact Time on the Recovery of Metals from the Mining Effluent of Lateritic Nickel by Chelating Resin Dowex XUS43605

Isadora Dias Perez, Jorge A. Soares Tenório and Denise C. Romano Espinosa

Abstract The generation of mining waste has been the subject of environmental, economic, and social concern. Thus, alternative and sustainable methods of metal treatment and recovery are desired. This paper focuses on the application of ion exchange technology for the recovery of metals from the mining effluent of lateritic nickel by chelating resin Dowex XUS43605. Chelating resin was chosen due to its ability to capture transition metals. 1 g of Dowex XUS43605 with 50 mL synthetic solution in 250 mL flask is shaken, in a speed of 200 rpm. The synthetic solution has nine types of metals, such as Al, Co, Cr, Cu, Fe^{3+}, Mg, Mn, Ni, and Zn. Batch technique was employed to examine the effects of contact time (1–7 h) when solution was adjusted at pH 1.5 at 25 °C. The present work demonstrates that the chelating resin shows negligibly higher selectivity for copper ions compared to the other metals. The metal ions (Al, Co, Cr, Mg, Mn, and Zn) present in the solution were not adsorbed.

Keywords Ion exchange · Batch adsorption · Transition metals

Introduction

The use of metals is related to the industrial production of military materials, batteries, building materials, and electronic equipment [1]. Copper, iron, and nickel are among the most widely used metals in the world due to their versatility. As a consequence, it is believed that demand for these metals, as well as their prices, will increase. In this way, the mining companies face the challenge of perfecting the production processes in order to meet future demand. In some cases, a completely new process may be proposed [2].

In mining limonite nickel ore, a liquor is generated after atmospheric leaching. This liquor is composed mainly of nickel, copper, and iron, as well as aluminum, cobalt, chromium, manganese, magnesium, and zinc [3]. There are separation tech-

I. D. Perez (✉) · J. A. Soares Tenório · D. C. Romano Espinosa
Chemical Engineering Department, Polytechnic School, University of São Paulo, 158 Av. Prof. Luciano Gualberto, Trav. 3, São Paulo, SP 05424-970, Brazil
e-mail: dpisadora@gmail.com

G. Gaustad et al. (eds.), *REWAS 2019*, The Minerals, Metals & Materials Series,
https://doi.org/10.1007/978-3-030-10386-6_45

niques that allow the recovery of metallic ions, e.g., chemical precipitation, reverse osmosis, solvent extraction, and ion exchange with solid adsorbents [4]. The choice of separation technique for a specific treatment involves the analysis of operating costs, environmental impacts, process selectivity, and effluent characteristics to be treated [5]. In this work, the ion exchange using chelating resin was chosen. Chelating resins present a degree of selectivity for some transition metals, especially divalent metal ions, which were not available for conventional (cationic and anionic) [6].

Commercial chelating resins have been widely used in the recovery study of metals, mainly nickel and copper. Commercial resins Dowex M4195, Amberlite IRC748, Ionac SR-5, and Purolite S930 may be mentioned [7]. Chelating resin chosen was Dowex XUS43605 with functional group: hydroxypropylpicolylamine (HPPA). HPPA is composed of two nitrogen atoms and one oxygen atom as electron donors [8].

In order to recover metals present in a synthetic solution, the present work intends to evaluate the influence of contact time on the adsorption capacity of resin.

Materials and Methods

Resin

The resin adopted for this research was the chelating resin Dowex XUS43605, which has hydroxypropylpicolylamine (HPPA) ligand supported by a polymeric backbone of styrene (St) cross-linked with divinylbenzene (DB) [9]. Table 1 shows the resin characteristics.

Prior to use, the resin was washed with acid solution (4 mol/L HCl) and deionized water at 60 °C, to remove all impurities. The procedure was repeated two times to ensure that the resin was in the protonated form and ready for use. Finally, the resin was dried in an oven at 60 °C for 2 h.

Synthetic Solution

Analytical grade reagents were used to prepare the multicomponent solution, and all metals concentrations of Al [as $Al_2(SO_4)_3 \cdot 17H_2O$], Co [as $CoSO_4 \cdot 7H_2O$], Cr [as $Cr_2(SO_4)_3 \cdot 8H_2O$], Cu [as $CuSO_4 \cdot 5H_2O$], Fe [as $Fe_2(SO_4)_3 \cdot 8H_2O$], Mg [as $MgSO_4 \cdot 7H_2O$], Mn [as $MnSO_4 \cdot H_2O$], Ni [as $NiSO_4 \cdot 6H_2O$], and Zn [as $ZnSO_4.7H_2O$] are equal 500 ppm. The metals were dissolved with distilled water.

Table 1 Characteristics of commercial resin Dowex XUS43605 [10]

Manufacturer	Dow Chemical Company
Matrix	Macro styrene
Size	320 μm
Functional groups	HPPA 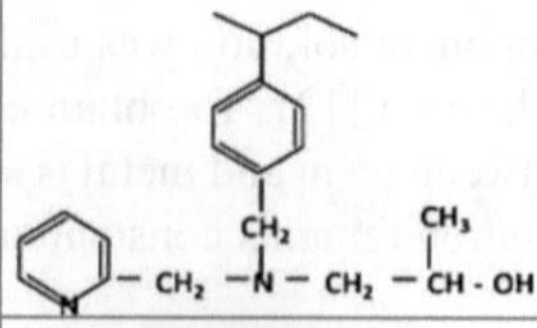
Visual aspect	

Contact Time Tests

A batch technique was employed to examine metals sorption from multicomponent solution. The contact time was determined through tests using synthetic solution. 50 mL of solution in pH 1.5 was agitated with 1.0 g resin, varying contact time in a range 1–7 h, in a speed of 200 rpm at 25 °C.

The adsorption capacity at equilibrium (q_e) was calculated by Eq. 1 [11]:

$$q_e = \frac{(C_o - C_e)}{M} \times V \qquad (1)$$

where q_e is the adsorption capacity at the equilibrium (mg/g), C_o and C_e are the initial and equilibrium metal concentration (mg/L), M is the amount of resin (g), and V is the volume (L) of metal solution used for sorption experiments, respectively.

The percentage removal of metal in equilibrium was calculated by Eq. 2 [5]:

$$\% = \frac{(C_o - C_e)}{C_o} \times 100 \qquad (2)$$

For all the experiments, the resin and solution separation was done by filtration through a millipore 20 μm filter paper. The remaining metals in the solution were quantified by fluorescence spectrometry energy dispersive (EDX).

Results and Discussion

Effect of Contact Time

The relationship between contact time and the adsorption capacity of the metals present in the multicomponent solution was evaluated in order to examine the time required to obtain equilibrium [12]. To obtain equilibrium reaction, it is necessary that the contact time between resin and metal is sufficient. At equilibrium, adsorbate concentration in the solution remains constant and no change in the rate adsorption capacity [13].

According to Veli and Alyüz [14], the contact time is considered as one of the most important factors affecting the adsorption efficiency. So, the contact time was evaluated and the samples were collected in the range of 1–7 h. The effect of contact time on adsorption capacity at equilibrium is shown in Fig. 1.

As shown in Fig. 1, it may be assumed that the aluminum, cobalt, chromium, manganese, magnesium, and zinc metals were not extracted by the Dowex XUS43605. The adsorbed metals were only copper, iron, and nickel. The adsorption of Cu, Fe^{3+}, and Ni is justified by the selectivity order of the resin: $Cu^{2+} >> Ni^{2+} > Fe^{3+} > Zn^{2+} > Co^{2+} > Cd^{2+} > Fe^{2+}$ to $pH < 2.0$ [15].

Adsorption of Al, Co, Cr, Mg, Mn, and Zn may not have occurred for two reasons: non-resin selectivity and concentration. It is known that adsorption capacity (q_e) is related to metal concentration in solution. Thus, the capacity value is proportional to the ion concentration. In the literature, q_e is reported due to the increase of metal concentration in solution [16]. According to Adebowale et al. [17], the relation between the metal concentration increase and adsorption capacity increase is justified by extra motive force, which is caused by higher metal concentration around the active resin sites. Therefore, the adsorption of Al, Co, Cr, Mg, Mn, and Zn was not favored due to concentration value. It is possible that for different experimental conditions, such as metal concentration, these metals are to be adsorbed.

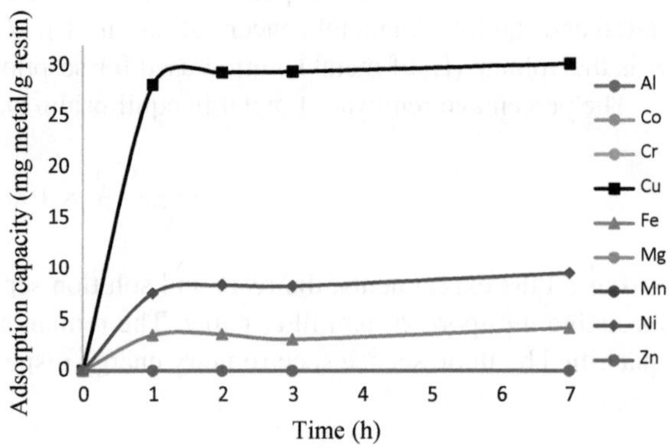

Fig. 1 Effect of contact time on adsorption capacity at equilibrium by Dowex XUS43605 chelating resin (conditions: 1.0 g of resin; 50 mL; pH 1.5; 25 °C)

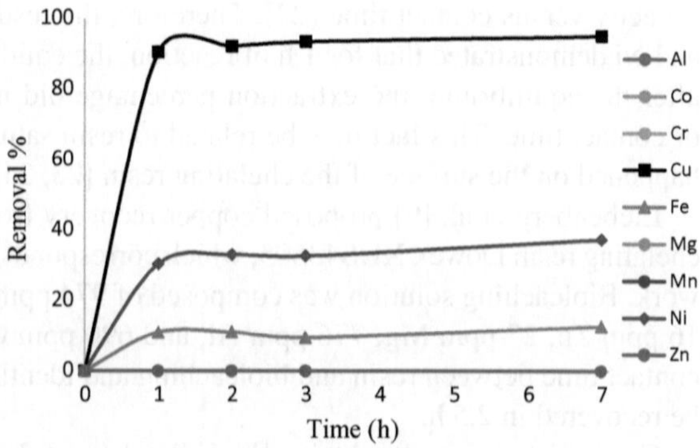

Fig. 2 Effect of contact time on percentage removal of metal in equilibrium by Dowex XUS43605 chelating resin (conditions: 1.0 g of resin; 50 mL; pH 1.5; 25 °C)

According to Fig. 1, the maximum adsorption for Cu, Fe^{3+}, and Ni occurred in the first hour of tests, reaching the equilibrium at 1 h. The rapid adsorption of these metals occurs due to the amount of available resin sites, since in the final phase of equilibrium, there is a reduction of the concentration of metal ions in solution and available resin sites [18]. Thus, it was verified that the saturation of resin by the Cu, Fe^{3+}, and Ni happened in 1 h.

It was verified that the adsorption capacity for copper was 30 mg of copper/g of resin, approximately, for all contact time studied. For iron and nickel, the values obtained were 4 mg of iron/g of resin and 9.6 mg of nickel/g of resin, respectively.

Observing these results and the mass of adsorbed metal per gram of resin, it was concluded that the selectivity order predicted by the literature is true and that the resin has more affinity for copper, followed by nickel and iron.

While Fig. 1 relates contact time with adsorption capacity, Fig. 2 indicates percentage removal of metal by versus contact time.

It was found that the maximum adsorption of the Cu, Fe^{3+}, and Ni by chelating resin was in 1 h, approximately, corresponding to 95%, 13%, and 37% extraction, respectively. After 1 h, the resin was saturated. So, the increase in the contact time had an irrelevant effect on the percentage of Cu, Fe^{3+}, and Ni removal, since from 1 h, the percentage adsorbed reached the equilibrium.

The reason for which copper has been more adsorbed by Dowex XUS43605 is related to its electronic configuration. Cu has incomplete 3d orbital in the electronic configuration and behaves as a Lewis acid. In this way, this metal is able to form chelate with a Lewis base, which donates electron pairs. In accordance with Wolowicz and Hubicki [19], copper has sorption affinity for chelating resins containing nitrogen atoms in the functional group and Dowex XUS43605 has two nitrogen and one oxygen as donor atoms [8]. Consequently, the rate of extraction of the copper will be higher than iron and nickel.

Observing Fig. 2, it is clear that there are two main stages. The first one is rapid and adsorbs more metal, and the second one is slow and adsorbs less metal [20, 21]. It is worth mentioning that the identification of the stages was reported by several authors in the literature considering only the visual behavior of the adsorption capacity versus contact time [22]. Therefore, the results obtained for the Cu, Fe^{3+}, and Ni demonstrated that for 1 h of reaction, the equilibrium was reached. Thereby, after the equilibrium, the extraction percentage did not increase with the increase of contact time. This fact may be related to resin saturation and that the adsorption happened on the surface of the chelating resin [23, 24].

Liebenberg et al. [9] proposed copper recovery from bioleaching solution using chelating resin Dowex XUS43605, which corresponds to the same one of the present work. Bioleaching solution was composed of 974 ppm Ni, 34 ppm Co, 267 ppm Fe, 16 ppm Zn, 27 ppm Mg, 776 ppm Al, and 690 ppm Cu. The authors evaluated the contact time between resin and bioleaching and identified that 80% of copper could be recovered in 2.5 h.

Comparing the results obtained by Liebenberg et al., it was verified that the equilibrium reached at the present work was 1.5 h faster. The difference between the results can be due to the concentration of each metal, since the solution of the present work has metals in the same concentration (500 ppm).

Conclusion

After the experiments, it was possible to conclude the following:

- Dowex XUS43605 was not able to recover all the metals present in the solution;
- Al, Co, Cr, Mg, Mn, and Zn remained in solution;
- Cu, Fe^{3+}, and Ni were the only adsorbed metals;
- The resin showed more affinity for Cu;
- Cu was the most extracted metal corresponding to 95%;
- The contact time investigated were sufficient to reach the reaction equilibrium of Cu, Fe^{3+}, and Ni.

Acknowledgement To the Counsel of Technological and Scientific Development (CNPq) for the financial support through doctorate grant.
To the Coordenação de Aperfeiçoamento de Pessoal de Nível Superior (CAPES) for the financial support through master grant.
To the Instituto Tecnológico Vale.

References

1. BNDES (2012) BNDES Setorial 36. Banco Nacional de Desenvolvimento Econômico e Social, Brasília
2. Luz AB, Sampaio JA, França SCA (2010) Tratamento de Minérios. CETEM, Rio de Janeiro
3. McDonald RG, Whittington BI (2008) Atmospheric acid leaching of nickel laterites review. Part II. Chloride and bio-technologies. Hydrometallurgy 91(1–4):56–69
4. Kentish SE, Stevens GW (2001) Innovations in separations technology for the recycling and re-use of liquid waste streams. Chem Eng J 84:149–159

5. Yadav S, Srivastava V, Banerjee S, Gode F, Sharma YC (2013) Studies on the removal of nickel from aqueous solutions using modified riverbed sand. Environ Sci Pollut Res 20(1):558–567
6. Ceglowski M, Schroeder G (2015) Preparation of porous resin with Schiff base chelating groups for removal of heavy metal ions from aqueous solutions. Chem Eng J 263:402–411
7. Deepatana A, Tang JA, Valix M (2006) Comparative study of chelating ion exchange resins for metal recovery from bioleaching of nickel laterite ores. Miner Eng 19(12):1280–1289
8. Marston CR, Rodgers M (2011) Process for separating copper and nickel from cobalt containing solutions. US Patente 20110290077A1, 1 Dec 2011
9. Liebenberg CJ, Dorfling C, Bradshaw SM, Akdogan GA, Eksteen JJ (2013) The recovery of copper from a pregnant sulphuric acid bioleach solution with developmental resin Dow XUS43605. J South African Inst Min Metall 113(5):389–397
10. Dow Chemical Company (2016) Ion exchange resins for chemical processing. http://msdssearch.dow.com/PublishedLiteratureDOWCOM/dh_07c8/0901b803807c8488.pdf?filepath=liquidseps/pdfs/noreg/177-02437.pdf&fromPage=GetDoc. Accessed 29 Aug 2018
11. Kumar R, Kumar M, Ahmad R, Barakat MA (2013) L-methionine modified Dowex-50 ion-exchanger of reduced size for the separation and removal of Cu(II) and Ni(II) from aqueous solution. Chem Eng J 218:32–38
12. Mendes FD, Martins AH (2004) Selective sorption of nickel and cobalt from sulphate solutions using chelating resins. Int J Miner Process 74(1–4):359–371
13. Zagorodni AA (2007) Ion exchange materials: properties and applications. Elsevier, Amsterdam
14. Veli S, Alyüz B (2007) Adsorption of copper and zinc from aqueous solutions by using natural clay. J Hazard Mater 149(1):226–233
15. Zaganiaris EJ (2013) Ion exchange resins and adsorbents in chemical processing. Books on Demand GmbH, Norderstedt
16. Jiang M, Jin X, Lu XQ, Chen Z (2010) Adsorption of Pb(II), Cd(II), Ni(II) and Cu(II) onto natural kaolinite clay. Desalination 252(1–3):33–39
17. Adebowale KO, Unuabonah IE, Olu-Owolabi BI (2006) The effect of some operating variables on the adsorption of lead and cadmium ions on kaolinite clay. J Hazard Mater 134(1–3):130–139
18. Gode F, Pehlivan E (2006) Removal of chromium(III) from aqueous solutions using Lewatit S 100: the effect of pH, time, metal concentration and temperature. J Hazard Mater 136(2):330–337
19. Wołowicz A, Hubicki Z (2012) The use of the chelating resin of a new generation Lewatit MonoPlus TP-220 with the bis-picolylamine functional groups in the removal of selected metal ions from acidic solutions. Chem Eng J 197:493–508
20. Han R, Han P, Cai Z, Zhao Z, Tang M (2008) Kinetics and isotherms of neutral red adsorption on peanut husk. J Environ Sci 20(9):1035–1041
21. El-Sayed GO, Dessouki HA, Ibrahiem SS (2011) Removal of Zn(II), Cd(II) and Mn(II) from aqueous solutions by adsorption on maize stalks. Malaysian J Anal Sci 15(1):8–21
22. Wang XS, Huang J, Hu HQ, Wang J, Qin Y (2007) Determination of kinetic and equilibrium parameters of the batch adsorption of Ni(II) from aqueous solutions by Na-mordenite. J Hazard Mater 142(1–2):468–476
23. Shaaban AF, Fadel DA, Mahmoud AA, Elkomy MA, Elbahy SM (2014) Synthesis of a new chelating resin bearing amidoxime group for adsorption of Cu(II), Ni(II) and Pb(II) by batch and fixed-bed column methods. J Environ Chem Eng 2(1):632–641
24. Ilaiyaraja P, Deb AKS, Ponraju D, Ali SM, Venkatraman B (2017) Surface engineering of PAMAM-SDB chelating resin with diglycolamic acid (DGA) functional group for efficient sorption of U(VI) and Th(IV) from aqueous medium. J Hazard Mater 328:1–11

Experimental Study on Phosphorus Vaporization for Converter Slag by SiC Reduction

Y. K. Xue, S. H. Wang, D. G. Zhao and C. X. Li

Abstract The phosphorus vaporization experiment of converter slag was carried out in laboratory with SiC as reducing agent. The results show that $2CaO{\cdot}SiO_2$ and $3CaO{\cdot}SiO_2$ are the main phases in the final slag. After phosphorus vaporization operation, the dendritic Fe phase and a large number of C phases were distributed in the phosphorus vaporization slag excepted the above phase. The experimental result shows that phosphorus vaporization rate decreased with increasing the basicity in the slag. Phosphorus vaporization rate first increased and then decreased with increasing FeO in the slag and reached the maximum when the content of FeO was 20–25%. Phosphorus vaporization rate showed an upward trend with the increase of nitrogen flow rate and tended to be slow when the nitrogen flow rate was 0.4 m^3/h. Phosphorus vaporization rate increased with increasing the reaction temperature, and the growth rate became gentle when the temperature was more than 1600 °C. Excessive addition of SiC will not obviously increase the phosphorus vaporization rate.

Keywords Converter · Final slag · Phosphorus vaporization

Introduction

Converter slag is a subsidiary product of steelmaking, and its output accounts for about 10–15% of converter steel output [1–4]. The annual output of converter slag is over 100 million tons if the annual output of converter steel is 700 million tons

Y. K. Xue · S. H. Wang (✉) · D. G. Zhao · C. X. Li
College of Metallurgy and Energy, North China University of Science and Technology, Tangshan 063009, Hebei, China
e-mail: 271093984@qq.com

Y. K. Xue
e-mail: xueyuekai965@163.com

Y. K. Xue · S. H. Wang · D. G. Zhao · C. X. Li
Tangshan City Special Metallurgy and Material Preparation Laboratory, Tangshan 063009, Hebei, China

G. Gaustad et al. (eds.), *REWAS 2019*, The Minerals, Metals & Materials Series,
https://doi.org/10.1007/978-3-030-10386-6_46

in China. In China, only Fe element in the converter slag was extracted and reused; other beneficial elements are treated as metallurgical solid waste [5–9]. It does not meet the concept of environmental protection and energy consumption saving. Most of the elements in the converter slag are beneficial components, and there are many applications in agriculture, construction, and environmental protection after treatment [9–11]. However, due to the treatment cost and related industry standards, the reuse rate of slag in the domestic converter is less than 30% [12]. The reuse of converter slag in the metallurgical field, especially in the converter, is an efficient and direct treatment method, and has become a key concern of steel enterprises and university researchers in recent years [13].

Professor Wang Xinhua first proposed a double-slag steelmaking process for converters in China, in which a final slag with low phosphorus and high basicity during is formed and can be recycled in the next furnace. Similarly, in order to obtain ultralow phosphorus steel, duplex process converter was adopted, and dephosphorization and decarburization operations are carried out in dephosphorization furnace and decarbonization furnace, respectively. In the process, the slag produced in the decarburization furnace can be recycled in the dephosphorization furnace and the amount of slagging agent was reduced by about 50% [14]. The final slag was recycled directly in some enterprises and it was harmful to steel quality.

The phosphorus vaporization process of converter slag proposed that the slag was stirred sufficiently during slag splashing process; if the slag was treated by reducing agent added before slag splashing, dephosphorization reaction proceeds sufficiently for the good kinetic conditions and the relative vacuum conditions. As the dephosphorization product P_2 exists in the form of gas, the process is called phosphorus vaporization process. Carbon is often used in the final slag reduction experiment as reducing agent [15]. Cheng Steel Co. has adopted the double production mode of "Vanadium-transfer converter + steel-making converter", and the semi-steel produced in the vanadium-refining converter was transported to steelmaking converter. The basicity of final slag in the steel-making converter is very high for the low Si content. To realize the recycling use of the slag in the steel-making converter, SiC is selected as the dephosphorization agent, and laboratory tests are carried out.

Experiment Scheme

Experimental Materials

The final slag in the experiment is taken from the 100-ton semi-steel converter of Cheng Steel Co., and its composition is shown in Table 1. CaO, SiO_2, MgO, P_2O_5, and other chemical reagents were added to the final slag to meet the composition requirement of experimental slag. The composition of MnO and FeO in experimental slag was adjusted by adding $MnCO_3$ and Fe_2O_3.

Table 1 Chemical composition of BOF slag in experiment

BOF slag	CaO (%)	SiO_2 (%)	MgO (%)	Al_2O_3 (%)	MnO (%)	P_2O_5 (%)	S (%)	FeO (%)	Total (%)
Initial compositions	43.2	12.09	8.06	1.49	2.47	2.51	0.11	26.66	96.59
Composition adjustment	44.73	12.69	8.34	1.54	2.56	2.60	0.11	27.42	100.00

Table 2 Chemical composition of SiC in experiment

Dephosphorization agents	SiC (%)	Free C (%)	SiO_2 (%)	H_2O (%)	S (%)	Others (%)	Total (%)
SiC	55	28.1	8.2	0.3	0.1	8.3	100

In the experiment, SiC powder was used as reducing agent and its composition is shown in Table 2. All of the materials above are broken down to below 100 mesh by jaw crusher.

Experimental Equipment

The main equipment used in this experiment is SL63-7CD-type 50 kW vacuum carbon tube resistance furnace and its maximum working temperature is 2273 K. This equipment is graphite heating body and vertical vacuum resistance furnace. The experiment was carried out under circulating flow nitrogen which was measured by rotor flowmeter. MgO crucible was used in this experiment. Slag and SiC powder were heated, melted, and dephosphorized in MgO crucible. Double-platinum rhodium thermocouple was used to measure temperature, and WT-702 temperature control instrument was used to control temperature. The temperature difference between the molten pool and the thermocouple is measured by the blank experiment. Then the temperature of the molten pool is controlled by the indicated temperature of the thermocouple. The temperature was controlled within 5 °C.

Experimental Methods

In the experiment, the holding time was fixed at 10 min. The main parameters of slag basicity, FeO content, nitrogen flow rate, reaction temperature, and SiC amount were changed to study the influence on phosphorus vaporization rate. The amount of SiC added into the slag is expressed in the form of equivalent multiple. One equivalent is the SiC amount which was required to reduce all Fe_2O_3 and P_2O_5 in the slag.

The experimental slag after adjustment was melted at high temperature and hold enough time before the experiment to ensure the homogeneous distribution of materials. After the experimental slag was prepared, phosphorus vaporization experiment was carried out in the laboratory. The experimental steps were listed as follows.

The experimental slag and SiC were put into the MgO crucible and stirring evenly, and the total weight was controlled in 120 mg. The system of the vacuum carbon tube resistance furnace was vacuum to about 100 Pa and the high-speed nitrogen flow was introduced into the furnace from the bottom of the resistance furnace until the nitrogen was filled with the furnace. Then, the nitrogen flow was stably controlled at the experimental level. The vacuum carbon tube resistance furnace was heated to experimental temperature under the program setting. The crucible was put into the vacuum carbon tube resistance furnace after preheating, and then kept for 10 min. The crucible is taken out from the furnace and water-cooled to room temperature to obtain phosphorus vaporization slag samples.

The compositions of the experimental slag and the phosphorus vaporization slag were tested by XRF analysis. Then, the sample of experimental slag and the phosphorus vaporization slag were prepared and determined by S-3400 N observation and energy spectrum analysis to observe mineral phase change. XRD ray diffraction was used to test the phase composition of slag.

Results and Discussion

Electron Image and XRD Results of Phosphorus Vaporization Slag

The backscattered electron images are shown in Fig. 1.

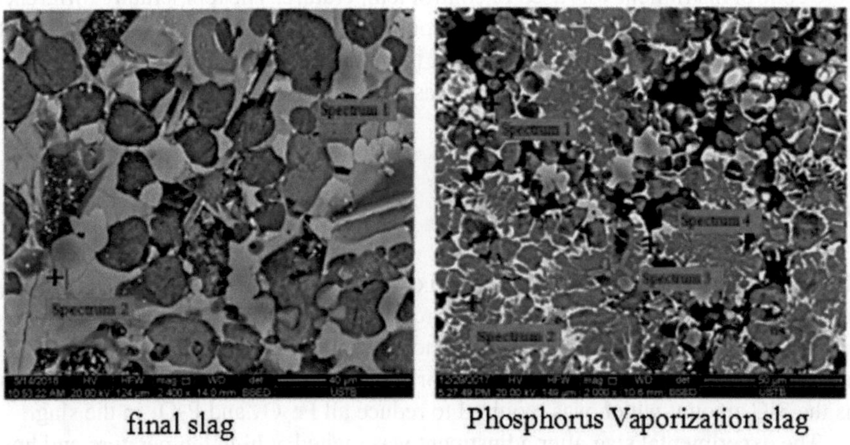

Fig. 1 Backscattered electron images of final slag and phosphorus vaporization slag

Table 3 EDS of each phase in converter slag

Final slag	C (%)	O (%)	Mg (%)	Si (%)	P (%)	Ca (%)	Fe (%)	Mn (%)	Rm = CaO/SiO_2
Spectrum 1	2.59	36.47	0.27	16.46	2.58	40.04	1.29	0.31	1.59
Spectrum 2	2.54	30.97	5.77	1.94	1.83	31.64	23.54	1.77	10.66
Phosphorus Vaporization slag	C	O	Mg	Si	P	Ca	Fe	Mn	Rm = CaO/SiO_2
Spectrum 1	11.53	23.94	0.57	15.7	2.98	42.48	2.23	0.56	1.77
Spectrum 2	7.46	24.96	0.95	1.17	0.39	33.15	30.94	0.98	18.51
Spectrum 3	8.77	23.95	9.3	8.74	0.34	9.5	34.31	5.09	0.71
Spectrum 4	70.4	20.34	4.11	0.59	0.42	2.61	1.12	0.41	2.89

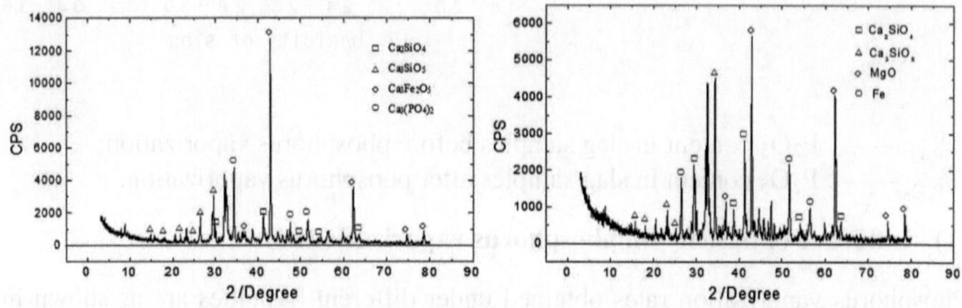

Fig. 2 X-ray diffraction patterns of final slag sample and phosphorus vaporization slag

In contrast to two electronic images, it is found that the dark gray phase and the light gray phase are the main phases in the experimental slag and the phosphorus vaporization slag. Combined with data in Table 3, it can be concluded that $2CaO{\cdot}SiO_2$ and $3CaO{\cdot}SiO_2$ were the main phase of dark gray and light gray. In the phosphorus vaporization slag, there was white iron phase containing carbon, and also the black C phase without reduction appeared.

The results of X-ray diffraction are shown in Fig. 2, and the results were consistent with the above conclusions.

Results and Analysis on Single Factor

The calculation method of dephosphorization rate is as follows:

$$\text{Dephosphorization rate} = \frac{\%^0_{P_2O_5} - \%^t_{P_2O_5}}{\%^0_{P_2O_5}} \times 100\% \quad (1)$$

in Formula (1):

Fig. 3 Influence of basicity on phosphorus vaporization rate

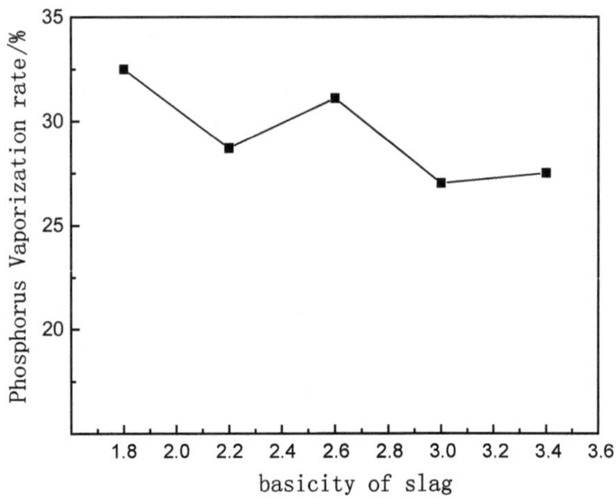

$\%^0_{P_2O_5}$—— P_2O_5 content in slag samples before phosphorus vaporization;
$\%^t_{P_2O_5}$—— P_2O_5 content in slag samples after phosphorus vaporization.

(1) **Influence of basicity on phosphorus vaporization rate**

Phosphorus vaporization rates obtained under different basicities are as shown in Fig. 3.

The linear relationship between basicity and dephosphorization rate in Fig. 3 shows that the phosphorus vaporization rate fluctuates with the increase of basicity but decreases as a whole. Therefore, it can be inferred that the optimum basicity for phosphorus vaporization of molten slag should be kept at no more than 2, and increasing basicity is not conducive to phosphorus vaporization. According to the research results of some scholars, P is mainly dissolved in the phosphorus-rich phase of $2CaO{\cdot}SiO_2$ and $3CaO{\cdot}SiO_2$ in the form of $3CaO{\cdot}PO_4$. Phosphorus needs to be transported from the inner of $2CaO{\cdot}SiO_2$ and $3CaO{\cdot}SiO_2$ to the surface and react with carbon. Too much phosphorus-rich phase is not conducive to dephosphorization.

(2) **Influence of FeO content on phosphorus vaporization rate**

Phosphorus vaporization rates obtained under different FeO contents are as shown in Fig. 4.

According to the curve in Fig. 4, the maximum phosphorus vaporization rate appeared when the FeO content was in the range of 22–25%. When FeO content was higher or lower than this value, the phosphorus vaporization rate showed a gradually decreasing trend. The melting point of slag was reduced with FeO increasing which acted as a flux and the mass transfer rate in phosphorus vaporization rate was promoted. However, FeO in the slag can react with SiC, and excessive FeO content in slag will compete with P_2O_5 for SiC, which is adverse to phosphorus vaporization reaction.

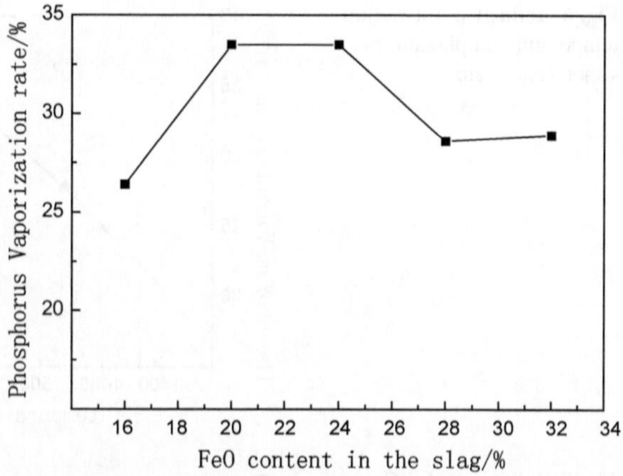

Fig. 4 Influence of FeO content on phosphorus vaporization rate

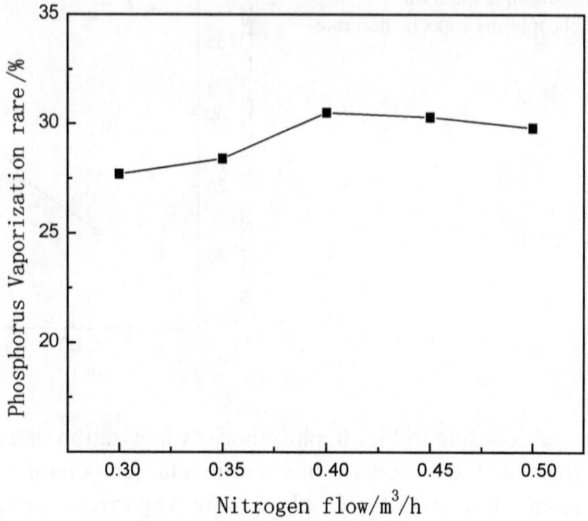

Fig. 5 Influence of nitrogen flow rate on phosphorus vaporization rate

(3) **Influence of nitrogen flow rate on phosphorus vaporization rate**

Phosphorus vaporization rates obtained with different nitrogen flow rates are shown in Fig. 5.

As shown in Fig. 5, increasing the flow rate of nitrogen was somewhat beneficial to phosphorus vaporization up to 0.4 m^3/h. Above this value, there was little influence of the nitrogen flow rate. The partial pressure of P_2 in the circulating nitrogen was almost zero and P_2 produced in the dephosphorization process injected into vacuum chamber of the circulating flow rate which promoted phosphorus vaporization reaction.

(4) **Influence of reaction temperature on Phosphorus Vaporization rate**

There were phosphorus vaporization rates under different reaction temperatures, as shown in Fig. 6.

Fig. 6 Influence of reaction temperature on phosphorus vaporization rate

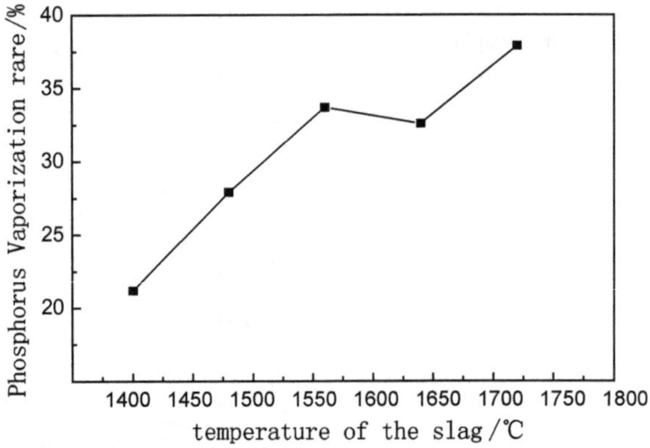

Fig. 7 Influence of SiC addition amount on phosphorus vaporization rate

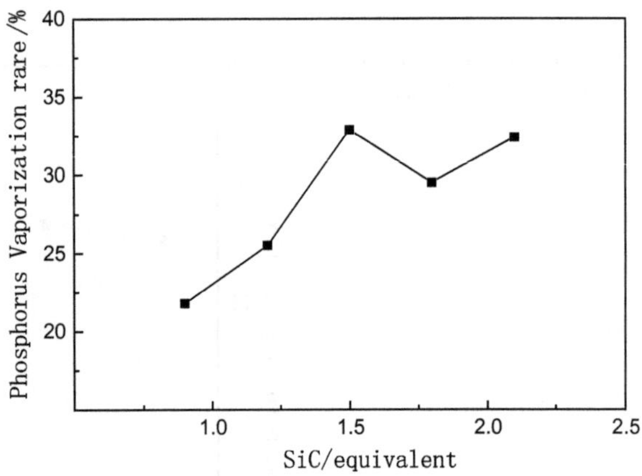

According to Fig. 6, phosphorus vaporization rate of the slag showed an upward trend with the increase of experimental temperature. When the temperature raised from 1400 to 1720 °C, phosphorus vaporization rate increased from 23 to 39%, raised by 16%. It is analyzed that the kinetic conditions of slag were improved for the increasing slag temperature and the reaction interface increased effectively. The phosphorus vaporization reaction is an interfacial reaction, and the uniformed mixture of slag and SiC promotes the phosphorus vaporization rate.

(5) **Influence of SiC addition amount on phosphorus vaporization rate**

There were phosphorus vaporization rates influenced by different SiC amounts, as shown in Fig. 7.

From Fig. 7, it can be seen that phosphorus vaporization rate of slag gasification increased with the SiC amount. When the SiC was 1.5 times the equivalent, phosphorus vaporization rate reached the maximum, and the continued increasing of SiC had no obvious effect on phosphorus vaporization rate. The reaction areas of dephosphorization were increased with the increase of SiC amount, but SiC existed as solid solution state for the high melting point which was unfavorable to phosphorus vaporization reaction.

Conclusion

(1) According to the analysis of electron microscopy and energy spectrum, $2CaOSiO_2$ and $3CaOSiO_2$ were the main phases in the final slag and the phosphorus vaporization slag reduced by SiC. After the phosphorus vaporization reaction with SiC, iron phase containing carbon and black C phase without reduction appeared in the phosphorus vaporization slag.

(2) The single factor test shows that the increase of basicity is not conducive to phosphorus vaporization reaction. The best FeO content for phosphorus vaporization reaction was 22–25%. Phosphorus vaporization rate which increased with nitrogen flow rate increasing remained stable when the nitrogen flow rate was 0.4 m^3/h. The increasing of temperature was beneficial for the increase of phosphorus vaporization rate. The increasing amount of SiC can increase phosphorus vaporization rate, and phosphorus vaporization rate tends to be stable when the amount of SiC reaches 1.5 times equivalent.

References

1. Ogawa Y, Yano M, Kitamura S, Hirata H (2001) Testu-to-Hagane 87:21–23
2. Mao MF, Cui YY, Wang DY, Min Y, Liu CJ (2013) J Iron Steel Res Int 20:17
3. Iwasaki M, Matsuo M (2011) Nippon Steel Technical Report 391:88–102
4. Jung SM, Do YJ (2006) Steel Res Int 77:312–317
5. Jung SM, Do YJ, Choi JH (2006) Steel Res Int 77:305–309
6. Turkdogan ET (2000) ISIJ Int 40:964–966
7. Li GQ, Zhang F, Zhang L, Sui ZT (2003) J Mater Metall 03:167–170
8. Wu XR, Nan JN, Chen RH (2010) J Anhui Univ Technol (Natl Sci) 27:24–27
9. Wang N, Liang ZG, Chen M, Zou ZS (2011) J Northeast Univ (Natl Sci) 06:814–818
10. Wang YC, Li HY, Li SW, Luo GP (2016) J Iron Steel Res 06:31–35
11. Wu M (2009) Steel Mak 02:16–19
12. Li CX, Li H, Zhou B (2015) China Metall 25:28–31
13. Ito M (1981) Iron Steel 67:126–129
14. Wu QF, Bao YP (2014) J Wuhan Univ Sci Technol (Natl Sci Ed) 06:411
15. Wang SH, Wu YQ, Liu XS, Xu ZR (2008) IRON STEEL 02:31–35

Research on Thermogravimetric-Differential Scanning Calorimeter of Spent Lithium Iron Phosphate Batteries Cathode Plate

Yafei Jie, Shenghai Yang, Yongming Chen, Zhiqiang Liu, Fang Hu, Nannan Liu and Yanqing Lai

Abstract The recovery of spent lithium iron phosphate batteries (LFPBs) has significant meaning in resource recycling and environmental protection. In order to investigate the effect of thermal treatment on the spent LFPBs cathode plate, in this paper, the thermogravimetric-differential scanning calorimetry (TG-DSC) of spent LFPBs cathode plate is researched. TG-DSC results indicate that two stages of weight losses and a stage of weight gain appear during the heating process with a weight change of −3.7, +1.0, and −2.4%. DSC curve showed two endothermic peaks at 165.6, 657.5 °C and two exothermic peaks at 475.6, 532.2 °C. XRD results indicate that $LiFePO_4$ is oxidized to $Li_3Fe_2(PO_4)_3$ and Fe_2O_3 during the heating process and the electrode material could be easily separated from aluminum foil due to the pyrolysis of the binder. SEM-EDS results indicate that the agglomeration degree of cathode powders decreased after the TG-DSC test, the mole fraction of C and F decreased from 23.98 and 7.03% to 1.06 and 0.32%, which was due to the pyrolysis of binders and conductive additive.

Keywords Spent LFPBs · Cathode sheets · TG-DSC test · XRD · SEM-ED

Y. Jie · S. Yang · Y. Chen (✉) · N. Liu · Y. Lai
School of Metallurgy and Environment, Central South University, Changsha 410083, China
e-mail: csuchenyongming@163.com

Z. Liu
Guangdong Research Institute of Rare Matals, Guangzhou 510651, China

Z. Liu
State Key Laboratory of Rare Metals Separation and Comprehensive Utilization, Guangzhou 510651, China

F. Hu
College of Chemistry and Chemical Engineering, Central South University, Changsha 410083, China

G. Gaustad et al. (eds.), *REWAS 2019*, The Minerals, Metals & Materials Series,
https://doi.org/10.1007/978-3-030-10386-6_47

Introduction

Rechargeable lithium-ion batteries (LIBs) have been commercialized for many years, due to their superior performance including high energy/power densities, long cycle life, memoryless effect and environmentally friendly property [1, 2]. Olivine-type lithium iron phosphate ($LiFePO_4$) and many other cathode materials have been applied to large electric vehicles, hybrid electric vehicles, and other power tools in recent years due to their inherent merits, including longer cycle life, high safety, and low cost [3]. However, the span of lithium iron phosphate batteries is about 3–5 years, when their effective capacity drops to 80% of the original capacity. An increase in consumption of LFPBs means increasing amounts of spent batteries [4]. According to statistics from China Automotive Technology and Research Center (CATARC), spent batteries from electric vehicles (EVs) will reach 120–170 thousand tons by 2020. $LiFePO_4$ batteries do not contain mercury, cadmium, lead or other toxic heavy metals, and thus they are often called eco-friendly green batteries. However, this does not mean that $LiFePO_4$ batteries are completely non-polluting products, as they contain metals, toxic electrolytes, organic chemicals and plastics, which can also lead to serious safety and environmental problems if not properly disposed of [6]. According to statistics, $LiFePO_4$ batteries typically contain 1.1% Li by weight. As a lithium-containing resource, recycling of spent $LiFePO_4$ batteries is important not only for the treatment of waste but also for the recovery of useful materials. Among the currently available recovery techniques for processing spent $LiFePO_4$ batteries, two main types have been widely deployed. One is the direct regeneration of the waste cathode material, the other is based upon strong acid leaching to recover the individual compound form [4, 5, 7]. The direct regeneration technology is sensitive to the presence of impurities and the direct regeneration of spent $LiFePO_4$ cathode materials is only possible with strict impurity control. Therefore, recovering the individual compound form is considered to be the most effective way to recycle spent $LiFePO_4$ batteries, especially the selective extraction of lithium salt, in which thermal pretreatment is critical. According to the melting points of PVDF and aluminum, thermal treatment techniques separate active materials from aluminum foil. Thermal analysis technology can be used to study how the material weight and thermal effect change as the temperature changes. Therefore, in this work, the TG-DSC techniques are performed to investigate the effect of thermal treatment on the spent LFPBs cathode plate.

Experimental

Materials

The spent LFPBs were discharged to avoid self-ignition by immersing them in 10 wt% NaCl solution for 24 h. Then those discharged batteries were further manually dismantled and separated into cathode plates anodes plate, separators and shells. Finally, the cathode plates were cut into small pieces with a size of approximately 3 * 3 mm, which were used as experimental materials for TG-DSC test.

TG-DSC Analysis

TG-DSC test was performed on STA449F3 analyzer (Netzsch, Germany). Samples (around 64.1 mg) were placed into the tray of the thermal analyzer, then heated from room temperature up to 850 °C at the heating rate of 10 °C/min. All measurements were carried out under the measured mixture of gas (20% O_2 + 80% Ar_2), with a gas flow of 100 ml/min.

Characterization

The crystalline phases of spent LFPBs electrode materials were determined by X-ray diffractometer (XRD, Rigaku D/max-2500). The surface morphology and qualitative constituent were examined by scanning electron microscopy (SEM, TESCAN MIRA3 LMU) equipped with an energy dispersive spectrometer (EDS, Oxford X-Max20).

Results and Discussion

TG-DSC Analysis of Spent Cathode Plate

As seen in Fig. 1, the TG curve shows two stages of weight losses and a stage of weight gain during the heating process with a weight change of −3.7, +1.0 and −2.4%. The weight change happens in three temperature ranges, 36–190 °C, 380–486.5 °C and 486.5–540 °C, respectively. The DTG curve shows six obvious mass loss drops in the low-temperature range, at 100.4 °C, 112.6 °C, 121.4 °C, 145.2 °C, 151.7 °C, and 159.6 °C, respectively. The DSC curve shows a small endothermic peak at 165.6 °C. Therefore, combined with physicochemical properties of organic and inorganic compounds added in the cathode plate [8], it could be concluded that the volatilization, decomposition, and hydrolysis of organic compounds and lithium salts mainly occur in the low-temperature range. In the high-temperature range, the DTG curve shows distinct weight gain peak at 486.5 °C, and the DSC curve shows an exothermic peak at 475.6 °C, which indicates the oxidation reaction of the $LiFePO_4$. This corresponds with previous work on the effect of heat treatment on the spent $LiFePO_4$ electrode [9] and the phase transformation during the heating process shown in Fig. 3. The

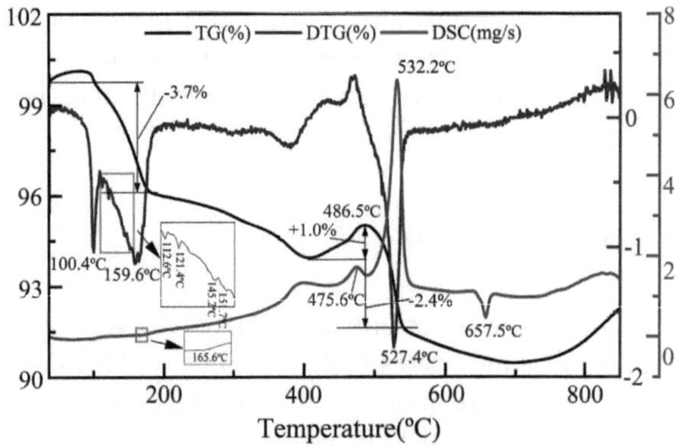

Fig. 1 TG-DSC-DTG analysis of cathode plate

DTG curve shows a distinct mass loss peak at 527.4 °C, and the DSC curve shows an exothermic peak at 532.2 °C, which indicates the oxidative decomposition of the binder. This correlates with previous work on the decomposition of PVDF [10, 11]. Finally, there is an endothermic peak around 657.5 °C attributed to the dissolution of aluminum foil, which is confirmed by the melting point of Al at 660 °C.

Characterization of Cathode Plate Before and After TG-DSC Test

The XRD patterns of spent LFPBs electrode materials are shown in Fig. 2. The composition of untreated electrode materials is $LiFePO_4$, and the dark grey electrode material (inset of Fig. 2) is coated on aluminum foil, making separation difficult. There is a significant phase change in the cathode sheets after the TG-DSC test as shown in Fig. 3. The electrode materials are composed of $Li_3Fe_2(PO_4)_3$ and Fe_2O_3, a change of color of the electrode material from dark grey to brick red suggests the oxidation of Fe^{2+} to Fe^{3+} during the TG-DSC test, thus the oxidation of the electrode materials during the heating process is expected to follow reaction Eq. (1). Besides, the electrode material could be easily separated from aluminum foil due to the pyrolysis of the binder.

$$12LiFePO_4 + 3O_2 \rightarrow 4Li_3Fe_2(PO_4)_3 + 2Fe_2O_3 \quad (1)$$

In order to further investigate the change of cathode plate during the heating process, its morphology and elemental compositions were studied by SEM-EDS. Figure 4a, b showed the SEM images under different magnifications of untreated electrode materials, the primary particles had a distribution size of 50–200 nm with an oval and circular shape. Due to the existence of binders, the cathode powder was agglomerated. From elemental mapping of Fe, P, O, C, F for untreated electrode materials, it can be seen that Fe, P, and O elements are uniformly distributed in the spent

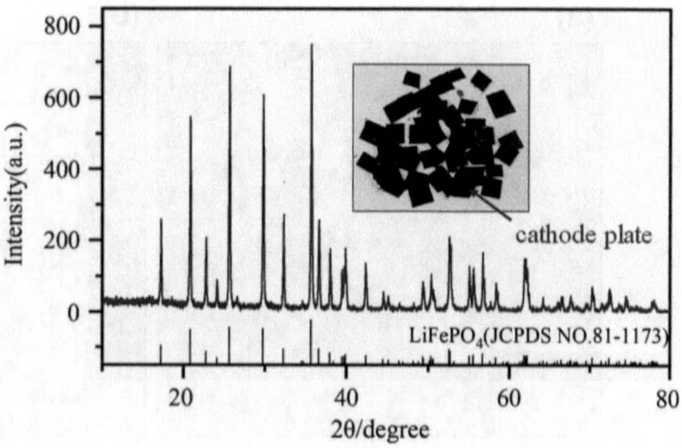

Fig. 2 XRD patterns of untreated spent LFPBs electrode materials

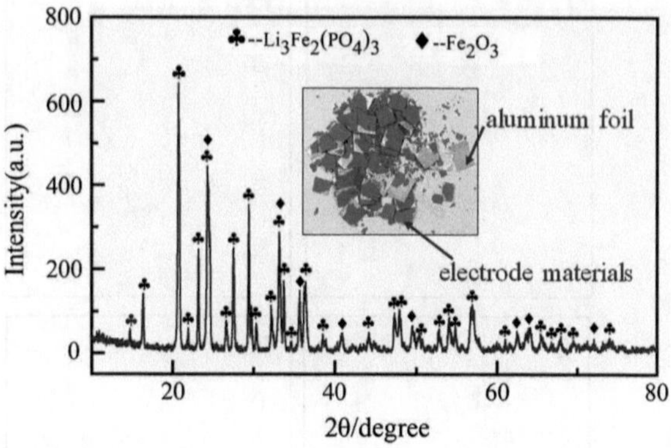

Fig. 3 XRD patterns of spent LFPBs electrode materials after TG-DSC test

LFPBs electrode materials, a small amount of C and F are irregularly distributed, and Al does not exist. Figure 5a, b shows the SEM images under different magnifications of electrode materials after TG-DSC test. When compared with untreated electrode materials, the agglomeration degree of cathode powders decreased significantly. Meanwhile, various sizes of the small irregular particle were observed on the large grains, which seemed to be rough. The elemental mapping of C, F and Al are also changed, in which C and F elements content decrease and almost disappear, and a small amount of Al element appears. Correspondingly, the EDS images of spent LFPBs electrode materials before and after TG-DSC test are shown in Fig. 6. The untreated cathode materials are mainly composed of P, Fe, O, C, and F. The mole fractions of C and F in Fig. 6a are 23.98 and 7.03%, which indicates that fluorocarbon materials, such as binders and conductive additive, make-up cathode materials. Figure 6b shows the elemental composition of the cathode powders after TG-DSC test are mainly O, P and Fe, the mole fractions of C and F are 1.06 and 0.32%, far less than those in untreated cathode materials. A possible explanation is due to the decomposition of PVDF during the thermal treatment. The mole fraction of Al is 0.14%, which may come from aluminum foil. It is corroded by fluorine-containing gases.

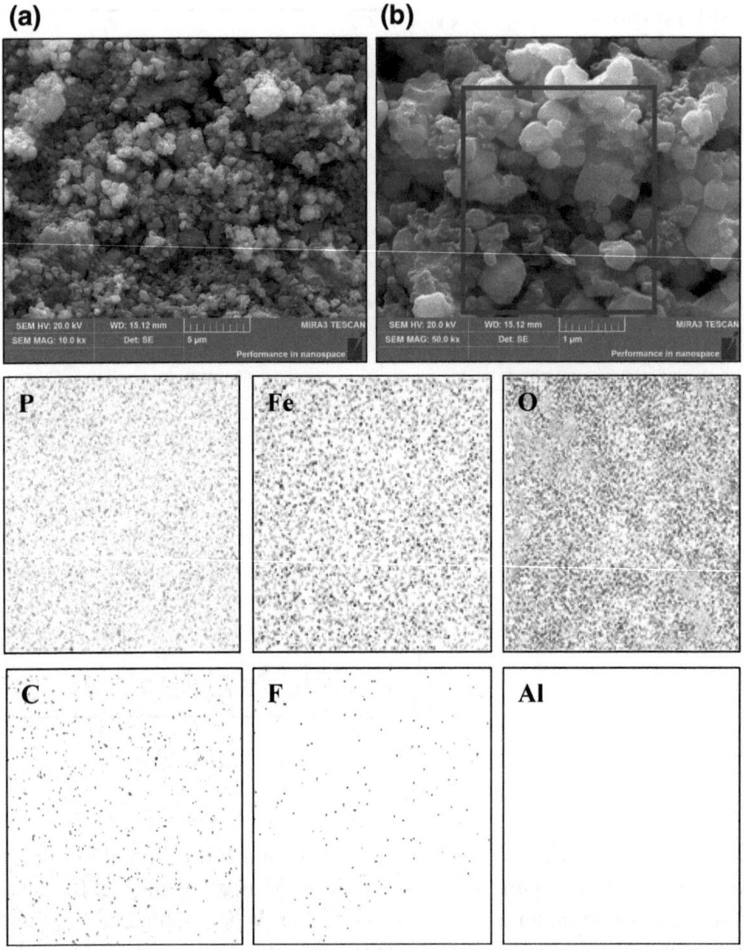

Fig. 4 SEM images of untreated spent LFPBs electrode materials, elemental mapping of Fe, P, O, C, F, Al for untreated electrode materials

Conclusion

1. In the range of 36–190 °C, the TG curve shows a stage of weight losses, which is due to the volatilization, decomposition, and hydrolysis of organic compounds and lithium salts. In the range of 380–486.5 °C, the TG curve shows a stage of weight gain, and the DSC curve shows an exothermic peak at 475.6 °C, which indicates the oxidation of the $LiFePO_4$. In the range of 486.5–540 °C, the TG curve shows a stage of weight losses, and the DSC curve shows an exothermic peak at 532.2 °C, which indicates the oxidative decomposition of the binder.

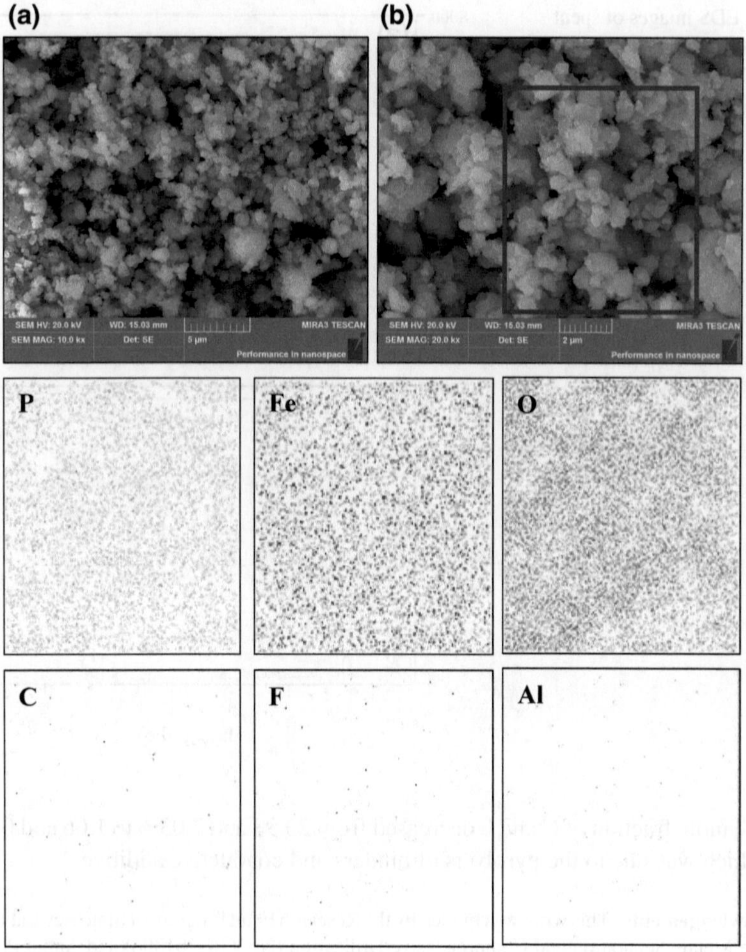

Fig. 5 SEM images of spent LFPBs electrode materials after TG-DSC test, elemental mapping of Fe, P, O, C, F, Al for electrode materials after TG-DSC test

There is an endothermic peak around 657.5 °C attributed to the dissolution of aluminum foil.

2. During the heating process, the $LiFePO_4$ is oxidized to $Li_3Fe_2(PO_4)_3$ and Fe_2O_3, as confirmed by XRD. The color of the electrode material changes from dark grey to brick red due to iron oxidation. The electrode material could be easily separated from aluminum foil by vibrating screen due to the pyrolysis of the binder.
3. The morphology and elemental composition analysis indicated that the agglomeration degree of cathode powders decreased obviously after the TG-DSC test,

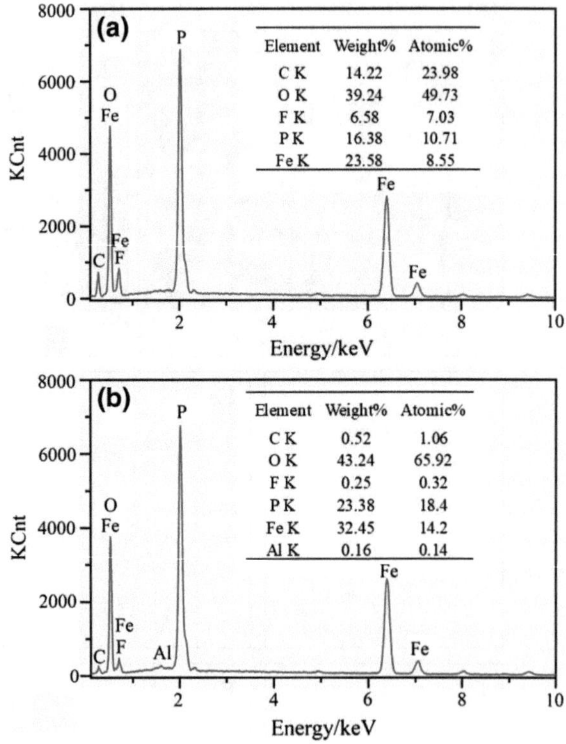

Fig. 6 EDS images of spent LFPBs electrode materials, **a** before TG-DSC test, **b** after TG-DSC test

the mole fraction of C and F decreased from 23.98 and 7.03% to 1.06 and 0.32%, which was due to the pyrolysis of binders and conductive additive.

Acknowledgements This work was funded by the Research Fund Program of State Key Laboratory of Rare Metals Separation and Comprehensive Utilization (No. GK-201806) and Anhui Province Research and Development Innovation Project for Automotive Power Battery Efficient Recycling System for which the authors are grateful.

References

1. Tu X, Zhou Y, Tian X et al (2016) Monodisperse $LiFePO_4$ microspheres embedded with well-dispersed nitrogen-doped carbon nanotubes as high-performance positive electrode material for lithium-ion batteries. Electrochim Acta 222:64–73
2. Aifantis KE, Hackney SA, Dempsey JP (2007) Design criteria for nanostructured Li-ion batteries. J Power Sources 165(2):874–879
3. Yang Y et al (2017) A closed-loop process for selective metal recovery from spent lithium iron phosphate batteries through mechanochemical activation. ACS Sustain Chem Eng 5(11):9972–9980
4. Bian D, Sun Y, Li S et al (2016) A novel process to recycle spent $LiFePO_4$ for synthesizing LiFePO4/C hierarchical microflowers. Electrochim Acta 190:134–140

5. Shin EJ, Kim S, Noh JK et al (2015) A green recycling process designed for LiFePO 4 cathode materials for Li-ion batteries. J Mater Chem A 3(21):11493–11502
6. Wang W, Wu Y (2017) An overview of recycling and treatment of spent $LiFePO_4$ batteries in China. Resour Conserv Recycl 127:233–243
7. Chen J, Li Q, Song J et al (2016) Environmentally friendly recycling and effective repairing of cathode powders from spent LiFePO 4 batteries. Green Chem 18(8):2500–2506
8. Liu YL, Wu JY, Li H (2014) Fundamental scientific aspects of lithium ion batteries (IX)—Nonaqueous electrolyte materials. Energy Storage Sci Technol 3(3):262–275
9. Zheng R, Zhao L, Wang W et al (2016) Optimized Li and Fe recovery from spent lithium-ion batteries via a solution-precipitation method. RSC Adv 6(49):43613–43625
10. Hanisch C, Loellhoeffel T, Diekmann J, Markley KJ, Haselrieder W, Kwade A (2015) J Clean Prod 108:301–311
11. Huang C, Yang H, Li Y et al (2015) Characterization of aluminum/poly (vinylidene fluoride) by thermogravimetric analysis, differential scanning calorimetry, and mass spectrometry. Anal Lett 48(13):2011–2021

Study of Precursor Preparation of Battery-Grade Lithium Iron Phosphate

Li-li Zhang, Wei-guang Zhang, Ting-an Zhang, Qiu-yue Zhao, Ying Zhang, Jing Liu and Kun Wang

Abstract In this paper, ferric sulfate was extracted from titanium white waste acid as the iron source of lithium iron phosphate precursor. The ferric sulfate obtained from titanium white waste acid, ammonium phosphate tribasic, and ammonia hydroxide were used as raw materials through liquid precipitation method to obtain iron phosphate as the precursor of lithium iron phosphate. Under the premise of ensuring the synthesis of $FePO_4{\cdot}2H_2O$, the effects of the pH, synthesis temperature and reaction time on the particle size of the resulting product were investigated. The results showed that high purity amorphous $FePO_4{\cdot}2H_2O$ with a median diameter of 38.4 μm was acquired through the condition of pH = 2, T = 25 °C, and t = 12 h, which meets the requirements for preparation of lithium iron phosphate and realizes high value-added utilization of discarded resources.

Keywords Iron phosphate · Precursor · Lithium iron phosphate · Median diameter

Introduction

Olivine-type $LiFePO_4$ has many advantages such as environmental friendliness, low price, excellent safety performance, thermal stability, and cycle performance and may be the most promising material for power battery and energy storage system [1–3]. $FePO_4$ as a precursor of $LiFePO_4$ has a similar structure to $LiFePO_4$. Therefore, it is only necessary to add a lithium source when synthesizing $LiFePO_4$. The process is simple and controllable, and the raw materials are green and environmentally friendly, so the effective preparation of $FePO_4$ has attracted wide attention of researchers [4, 5].

The synthetic methods of $FePO_4$ include precipitation method [6–11], solid phase method [12, 13], sol-gel method [14, 15], hydrothermal method [4, 16–20], microemulsion method and microwave radiation method [21]. However, these meth-

L. Zhang · W. Zhang · T. Zhang (✉) · Q. Zhao · Y. Zhang · J. Liu · K. Wang
Key Laboratory of Ecological Metallurgy of Multi-metal Intergrown Ores of Ministry of Education, Special Metallurgy and Process Engineering Institute, School of Metallurgy of Northeastern University, Shenyang 110819, China
e-mail: zta2000@163.net

G. Gaustad et al. (eds.), *REWAS 2019*, The Minerals, Metals & Materials Series,
https://doi.org/10.1007/978-3-030-10386-6_48

ods have high production costs, complicated process, and high equipment requirements. The particle size of $FePO_4$ prepared by some methods is too large and uneven. These factors restrict the large-scale production and application of $LiFePO_4$. It is of great significance to find a method for preparing the product with small particle size, which is economical in raw materials, simple in process, mild in production conditions and simple in operation.

In this paper, the micron-sized $FePO_4$ precursor with relatively uniform particle size was prepared by liquid phase precipitation method using $Fe_2(SO_4)_3$ which is separated and extracted from titanium white waste acid, $(NH_4)_3PO_4$ and $NH_3 \cdot H_2O$ as starting materials. Compared with other previous reports, the experimental procedure with low energy consumption is handy and requires low equipment. The obtained $FePO_4$ with relatively uniform distribution has a small median diameter. It proved that the product is completely suitable for the industrial production of $LiFePO_4$. In this reaction, the $Fe_2(SO_4)_3$ was used as the iron source and the $(NH_4)_3PO_4$ was used as the phosphorus source. In addition to the preparation of the micron-sized $FePO_4 \cdot 2H_2O$ precursor, it not only solves the environmental pollution problem caused by the titanium white waste acid, but also realizes the high-value utilization of the valuable metal in the titanium white waste acid. Under the condition of ensuring the synthesis of $FePO_4 \cdot 2H_2O$, the effects of the pH, the synthesis temperature and the reaction time on the particle size of the resulting product were investigated.

Experiments

Reagents and Apparatus

$(NH_4)_3PO_4$ (analytical grade, Sinopharm Chemical Reagent Co., Ltd., China) and $Fe_2(SO_4)_3$ (separated and extracted from titanium white waste acid) were used to prepare in this investigation. The pH of the reaction solution was adjusted by ammonium hydroxide (analytical grade, Sinopharm Chemical Reagent Co., Ltd., China) solutions. Water used in this work was deionized.

The concentration of iron ions in $Fe_2(SO_4)_3$ solution and the concentration of phosphate in $(NH_4)_3PO_4$ solution are measured by an inductively coupled plasma atomic emission spectrometry (ICP-AES, prodigy XP). The pH of the reaction solutions was measured by a digital pH meter (PHSJ-3F). The particle size of $FePO_4$ samples was obtained using laser particle size analyzer (Mastersizer 3000). The morphology of the samples was observed by a scanning electron microscopy (SEM, SU8010). The thermal characterization of the sample was measured by differential scanning calorimetry and thermal gravimetry (DSC-TG, STA449 F3). The thermal scans were performed with a heating rate of 10 K/min up to a temperature of 800 °C in an air stream. The phase of $FePO_4$ sample which was calcined at 500 °C for 6 h was determined by X-ray diffraction (XRD, PW3040/60) using Cu Kα radiation and 0.02° scanning step. The fired sample Fourier-transform infrared spectra of extraction was obtained by an infrared spectrometer (FT-IR, Nicolet iS50).

Experimental Procedures

Deionized water produced by an ultra-pure water production system was used to prepare all solutions. $(NH_4)_3PO_4$ (analytical grade) crystalline solids were dissolved in deionized water to make phosphate stock solutions for the production of $FePO_4$, $Fe_2(SO_4)_3$ solution (separated and extracted from titanium white waste acid) was added to deionized water to dilute solution to make iron stock solutions. The stock solutions were standardized by an inductively coupled plasma (ICP) emission spectrometry. Ammonium hydroxide (analytical grade, Sinopharm Chemical Reagent Co., Ltd., China) was diluted into a solution of ammonia hydroxide by water at a volume ratio of 1:4 and the dilute solution of ammonia hydroxide was used in this experiment.

Amorphous $FePO_4$ was synthesized by spontaneous precipitation from the solution of $Fe_2(SO_4)_3$ and $(NH_4)_3PO_4$, using the dilute solution of ammonia hydroxide. The solution of $Fe_2(SO_4)_3$ containing 0.625 mol/L iron and the solution of $(NH_4)_3PO_4$ containing 0.938 mol/L phosphorus were prepared. To the solution of $Fe_2(SO_4)$ containing 0.625 mol/L iron was added the solution of $(NH_4)_3PO_4$ containing 0.938 mol/L phosphorus, in the 1:1 volume ratio. The mixed solution of $Fe_2(SO_4)_3$ solution containing 0.625 mol/L iron and $(NH_4)_3PO_4$ solution containing 0.938 mol/L phosphorus was vigorously stirred at 600 r/min by magnetic stirring (T09-1S). Then the dilute solution of ammonia hydroxide was added to the mixed solution at 25 °C. After a while, precipitate started to gradually form after the addition of the dilute solution of ammonia hydroxide. The dilute solution of ammonia hydroxide is continuously added to the reaction solution so that the pH of the reaction solution measured by a digital pH meter is 1.8, 2.0, 2.2, 2.4, and 3.0, respectively. After the solution has reacted for a while, the precipitate was collected on a filter paper, washed several times and dried in an oven at 55 °C for 24 h. Then the reaction temperature and reaction time of the experiment were changed and the above experiments were repeated.

Results and Discussion

Effect of pH

The pH of the reaction solution was set to 1.8, 2.0, 2.2, 2.4, and 3, respectively. The precipitated product was obtained from the reactions which were carried out for 4 h

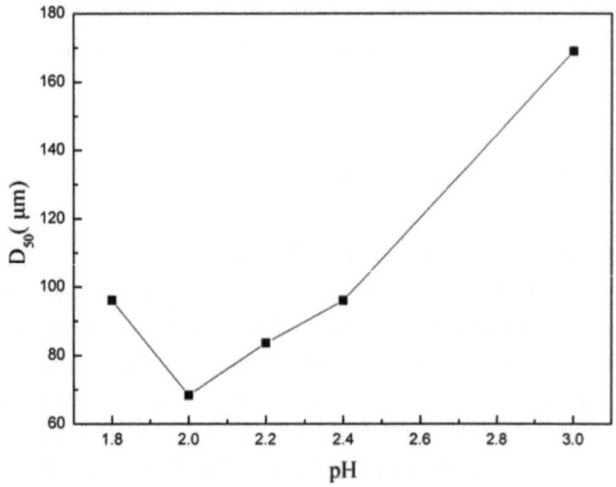

Fig. 1 Effect of pH on the particle size of the precipitated product

at 25 °C. Influence of pH on the particle size of the precipitated product was studied and the result is shown in Fig. 1.

It can be seen clearly from Fig. 1 that the median particle size of the product decreases as the pH increases in the range of pH 1.8–2.0 of the reaction solution. Compared with the median particle size of pH 1.8–2.0 of the reaction solution, the smallest particles of the product with a median diameter of 68.4 μm was acquired under the condition of pH = 2. It is not clear why the median particle size of the product decreases with the increase of pH in the range of 1.8–2.0. This needs further exploration and discussion. As the pH increases in the range of pH 2.0–3.0 of reaction solution, the median particle size of the product also increases. This is because the concentration of free Fe^{3+} and PO_4^{3-} in the reaction solution and the supersaturation of the reaction solution are determined by the pH of the reaction solution. With the increase of pH, the concentration of free Fe^{3+} and PO_4^{3-} increases, and the supersaturation of the reaction solution increases. A large number of crystal nuclei are formed instantaneously, and there is not enough time to grow, leading to the formation of finer precipitates. Primary particles with larger surface energy were easy to agglomerate into secondary particles. The size of secondary particles increases with the decrease of the size of primary particles. Therefore, it is necessary to set the pH of the reaction solution to 2.0 in order to synthesize precipitates with a smaller median size.

Effect of Temperature

Under experimental conditions of pH = 2, the reaction solution was placed at 25 °C, 40 °C, 50 °C, 60 °C, and 70 °C for 4 h, respectively. The effect of synthesis temperature on the particle size of the precipitated product obtained was studied. The result is shown in Fig. 2.

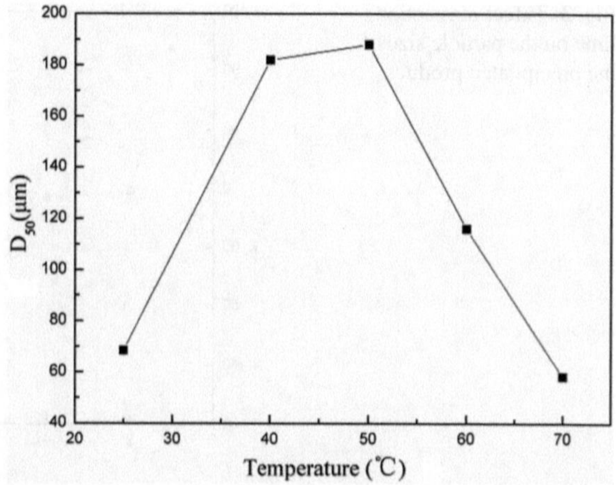

Fig. 2 Effect of synthesis temperature on the particle size of the precipitated product

Figure 2 compares the influence of the synthesis temperature on the particle size of the precipitated product. As can be seen from Fig. 2, the intermediate particle size of the precipitated product is smaller at 25 °C and increases with the increase of temperature. When the temperature reaches 50 °C, the median particle size of the precipitated product reached the maximum. Then, as the temperature continues to rise, the median diameter of the precipitates product decreases. This is because the synthesis temperature not only affects the nucleation rate and growth rate of the product, but also affects the interaction between primary particles of the precipitated product. When the temperature is relatively low, relatively few nuclei are formed, and these nuclei are easy to develop into large primary particles which are not easy to aggregate. Therefore, the secondary particles agglomerated by the primary particles have a smaller particle diameter. When the temperature rises initially, due to a large number of nuclei and quick rate of nucleation, the primary particles have no time to grow up causing the formation of small primary particles. Subsequently, these primary particles are clustered into larger secondary particles. When the temperature continues to increase, the high temperature accelerated the Brownian motion of molecules in solution so the adsorption effect between primary particles were destroyed resulting in a reduction of the median size of secondary particles formed by the aggregation of these primary particles. At the same time, the solubility product constant of $FePO_4$ increases with the increase of temperature making for the decrease of system oversaturation, but these factors are not conducive to the formation of a crystal. Considering the above factors, the reaction temperature of the following experiment was determined at 25 °C.

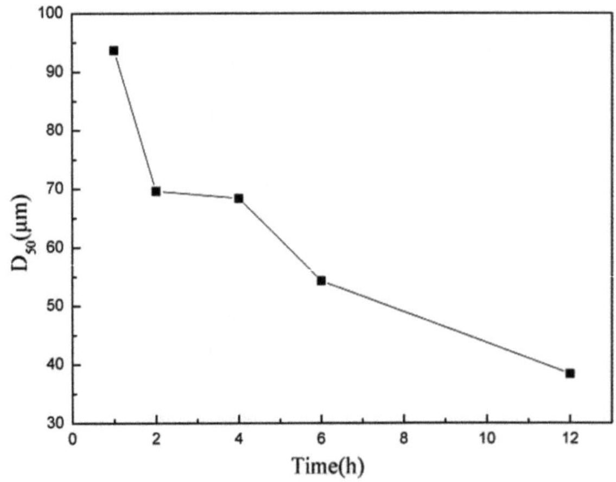

Fig. 3 Effect of reaction time on the particle size of the precipitated product

Effect of Time

The precipitated product was obtained from the reactions which were carried out for at 25 °C and pH = 2. Influence of the synthesis time ranging from 1 to 12 h on the particle size of the precipitated product was discussed and the result is shown in Fig. 3.

Figure 3 shows the effect of reaction time on the median particle size of the product. A conclusion can be drawn from Fig. 3 that the median particle size of the product decreases gradually with the extension of reaction time. In the range of 0–2 h, the median particle size of the product changes relatively and the change tends to be stable within 2–4 h, and finally the particle size shows a downward trend. The reason for these changes is that the reaction time affects the growth rate of the crystal and interaction between the small particles. Within 1–2 h, due to the delayed growth of the crystal nucleus in the short time, a large amount of fine precipitation was generated. The secondary particles formed by aggregation of smaller particles are larger. However, these larger secondary particles will rapidly become smaller secondary particles under stirring. Within 2–4 h, the crystal nucleus grow slowly with the prolongation of the reaction time. The effect of stirring on the secondary particles is not great, so the median size of the secondary particles of the product does not change much. In the 4–12 h period, with the further extension of reaction time, the primary particles grow rapidly. The surface energy of these primary particles decreases and the agglomeration decreases, so the secondary particle median size of the products decreases rapidly. Therefore, the reaction time is set to 12 h, which is suitable for the finer products.

Fig. 4 SEM micrograph of the sample obtained by pH = 2, 25 °C, 12 h

Table 1 The comparison of measured and theoretical values of iron phosphate

Name	P (%)	TFe (%)	P/Fe (mol)
Measurements of sample	16.27	29.92	0.98
Theoretical value of iron phosphate	16.59	29.97	1

The Analysis of Iron Phosphate Sample

This work further analyzed the iron phosphate sample which was obtained under pH = 2, 25 °C, 12 h. Figure 4 is the SEM image of the sample. It can be seen from Fig. 4 that the particles of iron phosphate sample are agglomerated from very small primary particles.

The mass fractions of phosphorus and iron element in the $FePO_4 \cdot 2H_2O$ obtained using the optimized experimental conditions were shown in Table 1. It can be seen from Table 1 that the mass fractions of iron and phosphorus are 29.92% and 16.27%, respectively, which are consistent with the theoretical values.

Figure 5 is the TG-DTA curve of $FePO_4 \cdot 2H_2O$ prepared experimentally. The DSC curve shows that there is an obvious endothermic peak between 130 and 210 °C, which is the endothermic peak of the sample losing the crystalline water and there is no obvious peak after that. The TG curve indicates that the mass loss rate of the sample is 19.256% from 30 to 210 °C, which is very close to the theoretical mass loss of 19.270% when $FePO_4 \cdot 2H_2O$ loses crystalline water. In summary, the sample obtained is $FePO_4 \cdot 2H_2O$.

After the sample was calcined in an air atmosphere at 500 °C for 6 h, the corresponding product was analyzed by XRD and infrared and the results were shown in Figs. 6 and 7. The XRD spectra are consistent with the standard spectra of $FePO_4$, indicating that the sample is free of impurities. It can be inferred from the infrared spectrum that the peak of 973.16 cm^{-1} is the stretching vibration between phos-

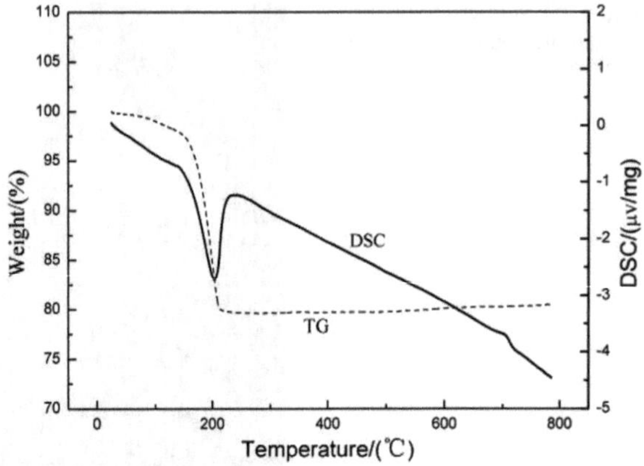

Fig. 5 TG/DSC curves of the $FePO_4{\cdot}2H_2O$ obtained by pH = 2, 25 °C, 12 h

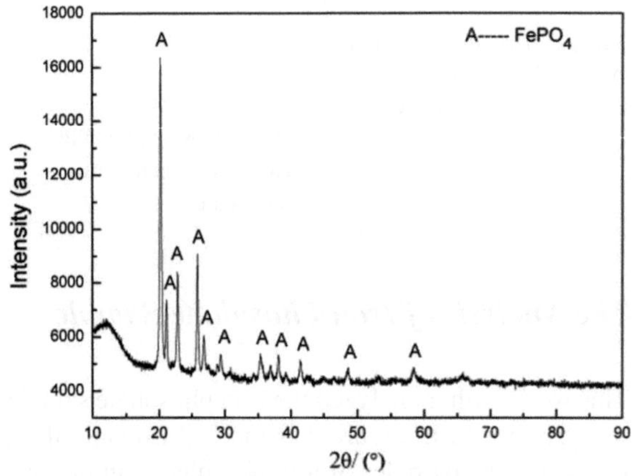

Fig. 6 XRD spectrum of crystalline $FePO_4$ obtained by heating the sample at 500 °C for 6 h under air atmosphere

phorus and oxygen bonds and the peaks of 633.90 and 590.60 cm^{-1} are symmetric bending vibrations between phosphorus and oxygen bonds. It shows that there is no impurity in the sample.

Conclusions

(1) In this paper, the micron-sized $FePO_4{\cdot}2H_2O$ precursor with relatively uniform particle size is prepared by liquid phase precipitation method using $Fe_2(SO_4)_3$ which is separated and extracted from titanium white waste acid, $(NH_4)_3PO_4$, and $NH_3{\cdot}H_2O$ as starting materials.

(2) Under the premise of ensuring the synthesis of $FePO_4{\cdot}2H_2O$, the effects of the pH, synthesis temperature and reaction time on the particle size of the resulting product were investigated. Compared with previous reports, the smallest amor-

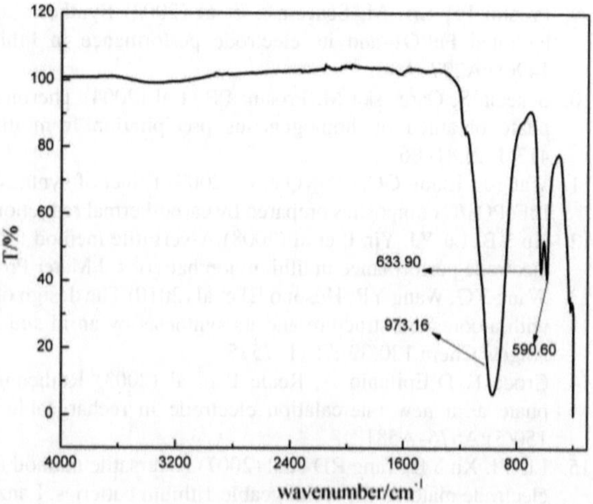

Fig. 7 Infrared spectrum of crystalline $FePO_4$ obtained by heating the sample at 500 °C for 6 h under air atmosphere

phous $FePO_4{\cdot}2H_2O$ with a median diameter of 38.4 μm was acquired through the condition of pH = 2, T = 25 °C and t = 12 h, respectively. The results show that $FePO_4{\cdot}2H_2O$ samples are suitable for further study as a potentially inexpensive material for the production of $LiFePO_4$.

Acknowledgements This work was financially supported by the National Natural Science Foundation of China (No. 51504059), the Fundamental Research Funds for the Central Universities (No. N162504016).

References

1. Padhi AK, Nanjundaswamy KS, Goodenough JB (1997) Phospho-olivines as positive-electrode materials for rechargeable lithium batteries. J Electrochem Soc 144(4):1188–1194
2. Wang DN, Wang HX, Yang JL et al (2016) Dynamic study of sub-micro sized $LiFePO_4$ cathodes by in-situ tender X-ray absorption near edge structure. J Power Sources 302:223–232
3. Gao SY, Su YF, Bao LY et al (2015) High-performance $LiFePO_4$/C electrode with polytetrafluoroethylene as an aqueous-based binder. J Power Sources 298:292–298
4. Song YN, Zavalij PY, Masatsugu Suzuki et al (2002) New iron (III) phosphate phases: crystal structure and electrochemical and magnetic properties. Inorg Chem 41(22):5778–5786
5. Wang X, Yang XH, Zheng HG et al (2005) Synthesis and electrochemical performance of amorphous hydrated iron phosphate nanoparticles. J Cryst Growth 274:214–217
6. Shi ZC, Attia A, Ye WL et al (2008) Synthesis, characterization and electrochemical performance of mesoporous $FePO_4$ as cathode material for rechargeable lithium batteries. Electrochim Acta 53(6):2665–2673
7. Yang SF, Song YN, Peter Y et al (2002) Reactivity, stability and electrochemical behavior of lithium iron phosphates. Electrochem Commun 4(3):239–244
8. Prosini PP, Carewska M, Scaccia S et al (2010) A new synthetic route for preparing LiFePO4 with enhanced electrochemical performance. Cheminiform 33(42):20–20

9. Prosini PP, Lisi M, Scaccia S et al (2002) Synthesis and characterization of amorphous hydrated $FePO_4$ and its electrode performance in lithium batteries. J Electrochem Soc 149(3):A297–A301
10. Seaecia S, Carewska M, Prosini PP et al (2004) Thermoanalytical study of iron (III) phosphate obtained by homogeneous precipitation from different media. Thermochim Acta 413(1–2):81–86
11. Wang L, Liang GC, Ou XQ et al (2009) Effect of synthesis temperature on the properties of $LiFePO_4$/C composites prepared by carbothermal reduction. J Power Sources 189(1):423–428
12. Xu YB, Lu YJ, Yin P et al (2008) A versatile method for preparing $FePO_4$ and study on its electrode performance in lithium ion batteries. J Mater Process Tech 204(1):513–519
13. Wang YG, Wang YR, Hosono EJ et al (2010) The design of a $LiFePO_4$/carbon nanocomposite with a coreshell structure and its synthesis by an in situ polymerization restriction method. Angew Chem 120(39):7571–7575
14. Croce F, D'Epifanio A, Reale P et al (2003) Ruthenium oxide-added quartz iron phosphate as a new intercalation electrode in rechargeable lithium cells. J Electrochem Soc 150(5):A576–A581
15. Lu YJ, Xu YB, Yang RD et al (2007) A versatile method for preparing $FePO_4$ as a promising electrode material for rechargeable 1ithium batteries. Lanzhou Univ (Nat Sci) 43(4):144–146
16. Guo XF, Ding WP, Wang XS et al (2001) Synthesis of a novel mesoporous iron phosphate. Chem Commun 8(8):709–710
17. Masquelier C, Reale P, Wurm C et al (2002) Hydrated iron phosphates $FePO_4{\cdot}nH_2O$ and $Fe_4(P_2O_7)_3{\cdot}nH_2O$ as 3 V positive electrodesin rechargeable lithium batteries. J Electrochem Soc 10(15):A1037–A1044
18. Kandori K, Nakashima H, Ishikawa T et al (2006) Control of size adsorptive properties of spherical ferric phosphate particles. J Colloid Interface Sci 300(1):225–231
19. Zaghiba K, Julien CM (2005) Structure and electrochemistry of $FePO_4{\cdot}2H_2O$ hydrate. J Power Sources 142(1):279–284
20. Mal NK, Bhaumik A, Matsukata M (2006) Syntheses of mesoporous hybrid iron oxophenyl phosphate, iron oxophosphate, and sulfonated oxophenyl phosphate. Ind Eng Chem Res 45(23):7748–7751
21. Okawa H, Yabuki J, Kawamura Y (2008) Synthesis of $FePO_4$ cathode material for lithium ion batteries by a sonochemical method. Mater Res Bull 43(5):1203–1208

Study on Vacuum Pyrolysis Process of Cathode Sheets from Spent Lithium-Ion Batteries

Weilun Li, Shenghai Yang, Nannan Liu, Yongming Chen, Yan Xi, Shuai Li, Yafei Jie and Fang Hu

Abstract Spent lithium-ion batteries(LIBs) contain lots of valuable metals such as nickel, cobalt, and lithium, together with organic solvents, binders, and other toxic materials. Therefore, recycling of spent LIBs is of great importance for comprehensive resource recovery and environmental protection. In this study, vacuum pyrolysis was used to dispose of the cathode sheets of LIBs. The effects of pyrolysis temperature and vacuum degree on the separation of cathode sheets and phase transition of valuable metal of cathode active powder were investigated in detail. The results showed that the effective separation of active powder and Al foil can be achieved under the optimized conditions of pyrolysis temperature of 600 °C and a vacuum degree of 1000 Pa, and the recovery rate of cathode active powder reached 98.04%. In the temperature range of 450–650 °C, with the increase of pyrolysis temperature, the XRD patterns of the cathode active powder showed that the characteristic peak of $Li[Ni_xCo_yMn_{1-x-y}]O_2$ gradually weakened and eventually disappeared.

Keywords Vacuum pyrolysis · Spent lithium ion batteries · Cathode sheets · Recovery rate · Phase transition

Introduction

The cathodes, anodes, and electrolytes are elements of LIBs rich in valuable metals. The most valuable element is high-quality cathode material composed of cobalt, lithium, nickel oxide. Taking the power ternary battery as an example, the average contents of lithium, nickel, and cobalt are 1.9%, 12.1%, and 2.3%, respectively. It is much higher than the corresponding metal grade in the original ore and has

W. Li · S. Yang · N. Liu · Y. Chen (✉) · Y. Xi · S. Li · Y. Jie
School of Metallurgy and Environment, Central South University, Changsha Hunan 410083, China
e-mail: csuchenyongming@163.com

F. Hu
College of Chemistry and Chemical Engineering, Central South University, Changsha Hunan 410083, China

G. Gaustad et al. (eds.), *REWAS 2019*, The Minerals, Metals & Materials Series,
https://doi.org/10.1007/978-3-030-10386-6_49

significant resource characteristics [1]. In addition, LIBs contain toxic materials such as separators, organic solvents, and binders, which may cause environmental risks. Therefore, the recycling of spent LIBs has the dual significance of comprehensive resource recovery and environmental protection.

There are mainly two routes for recovery of spent LIBs: pyrometallurgy and hydrometallurgy. Hydrometallurgy is widely adopted by many enterprises in China [2]. The disposal process generally includes four stages: pretreatment [3], leaching [4–6], purification and separation [7, 8], and product preparation [9–11]. In the hydrometallurgy recovery process, the components of spent LIBs are complicated, so it is necessary to pretreat the spent LIBs to separate and enrich the valuable materials. Pretreatment is mainly divided into mechanical crushing [12], solvent extraction [13, 14], alkaline dissolution [15] and heat treatment [16] according to different separation techniques. Li [17] combined mechanical crushing and ultrasonic cleaning in spent LIBs disposal process, and the recovery rate of element cobalt was 92%. But this strategy causes partial loss of element cobalt due to the presence of the binder. Contestabile et al. [18] used an organic solvent NMP to achieve effective separation of cathode material and Al foil. However, this method is susceptible to the type of binder and battery winding method, and organic solvents are expensive and volatile. Therefore, it would cause environmental pollution during the disposal process and would be challenging to implement an industrial scale. Chen et al. [19] treated the electrode active material with a NaOH solution to avoid the introduction of Al^{3+} in the next separation steps. It was found that 99.9% Al was dissolved under optimized conditions of 5 wt% NaOH solution and the liquid-to-solid ratio of 10:1 at room temperature for 4 h. However, this method increases the content of impurity ions in the solution and the amount of acid in the subsequent treatment and also causes difficulty in the treatment of the alkali solution.

Based on the above disadvantages, some scholars have proposed heat treatment methods to accomplish the separation of cathode active powder and Al foil. Sun and Qiu [20] used vacuum pyrolysis to treat cathode sheets of spent LIBs. The results demonstrated that the active powder on the cathode sheets can be effectively removed at 600 °C, vacuum degree of 1000 Pa for 30 min.

The above studies examined the effect of pyrolysis temperature on the separation effect of the cathode sheet but did not investigate in detail the effects of various heat treatment conditions on the separation effect of the cathode sheets, the phase transition of the cathode active powder. To investigate this topic, this study used a low-temperature vacuum tube furnace for the heat treatment of the cathode sheets under a hermetic vacuum atmosphere. The effects of pyrolysis temperature (450–700 °C) and vacuum degree (100–2000 Pa) on the separation of the cathode sheets and the phase transition of the valuable metal were systematically investigated.

Experimental

Experimental Materials

The experimental raw material spent ternary LIBs were provided by *Hunan Reshine New Material Co., Ltd*. All the solutions were prepared in deionized water.

Experimental Procedure

A flowsheet of the procedure was illustrated in Fig. 1.

The main experimental procedure steps are as follows:

(1) Dismantling and separation: To avoid the unsafe factors (self-ignition and short-circuiting), LIBs were discharged in the 10 wt% NaCl solution for 24 h. LIBs were disassembled into cathodes, anodes, organic separators and shells by manual dismantling, then the cathodes were cut into long sheets of length × width = 6 × 4.5 cm and which are stored in vacuum state.

(2) Vacuum pyrolysis: The experimental setup is shown in Fig. 2. Five pieces of cathode scraps were weighed and placed into the special corundum ark, then the ark was placed into a tubular furnace. The temperature of the tubular furnace was raised to a preset temperature with a heating rate of 10 °C/min in a preset vacuum degree. After the experiment, the Al foil and the cathode active powder were obtained by Taylor standard sieves (0.147 mm) vibrating and screening for 10 s. After weighing, the cathode active powder obtained by heat treatment was characterized, the crystalline phases were determined by X-ray diffractometer (XRD, Rigaku D/max-2500), the surface morphology and qualitative constituent were examined by scanning electron microscopy (SEM, TESCAN MIRA3 LMU) equipped with an energy dispersive spectrometer (EDS, Oxford X-Max20).

The formula for calculating the yield of the cathode active powder is as follows:

$$\text{cathode active powder yield} = \frac{m_1}{m\alpha_0} \times 100\% \tag{1}$$

Wherein, m_1 is the mass (g) of the cathode active powder of the sieve after screening, and m is the mass (g) of the cathode electrode before heat treatment. In addition, the cathode electrode of the raw material is uniformly coated during the experiment. So a current collector is selected as a standard set, and completely separated after the heat treatment to obtain the cathode electrode active material and the aluminum foil, and α_0 is the ratio (%) of the cathode electrode active powder to the cathode electrode sheet before heat treatment.

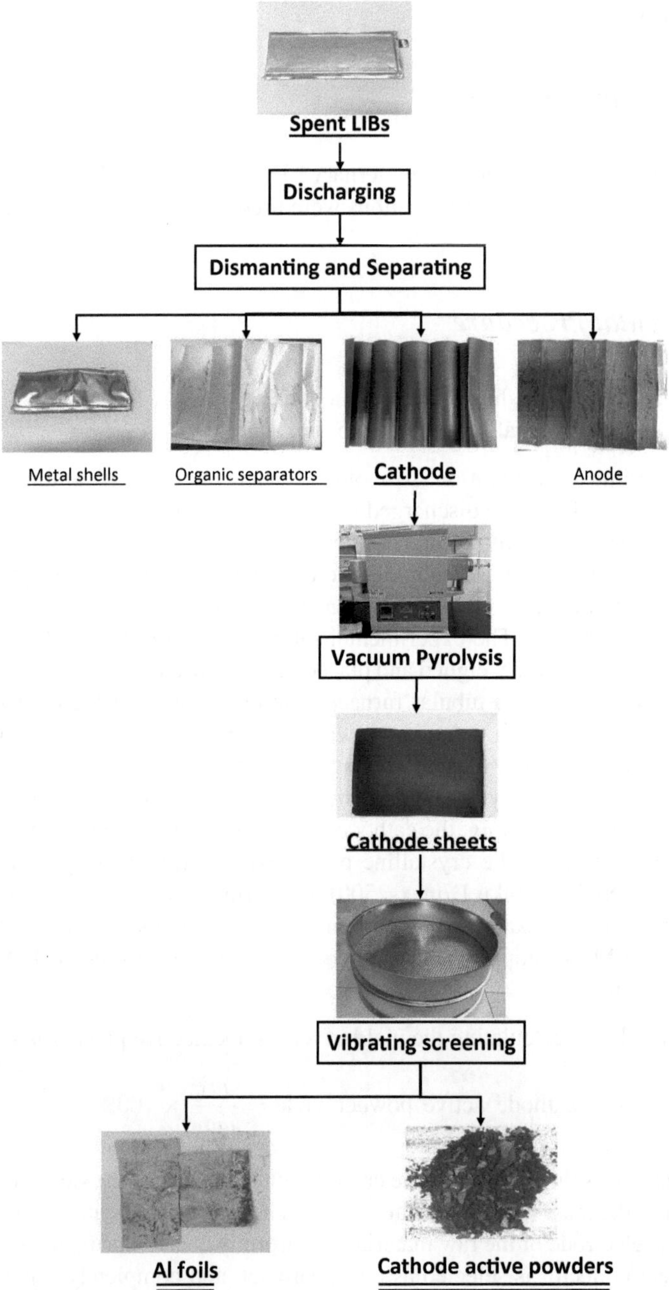

Fig. 1 The flowsheet of the experimental process

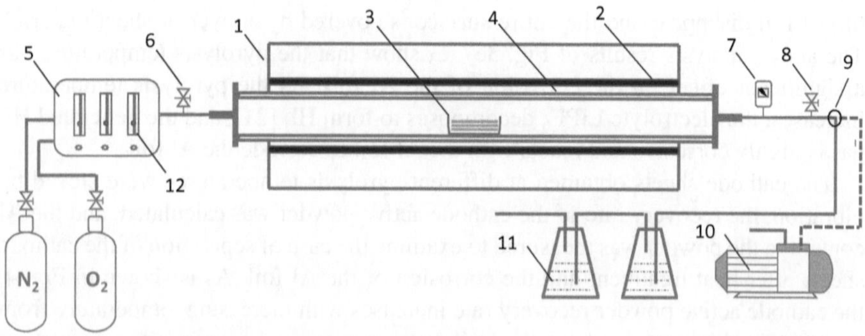

Fig. 2 Connection diagram of cathode scraps heat treatment equipment (1. small vacuum tube furnace; 2. heating rod; 3. corundum ark; 4. corundum tube; 5. mixing box; 6. intake valve; 7. vacuum gauge; 8. outlet valve; 9. two-way air conduction valve; 10. vacuum pump; 11. NaOH solution; 12. float flowmeter)

Results and Discussion

Separation of Cathode Active Powder and Al Foil

Effect of Pyrolysis Temperature

Under the pyrolysis vacuum degree of 1000 Pa and pyrolysis time of 30 min, the different pyrolysis temperatures(450–700 °C) for the separation of cathode active powder and Al foil were investigated.

The macroscopic morphology and SEM images of the cathode sheets and Al foils obtained at different pyrolysis temperatures are shown in Fig. 3. It can be seen from Fig. 3 that when the temperature rises to 500 °C, the cathode material on the cathode sheets begins to fall off. As the temperature further rises to 600 °C, the cathode sheets separation becomes easier, which indicates that the rise of temperature is favorable for removal of the binder in the cathode sheets. However, when the temperature rises to 650 °C, ablation at the current collector boundary occurs because the pyrolysis temperature is close to the melting temperature of the Al foil. The Al foil becomes brittle, which can be a barrier to the separation of the cathode active powder from the Al foil. When the temperature up to 700 °C, the cathode active powder and the Al foil are co-melted due to the melting of the Al foil, and the separation of the cathode active powder from the Al foil cannot be achieved. At the same time, the SEM was used to observe the morphology of the Al foil obtained at different pyrolysis temperatures. The results demonstrated that the macroscopic appearance of the Al foil looks flat, but the high-resolution SEM image shows that there are a large number of corrosion holes on the surface, and severe corrosion as the pyrolysis temperature increases. In addition, Fig. 3e_2 shows a large number of non-separation cathode material adhered to the Al foil, Fig. 3e_3 shows that the corrosion holes in the Al foil disappear and the entire surface is covered by a layer of sheet material. The above analysis results of Fig. 3e$_2$, e$_3$ show that the pyrolysis temperature has a significant effect on the corrosion of the Al foil. As the pyrolysis temperature increases, the electrolyte $LiPF_6$ decomposes to form HF [21], and the generated HF gas is highly corrosive and reacts with the Al foil to corrode the Al foil.

The cathode sheets obtained at different pyrolysis temperatures were sieved by vibration, the recovery rate of the cathode active powder was calculated, and the Al content in the powder was measured to examine the ease of separation of the cathode sheets after heat treatment and the corrosion of the Al foil. As is shown in Fig. 4, the cathode active powder recovery rate increases with increasing temperature from to 600 °C, which means the cathode sheets are more and more easily stripped off as the temperature increases. Subsequently, the recovery rate of the cathode active powder rapidly decreased from 98.04% at 600 °C to 67.24% when the temperature was increased to 650 °C, which indicates that the temperature was too high. The surface of the cathode sheets was ablated leading to the difficulty of the stripping

of the cathode sheets. Meanwhile, the Al content in the cathode active powder increased from 0.06% at 600 °C to 1.28% at 650 °C, indicating that the Al foil is heavily corroded, and resulting in a large amount of elemental Al entering the cathode active powder. The above results further confirm the conclusions drawn from Fig. 2. Therefore, the optimum pyrolysis temperature is 600 °C.

Effect of the Vacuum Degree

Under the pyrolysis temperature of 600 °C and pyrolysis time of 30 min, the effects of vacuum degree (100–2000 Pa) for the separation of cathode active powder and Al foil were investigated.

The macroscopic morphology of the cathode sheets and Al foil obtained under different vacuum degrees and the SEM images of the Al foil are shown in Fig. 5. It is known from Fig. 4a$_{1-2}$ that the black cathode active powder adheres to the aluminum foil. As can be seen from Fig. 5b$_{1-2}$, the cathode sheet is ablated in a large area, which is not conducive to the separation of the cathode active powder from the Al foil at 100 and 500 Pa. As can be seen from Fig. 5a, b, the black cathode active materials have not separated from Al foil. When the vacuum degree drops to 1000 Pa, the cathode sheet ablation phenomenon disappears and the surface of Al foil is clean after vibration screen. Then when the vacuum degree continues to be reduced to 2000 Pa, a small amount of cathode active material remains on the Al foil after screening and stripping. Above results indicate that excessively high and low vacuum degree is not favorable for the decomposition of the binder on the cathode sheet, which weakens the stripping effect of the cathode sheet. In addition, the SEM images (Fig. 5a$_3$–d$_3$) show that the chemical corrosion of the Al foil surface aggravated with the increase of the vacuum degree. As the vacuum degree decreases, the electrolyte decomposes more severely, releasing more HF, which intensifies the corrosion of aluminum foil.

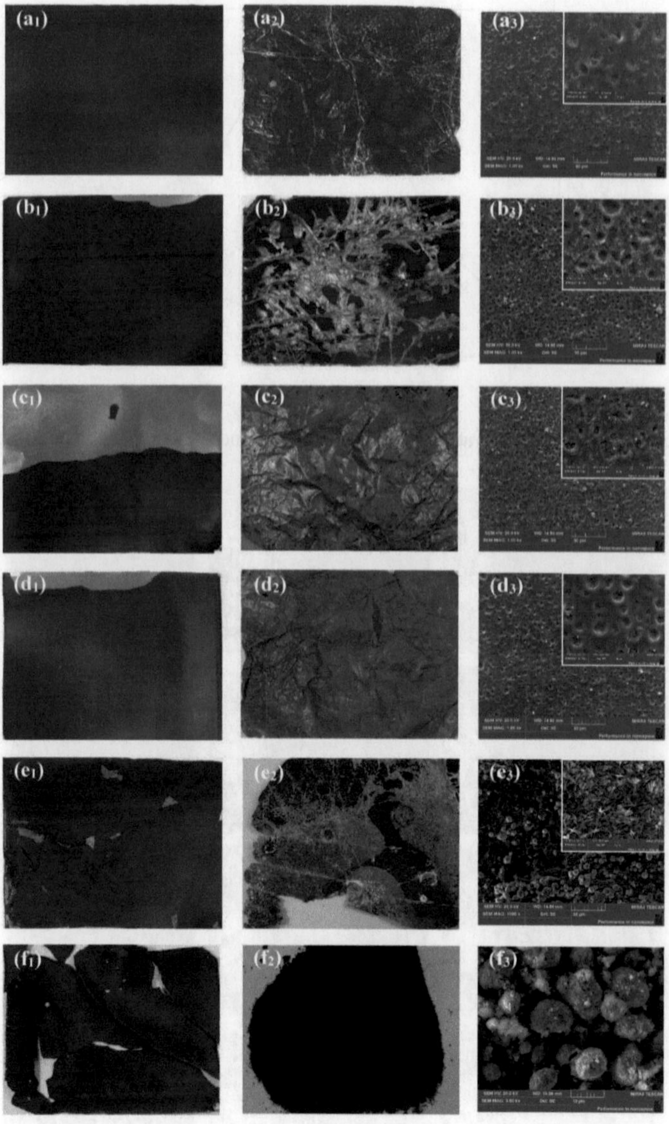

Fig. 3 Macroscopic and SEM micromorphology of cathode sheets and Al foils at different pyrolysis temperatures: 450 °C (**a**$_{1–3}$), 500 °C (**b**$_{1–3}$), 550 °C (**c**$_{1–3}$), 600 °C (**d**$_{1–3}$), 650 °C (**e**$_{1–3}$), 700 °C (**f**$_{1–3}$)

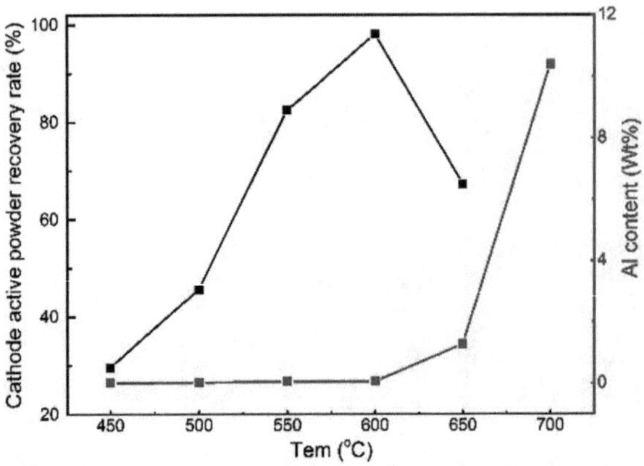

Fig. 4 Effect of pyrolysis temperature on the recovery rate and Al content of cathode active powder

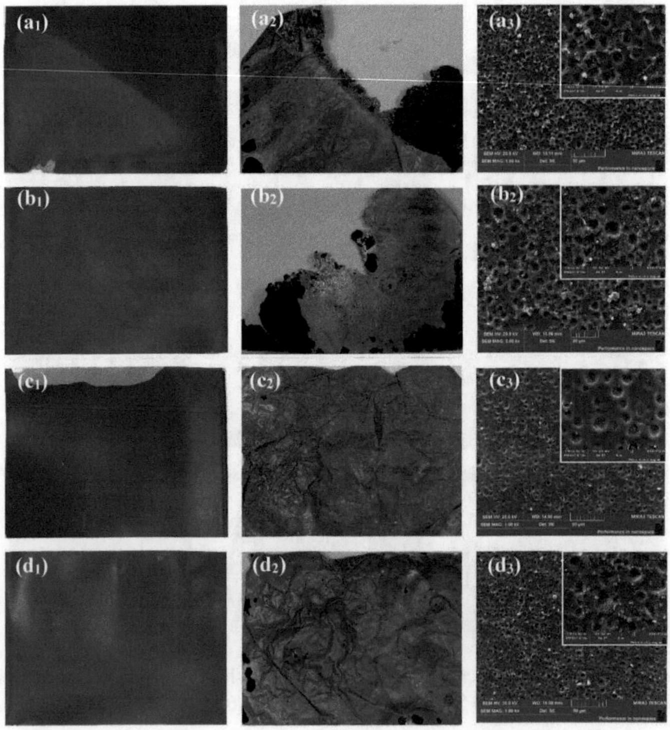

Fig. 5 Macroscopic morphology and SEM micromorphology of cathode sheets and Al foils under different vacuum conditions: 100 Pa ($\mathbf{a}_{1-3}$), 500 Pa ($\mathbf{b}_{1-3}$), 1000 Pa ($\mathbf{c}_{1-3}$), 2000 Pa ($\mathbf{d}_{1-3}$)

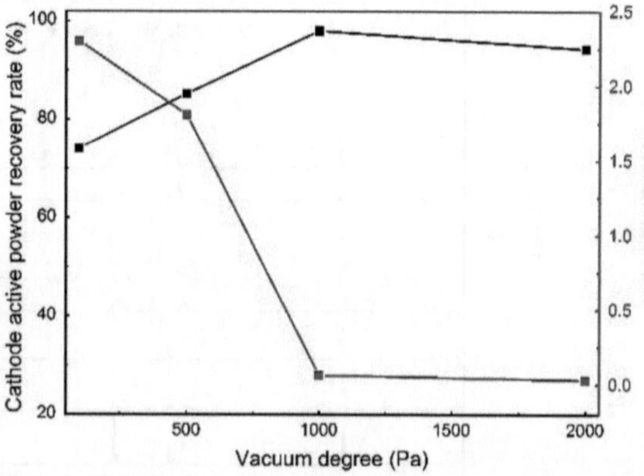

Fig. 6 Effect of vacuum degree on the recovery rate and Al content of cathode active powder

The cathode active powder recovery rate and Al content under different vacuum degrees are shown in Fig. 6. As can be seen from Fig. 6, as the vacuum degree is reduced from 100 to 1000 Pa, the recovery rate gradually increases, which indicates that a high vacuum degree does not favor the stripping of the cathode active powder. When the vacuum degree continues to decrease to 2000 Pa, the recovery rate decreases, which is caused by the low vacuum degree being unfavorable to the decomposition of the binder in the cathode sheets. The above results are consistent with the analysis results in Fig. 6. Therefore, a comprehensive selection of 1000 Pa is the optimal vacuum for the experiment.

Phase Transition of Valuable Metals in Cathode Active Powder

Effect of Pyrolysis Temperature

Under the pyrolysis vacuum degree of 1000 Pa and pyrolysis time of 30 min, the different pyrolysis temperatures(450–700 °C) for the phase transition of valuable metals in cathode active powder were investigated.

The results of XRD and SEM analyses on the cathode active powder and the untreated powder obtained at different pyrolysis temperatures are shown in Figs. 7 and 8. As can be seen from Fig. 7, the untreated cathode material phase is $LiNi_xCo_yMn_{1\text{-}x\text{-}y}O_2$. In the temperature range of 450–600 °C, the diffraction peaks of $LiNi_xCo_yMn_{1\text{-}x\text{-}y}O_2$ become weaker and weaker with the increase of temperature and the diffraction peaks of NiO and other phases appear at the same time. The diffraction peak of $Li[Ni_xCo_yMn_{1\text{-}x\text{-}y}]O_2$ disappears at 650 °C, but the diffraction peaks of NiO and other phases increase, which indicates the pyrolysis temperature has a significant effect on the phase transition of the valuable metal. As the temperature increases, the valuable metal in the cathode active powder is gradually reduced by a carbonaceous material such as acetylene black and binder. When the temperature continues to rise

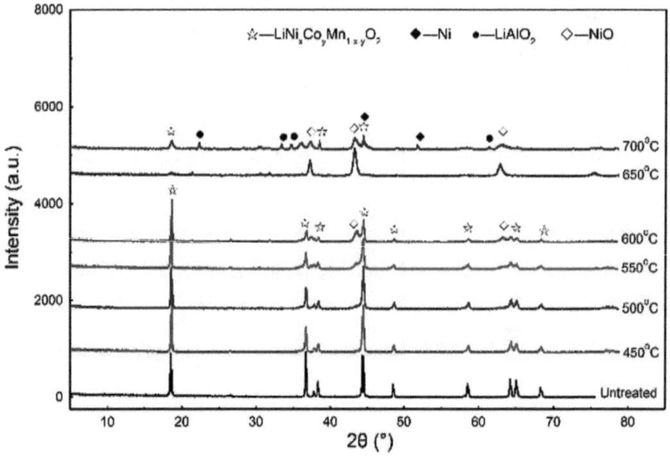

Fig. 7 XRD patterns of cathode active powder obtained from different pyrolysis temperatures

to 700 °C, the intensity of the diffraction peaks decreases, and the diffraction peaks of $LiNi_xCo_yMn_{1-x-y}O_2$, $LiAlO_2$, and Ni appear simultaneously. Because the pyrolysis temperature is too high, causing the Al foil to dissolve. The dissolved substances react with lithium in the cathode active powder to form $LiAlO_2$ adhering to the surface of the material, which affects the reduction of the valuable metal. The decomposition mechanism of the positive active material is complicated, and the decomposition is incomplete at 700 °C, so the diffraction peaks of $LiNi_xCo_yMn_{1-x-y}O_2$ are So, the diffraction peaks of the ternary appear again

Figure 8 shows the SEM images of the untreated raw material and the obtained cathode active powder at different pyrolysis temperatures. As can be seen from Fig. 8a, the powder crystal is agglomerated severely due to the presence of the binder, and a large number of small crystal grains are dispersed around the agglomerated crystal. Subsequently, the cathode sheets were subjected to vacuum pyrolysis at 450 °C. The obtained cathode active powder crystal was uniformly dispersed, but a large amount of floc organic matter was still present on the surface. When the temperature rises to 600 °C or more, the floc on the surface of the cathode active powder is significantly reduced, and the crystal surface is denser than that of 450 °C, which indicates that the increase of pyrolysis temperature is beneficial to the decomposition of the binder.

Finally, the backscattered and EDS scans of the pyrolysis products at of 700 °C were performed individually. The results are shown in Fig. 9. As can be seen from Fig. 9, two different brightness crystal forms exist in the SEM backscattering pattern. The results of the EDS surface scan show that nickel, cobalt, and manganese are distributed in the bright region of the backscattering image, and aluminum is distributed in another dark region. This indicates that the phase represented by the bright region in the backscattering pattern is $LiNi_xCo_yMn_{1-x-y}O_2$, the phase represented by the

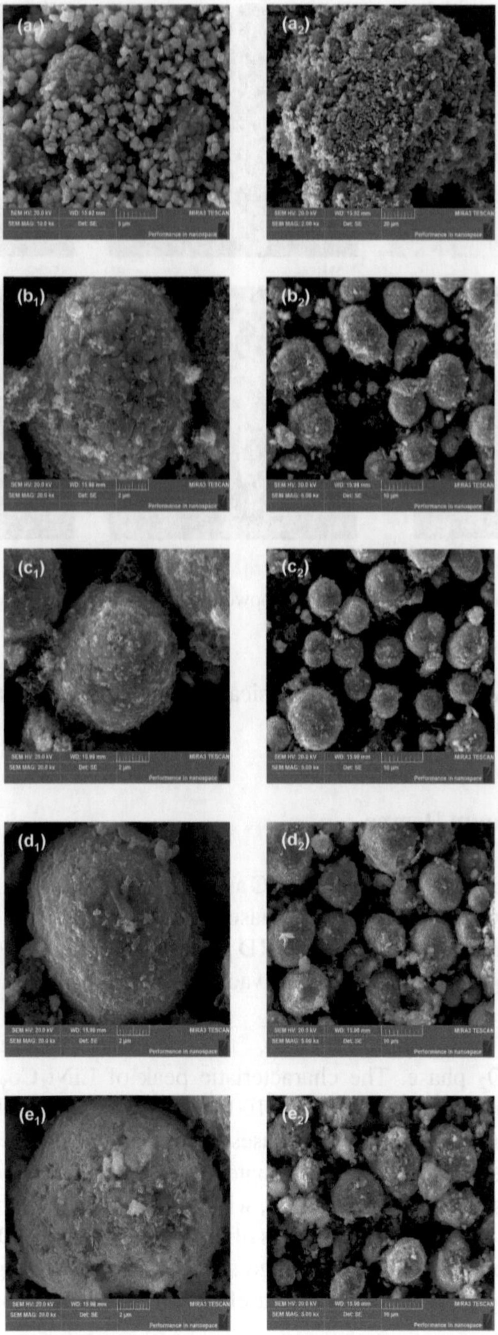

Fig. 8 SEM images of cathode active powder obtained from different pyrolysis temperatures: raw materials ($\mathbf{a}_{1\text{-}2}$), 450 °C ($\mathbf{b}_{1\text{–}2}$), 600 °C ($\mathbf{c}_{1\text{–}2}$), 650 °C ($\mathbf{d}_{1\text{–}2}$), 700 °C ($\mathbf{e}_{1\text{–}2}$)

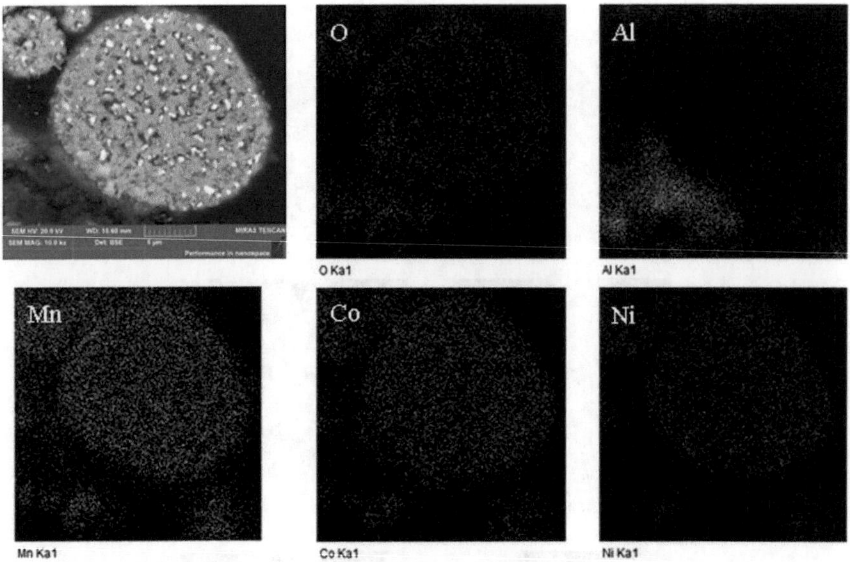

Fig. 9 SEM-EDS spectrum of cathode active powder obtained by vacuum pyrolysis at 700 °C

dark region is $LiAlO_2$, and $LiAlO_2$ is basically adhered to the surface of the cathode electrode material.

Effect of the Vacuum Degree

Under the pyrolysis temperature of 600 °C and pyrolysis time of 30 min, the effects of vacuum degree (100–2000 Pa) for the phase transition of valuable metals in cathode active powder were investigated. The XRD patterns and SEM images of the cathode active powders obtained under different vacuum degrees and the untreated powders are shown in Figs. 10 and 11.

As can be seen from Fig. 10, the untreated material contains the $LiNi_xCo_yMn_{1\text{-}x\text{-}y}O_2$ phase. The characteristic peak of $LiNi_xCo_yMn_{1\text{-}x\text{-}y}O_2$ in the cathode active powder is weakened at 100 Pa, and at the same time, characteristic peaks of intermediate reduction phases such as NiO appeared, which may be caused by the reduction of the cathode material by the carbonaceous organic matter during the pyrolysis process. In addition, with the continuous change of the vacuum degree, the characteristic peak intensities of the $LiNi_xCo_yMn_{1\text{-}x\text{-}y}O_2$ and NiO phases in the pyrolysis products did not change significantly, indicating that the change of the vacuum degree had no significant effect on the phase transformation of the pyrolysis products.

The SEM images of the cathode active powder under different vacuum degrees are shown in Fig. 11. Figure 11 shows that the grain morphology of 100 and 1000 Pa

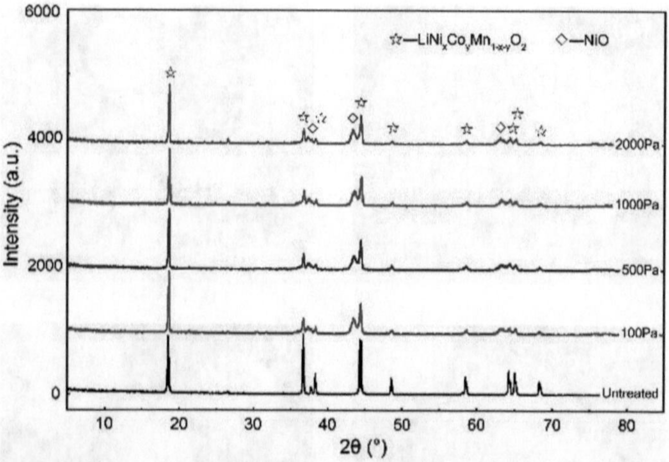

Fig. 10 XRD patterns of cathode active powder obtained from different vacuum degrees

is similar, and the spherical shape of the ternary cathode material is maintained. The floc is adhered to the surface of the particles, indicating that the binder, acetylene black and other organic substances are not all removed. The grain morphology of the products at 2000 Pa is slightly different but the spherical morphology of the ternary material is maintained.

Conclusion

(1) Under the optimized conditions of pyrolysis temperature of 600 °C and a vacuum degree of 1000 Pa, the binder was decomposed in large quantities to achieve efficient separation of the active powder and the Al foil current collector, and the cathode active powder recovery rate reached 98.04%.
(2) Pyrolysis temperature has a significant effect on the separation of cathode sheets and phase transition of valuable elements in cathode active material. In the temperature range of 450–650 °C, as the pyrolysis temperature increases, the characteristic peak of $LiNi_xCo_yMn_{1\text{-}x\text{-}y}O_2$ in the cathode active powder gradually weakens and disappears, and the intermediate NiO phase appears. When the pyrolysis temperature reaches at 700 °C, the $LiAlO_2$ phase is formed to coat the cathode electrode active material surface. At the same time, the results of SEM images show that with the pyrolysis temperature increases, the floccules on the surface of the cathode material are less and less, and the binder is removed more and more thoroughly.
(3) Vacuum degree has a significant effect on the separation of cathode sheets, but the effect on the phase transition of valuable elements in cathode active material is not obvious. Excessively high and low vacuum degrees are not favorable

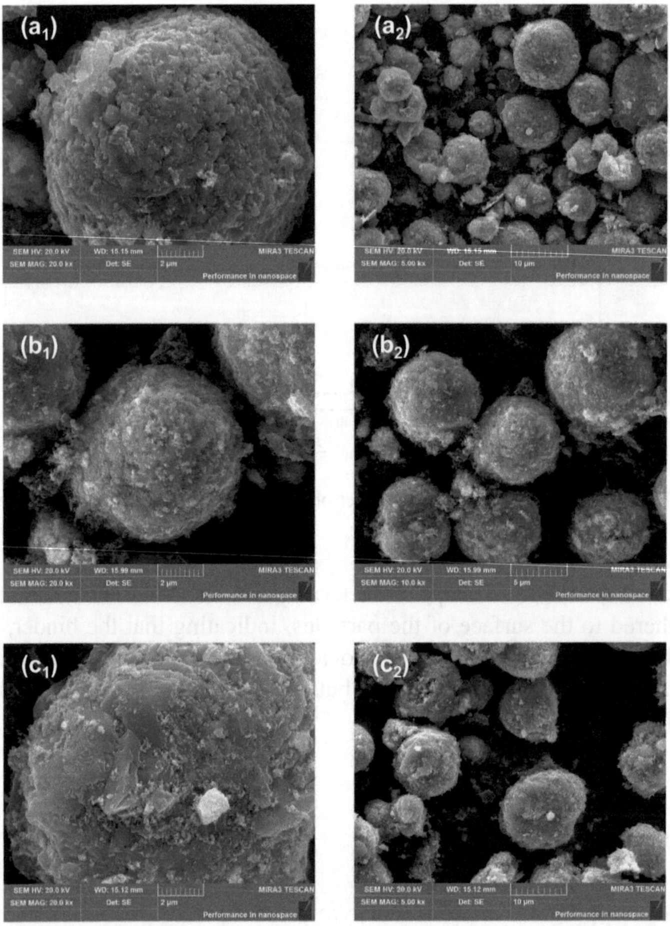

Fig. 11 SEM images of cathode active powder obtained by different vacuum degrees: 100 Pa (**a**$_{1-2}$), 1000 Pa (**b**$_{1-2}$), 2000 Pa (**c**$_{1-2}$)

for the decomposition of the binder on the cathode sheet. With the continuous change of the vacuum degree, the characteristic peak intensities of the $LiNi_xCo_yMn_{1-x-y}O_2$ and NiO phases in the pyrolysis products did not change significantly.

Acknowledgements We are grateful to Anhui Province Research and Development Innovation Project for Automotive Power Battery Efficient Recycling System and the Research Fund Program of State Key Laboratory of Rare Metals Separation and Comprehensive Utilization (No. GK-201806) for providing financial support.

References

1. Sinolink Securities. Deep research report on power lithium battery recycling industry [EB/OL]. Shanghai, 01 July 2017. http://www.sohu.com/a/153650710_355061
2. Chen Y, Liu N, Hu F, Ye L, Xi Y, Yang S (2018) Thermal treatment and ammoniacal leaching for the recovery of valuable metals from spent lithium-ion batteries. Waste Manag 75:469
3. Chagnes A, Pospiech B (2013) A brief review on hydrometallurgical technologies for recycling spent lithium-ion batteries. J Chem Technol Biotechnol 88(7):1191–1199
4. Meshram P, Pandey BD, Mankhand TR (2015) Recovery of valuable metals from cathodic active material of spent lithium ion batteries: leaching and kinetic aspects. Waste Manag 45:306–313
5. Li L, Dunn JB, Zhang XX, Gaines L, Chen RJ, Wu F et al (2013) Recovery of metals from spent lithium-ion batteries with organic acids as leaching reagents and environmental assessment. J Power Sources 233(233):180–189
6. Jha MK, Kumari A, Jha AK, Kumar V, Hait J, Pandey BD (2013) Recovery of lithium and cobalt from waste lithium ion batteries of mobile phone. Waste Manag 33(9):1890–1897
7. He J, Liu J, Li J, Lai Y, Wu X (2016) Enhanced ionic conductivity and electrochemical capacity of lithium ion battery based on PVDF-HFP/HDPE membrane. Mater Lett 170:126–129
8. Zhu SG, Wen-Zhi HE, Guang-Ming LI et al (2012) Recovery of Co and Li from spent lithium-ion batteries by combination method of acid leaching and chemical precipitation. Trans Nonferrous Metals Soc China 22(9):2274–2281
9. Li J, Zhao R, He X, Liu H (2009) Preparation of licoo 2, cathode materials from spent lithium–ion batteries. Ionics 15(1):111–113
10. Chen Y, Tian Q, Chen B, Shi X, Liao T (2011) Preparation of lithium carbonate from spodumene by a sodium carbonate autoclave process. Hydrometallurgy 109(1):43–46
11. Weng Y, Xu S, Huang G, Jiang C (2013) Synthesis and performance of $Li[(Ni_{1/3}Co_{1/3}Mn_{1/3})_{1-x}Mg_x]O_2$ prepared from spent lithium ion batteries. J Hazard Mater 246–247(4):163–172
12. Shin SM, Kim NH, Sohn JS, Yang DH, Kim YH (2005) Development of a metal recovery process from Li-ion battery wastes. Hydrometallurgy 79(3):172–181
13. Shuva MAH, Kurny A (2013) Hydrometallurgical recovery of value metals from spent lithium ion batteries. Am J Mater Eng Technol 1(1):8–12
14. Hanisch C, Haselrieder W, Kwade A (2011). Recovery of active materials from spent lithium-ion electrodes and electrode production rejects 85–89
15. Ferreira DA, Prados LMZ, Majuste D, Mansur MB (2009) Hydrometallurgical separation of aluminium, cobalt, copper and lithium from spent Li-ion batteries. J Power Sources 187(1):238–246
16. Paulino JF, Busnardo NG, Afonso JC (2008) Recovery of valuable elements from spent Li-batteries. J Hazard Mater 150(3):843–849
17. Zeng X, Li J (2014) Innovative application of ionic liquid to separate al and cathode materials from spent high-power lithium-ion batteries. J Hazard Mater 271(271):50–56
18. Contestabile M, Panero S, Scrosati B (2001) A laboratory-scale lithium-ion battery recycling process. J Power Sources 92(1):65–69
19. Chen L, Tang X, Zhang Y, Li L, Zeng Z, Zhang Y (2011) Process for the recovery of cobalt oxalate from spent lithium-ion batteries. Hydrometallurgy 108(1):80–86
20. Sun L, Qiu K (2011) Vacuum pyrolysis and hydrometallurgical process for the recovery of valuable metals from spent lithium-ion batteries. J Hazard Mater 194(11):378–384
21. Ravdel B, Abraham KM, Gitzendanner R, Dicarlo J, Lucht B, Campion C (2003) Thermal stability of lithium-ion battery electrolytes. J Power Sources 119:805–810

Waste Tire Rubber Powders Based Composite Materials

Carlos F. Revelo, Mauricio Correa, Claudio Aguilar and Henry A. Colorado

Abstract This investigation shows results for use of waste tire rubber powdered materials fabricated in composites by using polyurethane resin as a binder material. With this powders, processed at industrial scale for a recycling local company, rubber-based tiles were produced for several applications. The use of these materials is a solution that gives a valorization to the tires after use in many countries giving relief for an increasing problem worldwide with the tire use. Several concentrations and particle size distributions were investigated and tested. Tension and density tests were conducted in order to evaluate the flexible tiles for diverse applications. Scanning electron microscopy, Fourier-transform infrared spectroscopy, thermogravimetric analysis, and dynamic mechanical analyzer techniques were used to evaluate the microstructure of the samples. Weibull distribution analysis has been also included in order to analyze the variability of composite samples and thus characterize the manufacturing process at the industrial scale.

Keywords Waste tire rubber · Composites · Polyurethane · Mechanical characterization · Rubber tiles

C. F. Revelo · H. A. Colorado (✉)
CCComposites Laboratory, Universidad de Antioquia UdeA, Calle 70 No. 52-21, Medellín, Colombia
e-mail: henry.colorado@udea.edu.co

M. Correa
Grupo de Investigación En Ingeniería Y Gestión Ambiental, Universidad de Antioquia UdeA, Calle 70 No. 52-21, Medellín, Colombia

C. Aguilar
Metallurgical and Materials Department, Universidad Técnica Federico Santa María, Valparaíso, Chile

H. A. Colorado
Facultad de Ingenieria, Universidad de Antioquia, Bloque 20, Calle 67 No. 53-108, Medellín, Colombia

G. Gaustad et al. (eds.), *REWAS 2019*, The Minerals, Metals & Materials Series,
https://doi.org/10.1007/978-3-030-10386-6_50

Introduction

Tires are a complex part with several very different materials inside difficult to separate and energy consuming process has to be involved in order to recycle them. Because of the scale of the problem, many solutions and processes have to be developed in order to use a large amount of generated tires after the life cycle. This problem is even worst in developing countries were a combination of weak environmental policies, soft regulation enforcement, and lack of proper technology work together to worsen the situation [1].

Every year, around 1.4 billion of new tires are sold in the world, which in a short period of time will be as waste [2]. About 4 billion of waste tires are being uselessly stockpiled in landfills worldwide [3], which are associated now with health problems, and to adverse environmental and economic effects, associated with the contamination [3–5], which represent big challenges in engineering to provide the appropriate solutions via new products and waste minimization strategies driven by a green manufacturing. Among the applications for these wastes, gym rubber flooring [6], athletic rubber tracks [6], and asphalt-rubber pavement [7], concrete [8], mortars [9], combinations rubber-thermoplastics [10], and composites [11]. Piszczyk et al. [12] modified polyurethane foams with recycled rubber, for which the thermos-mechanical behavior has shown competitive results improving the density, thermal stability, and compressive strength.

The current investigation has been conducted over the development of flexible floors by using tire rubber waste from the tires sold in Colombia in combination with a flexible and inexpensive resin. These inexpensive and unutilized materials, such as rubber from waste tires in combination with resins, can function as additives [13, 14] and participate in the reactions to form new materials. The composite was fabricated using a polyurethane resin as a binder of the ground rubber particles in a hot press machine. The presented process is not only positive for the environment, but also good for a local Colombian company who is commercializing these materials.

Experimental

Waste tire rubber in four different granulometries (4 mm, 2 mm, No 20 mesh (0.841 mm), No 30 mesh (0.595 mm)) was supplied by Prismacaucho S.A.S from Colombia. Rubber was ground first processed in a line process containing a grinder for the particulate material. A primary and secondary grinding conducted to tailor different granulometry. The rubber grinding process is shown in Fig. 1. The polyurethane resin (pre-polymer) used does not require a catalyst to favor the bonding process of the rubber particles.

The preparation of the mixtures was conducted mechanically, with resin mixtures kept in a low proportion with respect to the rubber in order to increase maximize the

Fig. 1 Rubber grinding process. **a** Car tire waste, **b** primary and secondary grinding, **c** industrial linear vibrating sieve shaker, and **d** 30 mesh ground rubber

Table 1 Sample formulation fabricated using different waste contents and particle sizes

Ref.	Sample name	Formulation
S1	W20—2.5 mm	C/R 20—2.5 mm
S2	W10—2.5 mm	C/R 10—2.5 mm
S3	W20—4.0 mm	C/R 20—4 mm
S4	W10—4 mm	C/R 10—4 mm
S5	W20—No. 20	C/R 20—No. 20
S6	W10—No. 20	C/R 10—No. 20
S7	W20—No. 30	C/R 20—No. 30
S8	W10—No. 30	C/R 10—No. 30

rubber used, and to reduce the amount of polyurethane involved in the process. The formulations are shown in Table 1.

The mixtures were mechanically processed by making rubber sheets with a hot press machine, with a uniform material and temperature distribution on the surface, see Fig. 2. The main process parameters were the temperature for the upper and lower plates was 150 °C and 130 °C, respectively. The applied pressure was 1500 psi. Once the rubber-resin sheets were obtained, test pieces were cut with a die-cutter for tensile tests according to ASTM D638. Five test pieces were die-cut for each of

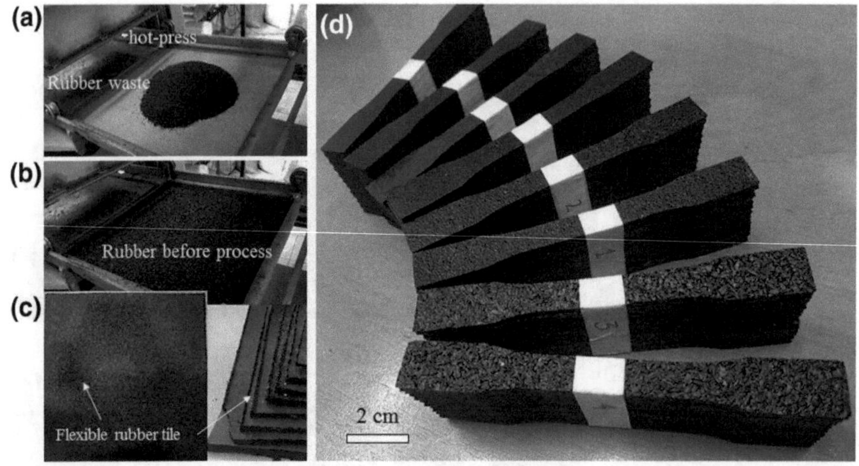

Fig. 2 Flexible tile fabrication, **a** hot press and rubber waste batch, **b** rubber particles with the binder on the mold, **c** flexible tiles after the processing, **d** samples ready for the tensile tests

the 8 different formulations mentioned, for a total of 40 specimens (see Fig. 1d), all thereafter subjected to tensile tests.

From the rubber sheets designed, cuts of square pieces of approximately 3 cm side were made for density tests in accordance with ASTM D792. Density tests were conducted for all samples based on the weight, dimensions and the buoyancy. For these tests, a high precision Mettler Toledo balance that followed Archimedes' principle was used [15]. In these calculations, the density of water was taken to be 1.0 g/cm^3. A set of 5 samples were tested for each composition. The two methods were used because in the traditional industry both are frequently used with the problem that they can show very different results.

Tensile tests (stress and deformation) were performed for the die-cut samples using a Shimadzu AG250KN universal testing machine at a heat speed of 50 mm/min. For this analysis, 5 samples for each formulation were tested. At the end of this test, a photographic record of the fractures of each test tube was made as shown in Fig. 6.

The microstructure of samples and raw materials was also investigated using scanning electron microscopy (SEM). Samples were also analyzed with energy dispersed spectroscopy (EDS). All samples after the drying process were mounted on a carbon tape and gold sputtered with a Hummer 6.2 system, at conditions of 15 mA AC for 30 s, in order to make a thin film of Au of about 2 nm. The SEM used was a JEOL JSM 6700R in a high vacuum mode.

Analysis and Results

Density results obtained by the measurement of dimensions-weight (dimensional method) and also by the Archimedes balance (Archimedes method) are presented in Fig. 3. In the dimensional method (continuous line), the density of all the specimens had a similar behavior obtaining values close to 1 g/cm^3. It is known that this macro method idealizes the material as a material free of irregularities, taking the dimensions of the piece of the maximum volume it reached in the process. On the other hand, the Archimedes method (dotted line) shows an increase in density for all formulations compared to the dimensional method, giving values close to 3 g/cm^3. The formulation with the highest density value was M4 (4.89 g/cm^3). Statistically, this formulation does not present the same trend as the other samples.

Figure 4 shows the results of the tensile strength. The M3 formulation is the one with the highest stress (1.70 MPa). Other formulations with higher values of stress were M1 and M2, with values of 1.49 MPa and 0.70 MPa respectively. These formations are characterized by having a larger grain size in their formulation, which contributes to develop greater tensile stress. On the contrary, as the grain size decreases, the stress tends to decrease as it was observed in the other formulations (M4, M5 M6, M7, and M8). The maximum values for the tension stress were summarized in Fig. 5a.

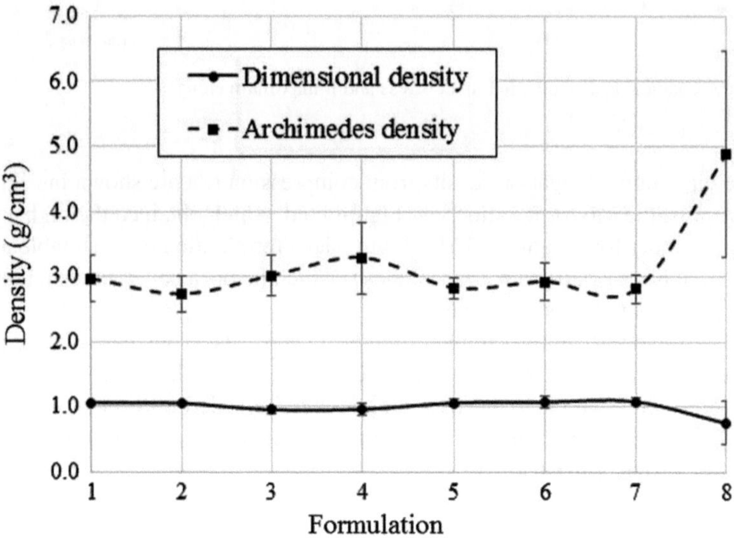

Fig. 3 Density results

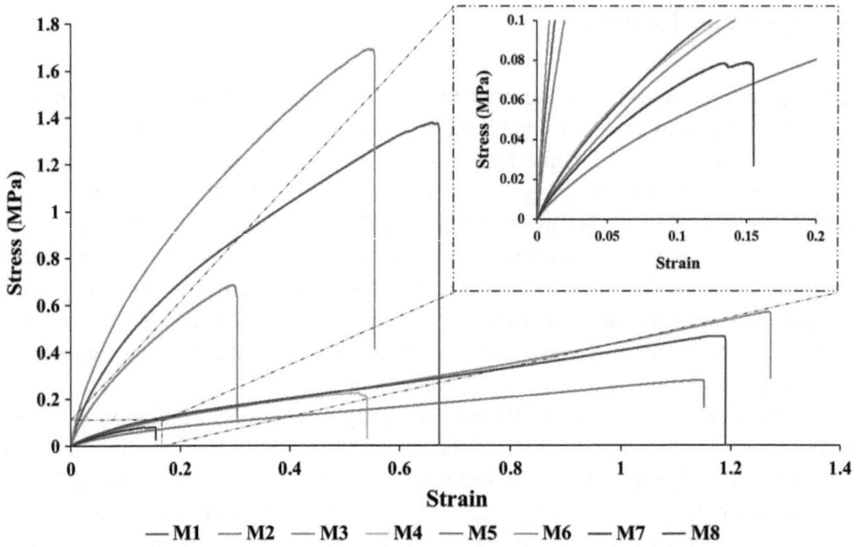

Fig. 4 Stress-strain curve

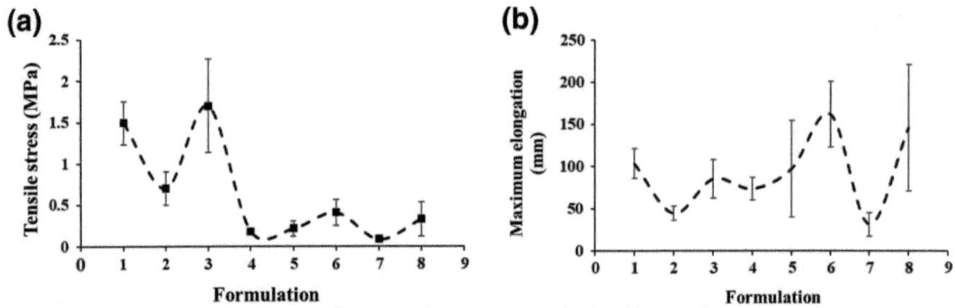

Fig. 5 Mechanical test results. **a** Tensile stress and **b** maximum elongation

The maximum elongation results from compression test are shown in Fig. 5b. In these results, the M6 formulation was highlighted, which obtained the highest value among the other formulations (161.82 mm). This formulation has a suitable amount of rubber, it has a small grain size (20 mesh) which favors a better adhesion between rubber particles to the polyurethane resin.

Figure 6 summarizes the samples after the tensile tests. From these images is clear that formulations M7 and M8 have multiple fractures, which has been associated with the material combination, as these formulations contained a low level of polyurethane resin. Thus, the rubber composite did not have a good amount of binder and thus leading in a weaker material.

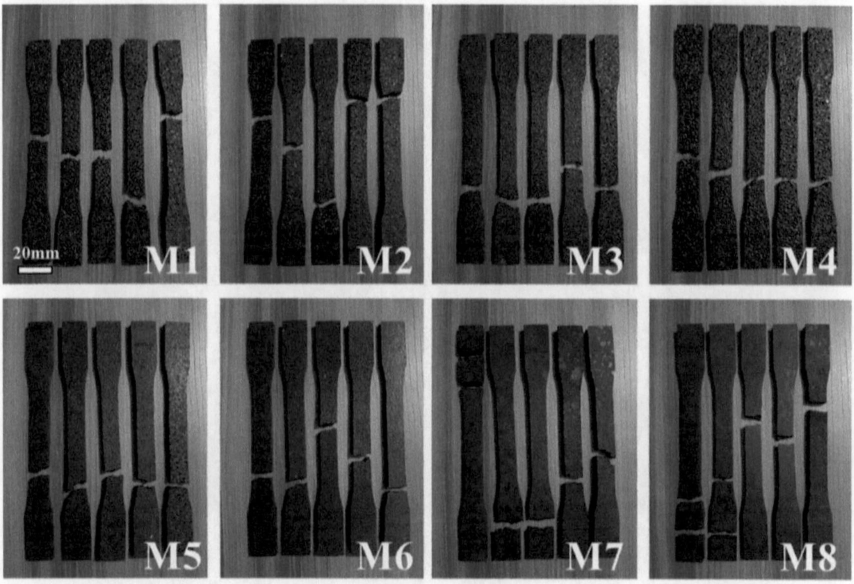

Fig. 6 Fracture of specimens after tensile stress

In general, all the SEM images did not reveal macro defects, therefore, confirming an adequate process. Homogeneous surfaces are observed with the presence of some white spots, characteristic of the polyurethane resin and some impurities, and a minimum number of pores on the surfaces. It is highlighted the M7 formulation shows the presence of some irregular shapes agglomerated in the surface. The good adhesion and impregnation of the resin to the rubber is reflected in the homogeneity of the surface as well (Fig. 7).

Conclusion

The experiments conducted in this research proved a very scalable and simple process able to use a large amount of waste tire rubber. The flexible properties of the composite enable this to be used in multiple application where impact loads are common, such a floor tiles used in gyms or kinder gardens, very versatile rubber-based tiles easy to install and tailor its mechanical behavior by using high performance resins [16, 17] with fibers [18, 19] or particles [17]. Nanocomposites are also a promising field to explore with this composites particularly as structural material [20, 21]. From the University–Industry collaboration, it was really positive the experience as the use of combined facilities to develop products benefit the production sector in the country.

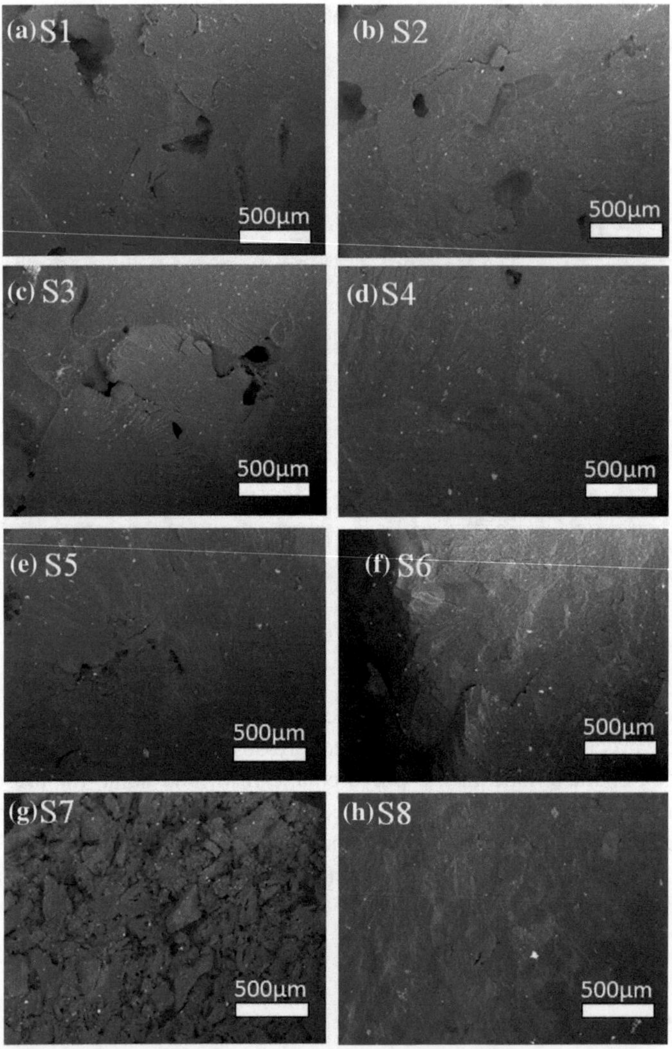

Fig. 7 SEM images for all rubber-resin formulations

Acknowledgements The authors wish to thank the engineer Juan C. Salazar from Prismacaucho S.A.S. for his support and useful suggestions in this investigation.

References

1. Yilmaz A, Degirmenci N (2009) Possibility of using waste tire rubber and fly ash with Portland cement as construction materials. Waste Manag 29(5):1541–1546
2. Uruburu Á, Ponce-Cueto E, Cobo-Benita JR, Ordieres-Meré J (2013) The new challenges of end-of-life tyres management systems: a Spanish case study. Waste Manag 33(3):679–688
3. Sienkiewicz M, Kucinska-Lipka J, Janik H, Balas A (2012) Progress in used tyres management in the European Union: a review. Waste Manag 32(10):1742–1751
4. Dehghani M et al (2017) The effects of air pollutants on the mortality rate of lung cancer and leukemia 15
5. Leung DYC, Wang CL (1998) Kinetic study of scrap tyre pyrolysis and combustion. J Anal Appl Pyrolysis 45(2):153–169
6. Di Mundo R, Petrella A, Notarnicola M (2018) Surface and bulk hydrophobic cement composites by tyre rubber addition. Constr Build Mater 172:176–184
7. Lo Presti D (2013) Recycled tyre rubber modified bitumens for road asphalt mixtures: a literature review. Constr Build Mater 49:863–881
8. Thomas BS, Gupta RC, John Panicker V (2015) Experimental and modelling studies on high strength concrete containing waste tire rubber. Sustain Cities Soc 19:68–73
9. Popovici A et al (2015) Modern mortars with electronic waste scraps (glass and plastic). Mater Plast 52(4):588–592
10. Zhang SL, Zhang ZX, Pal K, Xin ZX, Suh J, Kim JK (2010) Prediction of mechanical properties of waste polypropylene/waste ground rubber tire powder blends using artificial neural networks. Mater Des 31(8):3624–3629
11. Xu M, Li J (2012) Effect of adding rubber powder to poplar particles on composite properties. Bioresour Technol 118:56–60
12. Piszczyk Ł, Hejna A, Formela K, Danowska M, Strankowski M (2015) Effect of ground tire rubber on structural, mechanical and thermal properties of flexible polyurethane foams. Iran Polym J 24(1):75–84
13. Colorado HA, Singh D (2014) High-sodium waste streams stabilized with inorganic acid–base phosphate ceramics fabricated at room temperature. Ceram Int, Part B 40(7):10621–10631
14. Colorado HA, Colorado SA (2016) Portland cement with battery waste contents. In: REWAS 2016. Springer, pp 57–63
15. Teh E-J, Leong YK, Liu Y, Fourie AB, Fahey M (2009) Differences in the rheology and surface chemistry of kaolin clay slurries: the source of the variations. Chem Eng Sci 64(17):3817–3825
16. Zhang X et al (2014) Vinyl ester resin: rheological behaviors, curing kinetics, thermomechanical, and tensile properties. AIChE J 60(1):266–274
17. Colorado HA, Yuan W, Guo Z, Juanri J, Yang J-M (2014) Poly-dicyclopentadiene-wollastonite composites toward structural applications. J Compos Mater 48(16):2023–2031
18. Neves Monteiro S et al (2018) Fique fabric: a promising reinforcement for polymer composites. Polymers (Basel) 10(3):246
19. Teles MCA, Altoé GR, Amoy Netto P, Colorado H, Margem FM, Monteiro SN (2015) Fique fiber tensile elastic modulus dependence with diameter using the Weibull statistical analysis. Mater Res 18:193–199
20. Colorado HA, Nino JC, Restrepo O (2018) Applications and opportunities of nanomaterials in construction and infrastructure. In: TMS annual meeting & exhibition, pp 437–452
21. Wang Z et al (2012) Effective functionalization of carbon nanotubes for bisphenol F epoxy matrix composites. Mater Res 15(4):510–516

Author Index

G. Gaustad et al. (eds.), *REWAS 2019*, The Minerals, Metals & Materials Series,
https://doi.org/10.1007/978-3-030-10386-6

Subject Index

G. Gaustad et al. (eds.), *REWAS 2019*, The Minerals, Metals & Materials Series,
https://doi.org/10.1007/978-3-030-10386-6

MIX
Papier aus verantwortungsvollen Quellen
Paper from responsible sources
FSC® C105338

If you have any concerns about our products,
you can contact us on
ProductSafety@springernature.com

In case Publisher is established outside the EU,
the EU authorized representative is:
Springer Nature Customer Service Center GmbH
Europaplatz 3, 69115 Heidelberg, Germany

Printed by Libri Plureos GmbH
in Hamburg, Germany